Performance Evolution and Control for Engineering Structures

— Proceedings of the 9th Asia-Pacific Young Researchers and Graduates Symposium（YRGS 2019）

工程结构性能演化与控制

——第九届亚太地区青年学者和研究生论坛

Editors CHEN Jianbing YU Qian-Qian PENG Yongbo
陈建兵 余倩倩 彭勇波 主编

内 容 提 要

工程结构在服役过程中经历反复荷载和环境介质的长期作用，不可避免地会产生损伤累积、性能退化、风貌丧失等问题，需要对其进行性能评估与修复加固。工程结构性能演化与控制已成为当前国际学术研究的前沿和热点。本书是第九届亚太地区青年学者和研究生论坛（the 9th Asia-Pacific Young Researchers and Graduates Symposium YRGS 2019—Performance Evolution and Control for Engineering Structures）的论文集，收录了来自丹麦、韩国、美国、意大利、印度、日本和中国的各位作者的 76 篇论文和扩展摘要，反映了近年来工程结构性能演化与控制的国际国内最新进展。本书可供土木工程专业的教师、研究人员、研究生和高年级本科生及工程师参考。

图书在版编目（CIP）数据

工程结构性能演化与控制：第九届亚太地区青年学者和研究生论坛 = Performance Evolution and Control for Engineering Structures — Proceedings of the 9th Asia-Pacific Young Researchers and Graduates Symposium（YRGS 2019）：英文/陈建兵，余倩倩，彭勇波主编. —上海：同济大学出版社，2019.12

ISBN 978-7-5608-8854-5

Ⅰ.①工… Ⅱ.①陈… ②余… ③彭… Ⅲ.①工程结构-结构性能-国际学术会议-文集-英文 Ⅳ.①TU31-53

中国版本图书馆 CIP 数据核字（2019）第 259109 号

Performance Evolution and Control for Engineering Structures — Proceedings of the 9th Asia-Pacific Young Researchers and Graduates Symposium (YRGS 2019)

工程结构性能演化与控制——第九届亚太地区青年学者和研究生论坛

Editors CHEN Jianbing YU Qian-Qian PENG Yongbo

陈建兵 余倩倩 彭勇波 主编

责任编辑 陆克丽霞 李 杰 **责任校对** 徐春莲 **封面设计** 陈益平

出版发行 同济大学出版社 www.tongjipress.com.cn

（地址：上海市四平路 1239 号 邮编：200092 电话：021-65985622）

经 销 全国各地新华书店、建筑书店、网络书店

排 版 南京文脉图文设计制作有限公司

印 刷 当纳利（上海）信息技术有限公司

开 本 787 mm×1092 mm 1/16

印 张 17

字 数 424 000

版 次 2019 年 12 月第 1 版 2019 年 12 月第 1 次印刷

书 号 ISBN 978-7-5608-8854-5

定 价 98.00 元

Preface

The Asia-Pacific Young Researchers and Graduates Symposium (YRGS) is primarily a platform for early-stage structural engineering professors, research scientists, professional engineers, postdoctoral fellows and postgraduate students to present their latest findings within the context of wide structural engineering discipline. The symposium features oral presentations predominantly from early-career structural engineering people but also includes talks from leading figures in the field. It provides an opportunity for learning about future career paths and networking with fellow researchers. The previous YRGSs has been held in Kunsan, Korea since 2009, then in Hangzhou (China), Taipei (China), Hong Kong (China), Jaipur (India), Bangkok (Thailand), Kuala Lumpur (Malaysia), Tokyo (Japan) from 2010 to 2017. Up to now, it has been successfully organized for 8 times. With 10 years' development, the YRGS has become attractive to many elites and young scholars all over the world with its distinct characteristics and features.

The 9th Asia-Pacific Young Researchers and Graduates Symposium-Performance Evolution and Control for Engineering Structures (YRGS 2019) will be held in Shanghai on 19 - 20 December 2019. It is sponsored by College of Civil Engineering, Tongji University, Key Laboratory of Performance Evolution and Control for Engineering Structures of Ministry of Education of China and Asia Concrete Federation (ACF), to promote close international communication and cooperation, and to figure out the future development of structural engineering. The objectives of the upcoming symposium (YRGS 2019) are again to provide a forum to deal with the state of the art as well as emerging concept and technology, e.g., multi-scale and multi-field methodologies, super-computing, big-data and AI, etc., related to research and practice, particularly in the theme of performance evolution and control for engineering structures, which has been receiving increasing attention due to its important role in sustainability-oriented life-cycle performance civil engineering and resilience of city and countryside.

Over 80 abstracts from 9 countries and regions were submitted to the symposium and accepted for publication in the abstract proceedings of YRGS 2019. The authors are from China (including Hong Kong and Taiwan), Denmark, India, Italy, Japan, South Korea, and USA. On behalf of the YRGS 2019 organizing committee and Tongji University, the chairs of the Symposium would like to cordially welcome all the authors and participants. We also take this opportunity to express our sincere thanks to the distinguished keynote lecturers—and to all the members of the Advisory Committee, International Scientific Committee and International Steering Committee.

We believe with your contribution and participation, the YRGS 2019 will be a successful event. We earnestly hope that all the participants will enjoy their stay and have a great time in Shanghai.

CHEN Jianbing   YU Qian-Qian   PENG Yongbo
Chairs, YRGS 2019
Department of Structural Engineering, Tongji University, Shanghai, China

Contents

Keynote Lectures

Life-cycle management of concrete structures based on sustainability framework

H. Yokota

Faculty of Engineering, Hokkaido University, Sapporo, Hokkaido 060-0808128, Japan

Abstract

A concrete structure has to be thoroughly planned, designed, executed and maintained to keep its performance over its corresponding requirements throughout the life-cycle. However, structures suffering from serious damages in structural members have been often found due to various reasons. One of the reasons is lack of total management of the structure. The life-cycle management is an organized system to support engineering-based decision making for ensuring performance of a structure at the design, execution, maintenance, and all related work during its life-cycle. The life-cycle management is implemented according to the life-cycle management scenario in which balance of several sustainability indicators would be considered with ensuring overall sustainability. This paper presents the concept and framework of the life-cycle management of concrete structures to ensure sustainability during the life-cycle of the structure.

1. Introduction

A concrete structure is constructed with its own purposes such as supporting socio-economic activities, protecting people from disasters, and ensuring a comfortable and safe life. The structure is required to maintain its function and performance to achieve these purposes during its design service life. However, serious damages have been sometimes found, which may provoke performance degradation, and even structural collapse may be consequences. The life of a structure is made up of all the activities including planning, basic and detailed designs, execution, maintenance and intervention, and decommissioning. When coordination of those activities is not ensured, such damages may be found. Therefore, it is very important to coordinate these activities sufficiently. The life-cycle management (LCM) is the overall strategy with the aim of ensuring that the structure meets the associated performance requirements. The strategy is embodied by an LCM scenario. The scenario is formulated at the time of planning and design and may be subsequently modified at each stage of structure's life-cycle. LCM also contributes to realize a sustainable society through structures. Sustainability is defined in terms of environmental, economic, and social aspects. During the life-cycle of structures, sustainability is generally considered with one or a few sustainability indicators. The introduction of LCM for a structure would contribute to all aspects of sustainability while maintaining the function and performance. This paper introduces the principles and framework of LCM for concrete structures. This paper has been prepared by editing the parts of the contents of author's previously published papers[1, 2].

2. Framework of Life-Cycle Management

A concrete structure passes through different stages during its life: from the planning, design, execution, use, and to the end-of-life stages. Due to its long life, it involves different parties at each stage. This implies that it is essential to coordinate all the stages with transferring important information from one stage to another in an appropriate form. LCM is systematic and coordinated activity and practice through which a structure is appropriately managed over its life cycle.

The overall framework of the LCM is summarized in Figure 1. LCM is implemented according to the LCM scenario in which balance of several sustainability indicators should be considered with ensuring performance requirements. The sustainability indicators will be determined from the social, environmental and economic points of view. The scenario should be regularly reviewed and evaluated based on the PDCA cycle[3] and be updated if necessary. As shown in the Figure 1, the LCM is an integrated concept to assist in activities managing the total life-cycle of structure based on managements of each stage to ensure structural functions and performance and to achieve sustainability. As a platform to share the information, BIM has a big potential for use[4].

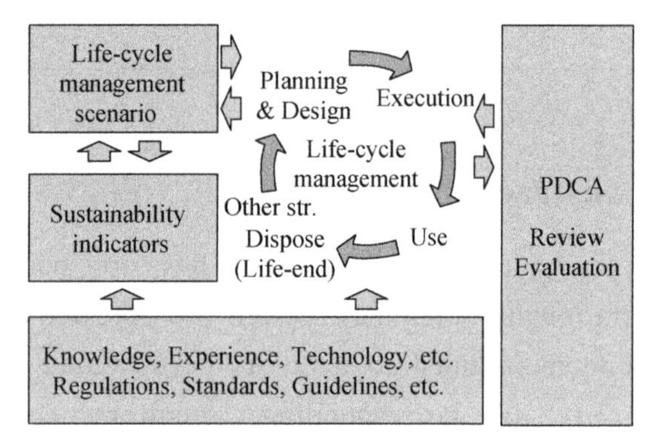

Figure 1 Framework of the life-cycle management

3. Procedure of Life-Cycle Management

Figure 2 shows the standard procedure of LCM. For a new structure, an LCM scenario should be formulated during or after the planning stage of the structure. The scenario includes the fundamental strategy on how the structure will be managed in terms of structural performance and sustainability aspects. The structure is generally designed to keep its structural function and performance without major interventions; however, planned interventions can be included in the scenario if they are required. The scenario mediates among the stages of the structural life-cycle. Design will be carried out to satisfy the scenario initially formulated. When the design outputs do not satisfy the scenario, either the scenario is modified to be consistent with design outputs and/or design is carried out again. After the execution, initial assessment is carried out to check the conditions of the structure. When any defect is found from the assessment, interventions should be taken as required. Then, it will be judged whether the scenario is suitable for the subsequent life-cycle of the structure or not. When the scenario is found to be unsuitable, the scenario should be updated. During the use stage, the structure is periodically assessed its conditions and performance possessed, and the above procedure should be repeated. When the scenario has been updated, the updated scenario should be reflected on subsequent management. If it is judged that interventions should not be taken from the sustainability evaluation mentioned later, the structure goes to the end-of-life stage.

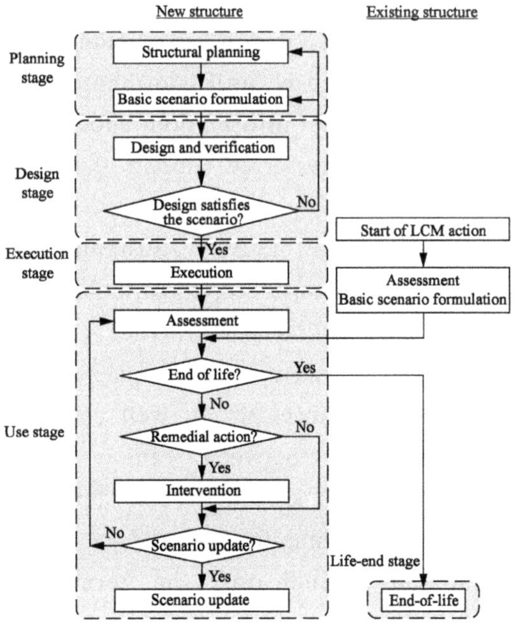

Figure 2 Standard procedure of life-cycle management

For an existing structure, the assessment should be carried out before starting the LCM procedure. The scenario is formulated according to the result of the first assessment and documents even they may not be enough. When the assessment results conclude that interventions are difficult to take to recover structural performance, the structure goes to the life-of-end stage; otherwise, the same procedure as that for a new structure can be followed.

4. Sustainability Indicators

It is necessary to choose suitable indicators to objectively evaluate LCM scenarios and make decisions. The author proposes to use sustainability indicators for this purpose. Sustainability is defined as a concept based on the environmental, economic and social aspects, and is one of the key issues in a construction sector to be well considered in the 21st century[5].

It is easy to understand that collapse of structure impairs the sustainability because the treatment of debris produced by destruction of structures needs huge energy and reconstruction of structures requires an additional amount of resources and energy. Many people might be killed or injured, and employment and production bases would be temporally unavailable. Engineers keep it in mind what might happens by phenomena that are not covered by the design. Thus, the safety margin or safety redundancy that represents the resistance of structure directly links the social sustainability. As the environmental aspect of sustainability, appropriate indicators are set for environmental impacts in the execution and use stages of the structure, such as resource consumption, greenhouse gas emission, and impacts on the ambient environment. As the economic aspect, all the direct and indirect costs during execution and the use and at the end of life of the structure, as well as the benefits and values provided by the structure, can be set as indicators[6].

It is not so easy to find the best solution among alternatives because no comprehensive indicator exists. For example, when the margin of safety (safety redundancy) is taken more, more resources and energy may be needed for construction and higher construction cost will be consequences, as indicated in Figure 3[7]. This is the collision among the sustainability indicators. Therefore, the sufficient balance among each sustainability indicator should be achieved.

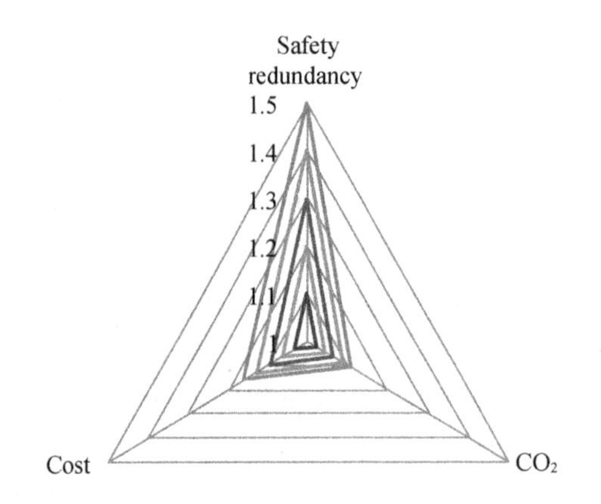

Figure 3 Balance between social (safety redundancy), environmental (CO_2 emission) and economic (cost) indicators[7]

Sustainability indicators should be well considered in the life-cycle of structure. Durability is directly related to structural performance such as safety and serviceability, while resilience and robustness are related to the safety margin and the mechanism of failure. Sustainability is systemized at each stage of life-cycle to consider in a comprehensive manner, safety and serviceability under the social aspect, cost under the economic aspect, and resources and energy under the environmental aspect. LCM allows for designers to find a good balance between social, economic and environmental indicators.

5. Conclusions

LCM is an integrated concept to assist in activities managing the total life-cycle of structures to realize sustainability. The following are concluding remarks in this paper:

(1) For doing infrastructure management, planning, design, execution, and use stages should be well coordinated, in which necessary information should be shared and transferred among the life-cycle stages.

(2) During LCM, sustainability should be well considered to formulate the scenario. Structural performance can be included in the sustainability concept.

(3) It is necessary to well consider the balance between the safety redundancy that should be considered as social sustainability and other sustainability indicators in terms of economic and environmental aspects.

(4) A concrete structure inherently has a long life when it is well designed, executed, and maintained. It can achieve

longer life with proper LCM, which will result in realizing sustainability.

Based on the concept and framework of the life-cycle management presented in this paper, International Standard, ISO/DIS 22040 *Life cycle management of concrete structures* is now in progress of development at ISO Technical Committee 71 *Concrete, reinforced and pre-stressed concrete*. It is expected to be published as ISO in 2021.

References

[1] YOKOTA H. Considerations for life-cycle of concrete structures[C]// Proceedings of the 8th International Conference of Asian Concrete Federation. Fuzhou, China, 2018: 1-8.

[2] YOKOTA H. Practical application of life-cycle management system for shore protection facilities[J]. Journal of Structure and Infrastructure Engineering, 2016, 13(1): 34-43.

[3] YOKOTA H, NAGAI K, MATSUMOTO K, et al. Prospect for implementation of road infrastructure asset management[J]. Advanced Engineering Forum, 2017, 21: 366-371.

[4] ZHANG Y, YOKOTA H, ZHU Y. Sensitivity analyses on chloride ion penetration into undersea tunnel concrete [J]. Journal of Advanced Concrete Technology, 2019, 17(10): 592-602.

[5] SAKAI K. Sustainability design for innovations of concrete technologies[C]// Proceedings of the 2nd ACF Symposium on Innovations for Sustainable Concrete Infrastructures. Chiang Mai, Thailand: Asian Concrete Federation, 2017.

[6] KAWABATA Y, KATO E, YOKOTA H, et al. Net present value as an effective indicator leading to preventive maintenance of port mooring facilities[J]. Journal of Structure and Infrastructure Engineering, 2019: 1-12.

[7] YOKOTA H, GOTO S, SAKAI K. Parametric analyses on sustainability indicators for design, execution and maintenance of concrete structures[C]// Proceedings of the 2nd International Conference on Concrete Sustainability (ICCS 16), Madrid, Spain, 2016: 1046-1053.

Fracture energy and tension-softening behavior of slag-fly ash blended geopolymer concrete

J. G. Dai

Department of Civil and Environmental Engineering, The Hong Kong Polytechnic University, Hung Hom, Kowloon, Hong Hong, China

Y. Ding

College of Civil Engineering, Chongqing University, Chongqing 400044, China

Abstract

Recently, geopolymer cement (GC) concrete has been extensively researched as a potential alternative to Portland cement (PC) concrete owning to its environmental friendliness. Similar to PC concrete, GC concrete is a brittle material in nature. However, little has been understood on its fracture properties in terms of the fracture energy and the tension-softening curve, both of which are important for the structural analysis and safe application of GC concrete in engineering practice. This paper presents the research findings arisen from a recent PhD project on the static fracture properties of GC concrete as compared to those of PC concrete. The fracture energy and tension-softening behaviors are compared between GC and PC mortar/concrete through extensive three-point bending tests, in considerations of the effects of various material compositions including the fly ash/slag ratio, the modulus ratio, the alkali concentration and water/binder ratio.

1. Introduction

Portland cement (PC) concrete is the most widely used construction material in the world. However, the increasing emphasis on the sustainability has highlighted the adverse effects of PC production on the environment and motivated researchers to explore new cementitious materials like geopolymer cement (GC) as a partial or complete alternative to PC for two main reasons: (1) through chemical activation, industry waste such as fly ash (FA) and ground granulated blast furnace slag (GGBFS) can be utilized[1]; (2) the carbon footprint of GC is significantly lower than that of PC, making it a more environmentally-friendly choice[2]. However, it should be noted that similar to PC concrete (PCC), the GC concrete (GCC) still has a brittle nature. Therefore, an in-depth understanding of its fracture properties, such as the fracture energy and tension softening behavior[3] is a prerequisite for the safe application of GCC in practical structural applications.

The properties of GCC largely depend on the used raw materials. Initially, the GC was strictly defined as an alkali-activated aluminosilicate (e. g. metakaolin as a precursor) without (or with little) other components[4]. As a byproduct of coal power plants, low-calcium content FA can be also activated to form geopolymer, leading to a major hydration product of N-A-S-H gel with a three-dimensional spatial structure[5]. However, high alkali concentration and heat curing are always needed to achieve a reasonable compressive strength due to its high activation energy[6]. The heat curing methods are acceptable in precast geopolymer industry, yet very challenging to be implemented in in-situ constructional operations[7]. Therefore, it is imperative to develop other chemical processes for room-temperature hardening of GCC. Therefore, in alkali-activated systems, calcium-containing solid aluminosilicate sources, such as ground granulated blast-furnace slag (GGBS) or high-calcium fly ash (FA) were usually introduced under high alkaline environment to form C-(A)-S-H gels[1]. Such chemical product shows some similar features to that of C-S-H gel that is dominant in the hydraulic cement. The big advantage of slag/fly ash blended geopolymer concrete lies its ambient temperature curing capacity. The compressive strength of slag/FA blended geopolymer concrete usually increases with the amount of slag[8, 9].

Up to now limited research has been conducted on the fracture properties of GC concrete in consideration of different material compositions. A recent PhD project was completed at the Hong Kong Polytechnic University on the static fracture and dynamic mechanical performance of GC concrete[8, 10-12]. This paper presents a summary of the research findings arisen from this project on the static fracture properties of GC concrete.

2. Experimental Program

2.1 Raw Materials

The ground granulated blast furnace slag (GGBFS) used in the research was produced by Huaxin Cement Co. in Hunan, China. The fly ash (FA) was produced by Lianyuan Power Plant in Hunan, China. Their chemical compositions obtained from the X-ray fluorescence (XRF) analysis are presented in Table 1. Figure 1 shows the particle size distributions of the GGBFS and FA. The former was mainly in the range from 2 μm to 50 μm while the latter had a wider range from 1 μm to 300 μm. PC produced by the Anhui Conch Cement Company, China with a specific surface area of 336 m^2/kg and loss on ignition of 1.62% was used in this research. Its chemical compositions are also indicated in Table 1. The morphology of GGBFS and FA particles observed by the scanning electron microscope (SEM) is shown in Figure 2, which illustrates that GGBFS particles are predominately of anomalous shape with clear edges and angles while most FA particles are solid spheres. The alkali activator liquid used was a combination of sodium silicate solution and sodium hydroxide. The sodium silicate solution was a commercially available product with a water content of 59% (by mass) and a modulus (the mole ratio of SiO_2 to Na_2O) of 3.7. The modulus of the alkali activator was further adjusted to the designed values by adding sodium hydroxide (NaOH) flakes with 99% purity, which was purchased from the Tianjin Bohai Chemical Industry, China. Gravel from local river were used as the coarse aggregate whose bulk specific density was 2 530 kg/m^3 with a maximum size of 10 mm. The water absorption of the coarse aggregate was 1.83%. The fine aggregate used was natural, uncrushed river sand. Its specific density, absorption, and fineness modulus were 2 340 kg/m^3, 2.75% and 2.47, respectively.

Table 1 Chemical composition of GGBFS, FA and PC (% by mass)

	CaO	Al_2O_3	SiO_2	SO_3	P_2O_5	MgO	Na_2O	K_2O	TiO_2
Slag	38.8	14.8	33.8	2.49	0.05	7.09	0.25	0.44	1.23
FA	1.32	34.1	52.6	0.33	0.17	0.50	0.34	1.37	1.72
PC	65.1	4.81	21.9	0.51	—	—	0.65	—	1.62

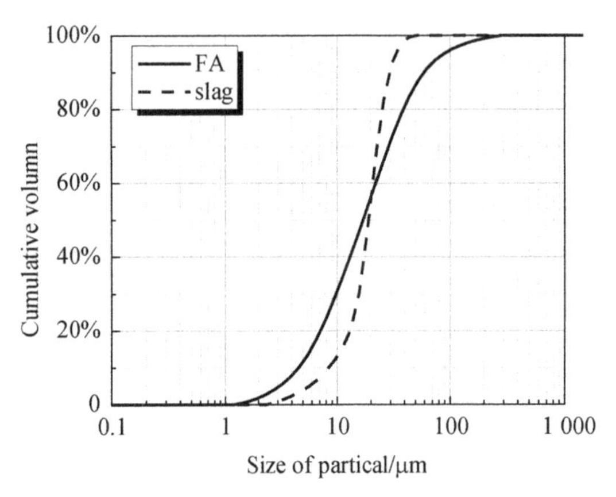

Figure 1 Particle size distributions of GGBFS and FA

(a) FA

(b) GGBFS

Figure 2 Morphology of FA and GGBFS particles

2.2 Material Mix Proportions

Three batches of specimens were prepared, including PC mortar and concrete, alkali-activated slag (AAS) geopolymer mortar and gepolymer concrete, and alkali-activated slag-fly ash (AASF) blended geopolymer concrete. The mixing proportions for these three batches of specimens are summarized in Table 2, Table 3 and Table 4, respectively.

Powder polycarboxylate superplasticizer was used here and its dosage was by the mass fraction of cement. It should be noted that the first batch and second batch specimens were prepared for comparison between PC mortar/concrete and AAS geopolymer mortar/concrete at three different target strength grades (30, 50 and 70 MPa at 28 d). The third batch of specimens were prepared to observe the effect of material parameters on the fracture properties of geopolymer concrete. The test parameters included slag/fly ash ratio, water/binder ratio, alkali concentration (the percentage of Na_2O by mass of slag, n) and modulus of the alkali activator (i.e. the mole ratio of SiO_2 to Na_2O, M_s), all of which are thought to have significant effects on the compressive strength of the GC concrete. The total water used to calculate the water/binder ratio included the water added and the water contained in the sodium silicate solution. In the tables, PCM and PCC represent ordinary Portland cement mortar and concrete, respectively; AASM and AASC represent the alkali-activated slag geopolymer mortar and concrete, respectively, for which slag was used as the sole precursor material; AASFC represents alkali-activated slag/FA blended geopolymer concrete.

Table 2 Mix proportions of PCM and PCC

	Cement /(kg · m^{-3})	Sand /(kg · m^{-3})	Stone /(kg · m^{-3})	Water /(kg · m^{-3})	w/c	Super plasticizer	Sand ratio
PCM30	600	1 200	—	300	0.50	—	—
PCM50	700	1 155	—	245	0.35	0.09%	—
PCM70	850	1 010	—	240	0.30	0.16%	—
PCC30	350	776	1 164	210	0.60	—	0.40
PCC50	380	795	1 192	133	0.35	0.42%	0.40
PCC70	420	782	1 172	126	0.30	0.50%	0.40

Table 3 Mix proportions of AASM and AASC

	n/%	M_s	Slag /(kg · m^{-3})	Sand /(kg · m^{-3})	Stone /(kg · m^{-3})	Water /(kg · m^{-3})	Alkali activator: Sodium silicate solution/ (kg · m^{-3})	Alkali activator: Sodium hydroxide/ (kg · m^{-3})	w/b	Sand ratio
AASM30	3.0	1.5	783	1 174	—	276	109	18	0.44	—
AASM50	4.0	1.5	783	1 174	—	253	145	24	0.44	—
AASM70	5.0	1.5	783	1 174	—	254	182	30	0.44	—
AASC30	3.0	1.5	350	746	1 120	127	49	8	0.45	0.40
AASC50	4.0	1.5	380	724	1 087	127	71	11	0.45	0.40
AASC70	4.5	2.0	420	694	1 041	117	117	11	0.45	0.40

Note: "n" means the alkali concentration and "M_s" means the modulus of alkali activator.

Table 4 Mix proportions of AASFC

	n/%	M_s	Slag /(kg·m^{-3})	Fly ash /(kg·m^{-3})	Sand /(kg·m^{-3})	Stone /(kg·m^{-3})	Water /(kg·m^{-3})	Alkali activator: Sodium silicate solution /(kg·m^{-3})	Alkali activator: Sodium hydroxide /(kg·m^{-3})	w/b
AASFC-1	3	1.5	200	200	716	1 074	145	56	9	0.45
AASFC-2	4	1.5	200	200	712	1 068	133	74	12	0.45
AASFC-3	5	1.5	200	200	708	1 062	122	93	15	0.45
AASFC-4	5	1.0	200	200	712	1 068	139	62	19	0.45
AASFC-5	5	2.0	200	200	704	1 056	104	124	12	0.45
AASFC-6	5	1.5	200	200	716	1 074	102	93	15	0.40
AASFC-7	5	1.5	200	200	700	1 050	142	93	15	0.50
AASFC-8	4	1.5	300	100	712	1 068	133	74	12	0.45
AASC-9	4	1.5	400	0	712	1 068	133	74	12	0.45

Note: "n" means the alkali concentration and "M_s" means the modulus of alkali activator.

2.3 Specimen Preparation Procedures

For all geopolymer mortar/concrete specimens, the weighted sodium hydroxide, sodium silicate solution and water were firstly mixed to form an alkali activator solution. A large amount of heat was released when sodium hydroxide was dissolved in water, so the alkali activator solution was prepared one day before casting of the specimen, to ensure that the solution was cooled down to ambient temperature. The weighted GGBFS, fly ash, sand and coarse aggregate were added into a mixer and dry-mixed for 2 minutes. After the dry materials were uniformly mixed, the alkali activator solution was added and mixed with the solid fraction for another 2 minutes. The final mixture was then poured into the prepared molds. The mixing process of PC mortar and PC concrete was simpler. Weighted solid fractions were dry-mixed for 2 minutes then water was added and mixed with the solid fractions for another 2 minutes. The curing conditions of the PC and GC mortar/concrete specimens were the same. All the specimens were demolded 24 hours after the casting and cured in an environmental chamber with a constant temperature of (21 ± 1) ℃ and a related humidity of ≥95% for 28 days.

2.4 Three Point Bending Tests

Three-point bending tests (TPB) were used to evaluate the fracture properties of PC and GC mortar/concrete beams. The beam dimensions were 100 mm × 100 mm × 515 mm with a span-to-depth ratio of 4.0. Each beam specimen was precut with a notch in the middle, 40 mm deep and 3 mm wide. The configuration of the beam specimens is illustrated in Figure 3. Three identical specimens were prepared for each type of mix, thus a total of 63 beams were tested.

The crack mouth opening displacement (CMOD) and crack tip opening displacement (CTOD) were measured using clip gauges clamped at both the mouth and the tip of the precut notch. The mid-span displacement (δ) of the beam was detected by two linear variable differential transformers (LVDTs). The displacements of the two supports were also measured by the LVDTs to remove the effect of support settlement on the mid-span displacement. To obtain the complete load-displacement ($P-\delta$, $P-$CMOD, P-CTOD) curves, a closed-loop servo controlled hydraulic jack of 100 kN capacity was used and operated at a stable and slow loading rate of 0.02 mm/min.

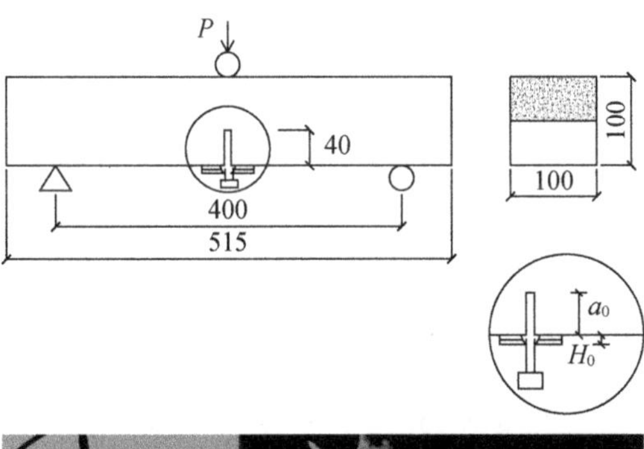

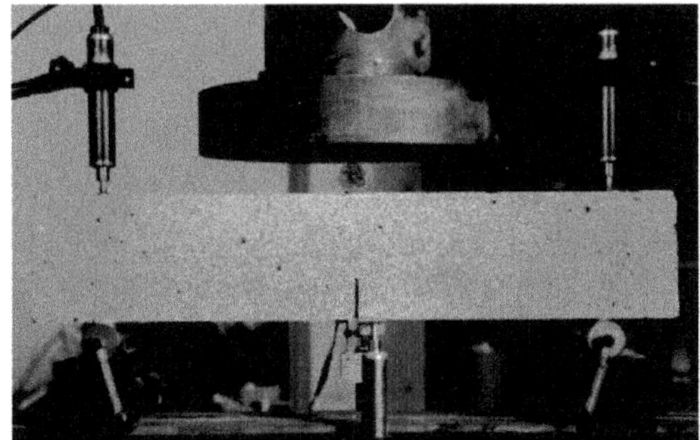

Figure 3　Configuration of three-point bending test beam (unit: mm)

3. Results and Discussion

3.1 Strength and Elastic Modulus

Figure 4(a) presents the average values and variations of the compressive strengths of PC and AAS geopolymer mortar (M30, M50 and M70) and geopolymer concrete (C30, C50 and C70). It is clear that both PC and AAS geopolymer mortar and concrete achieved very similar 28-day compressive strengths, facilitating a sound basis for their comparisons. The average compressive strengths were obtained from the testing results of three identical specimens in each group. The maximum strength deviation was ±6.4%, indicating a good quality control. Figure 4(b) depicts the modulus of elasticity of PC and AAS geopolymer mortar and concrete. The PC concrete clearly had an elastic modulus 25%～35% higher than AAS geopolymer concrete, and for the PCM, it was also 25%～35% higher than AASM given the same compressive strength grade. With C70, the elastic modulus of PC concrete was (35.2 ±1.1) GPa while the elastic modulus of AAS gepolymer concrete was (23.6 ± 1.2) GPa. The average value of the latter was 33% lower than that of the former.

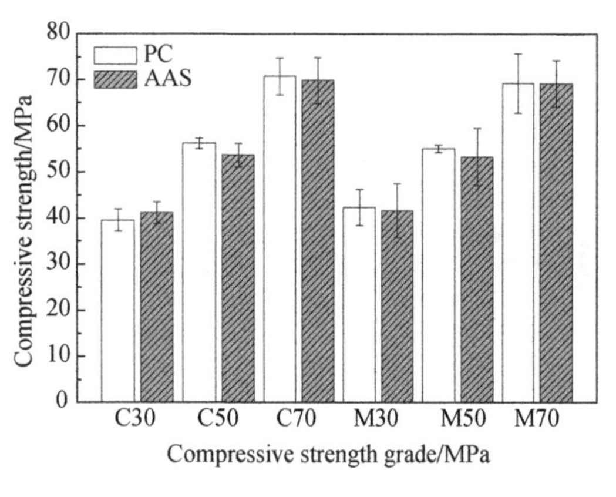

(a) Compressive strength

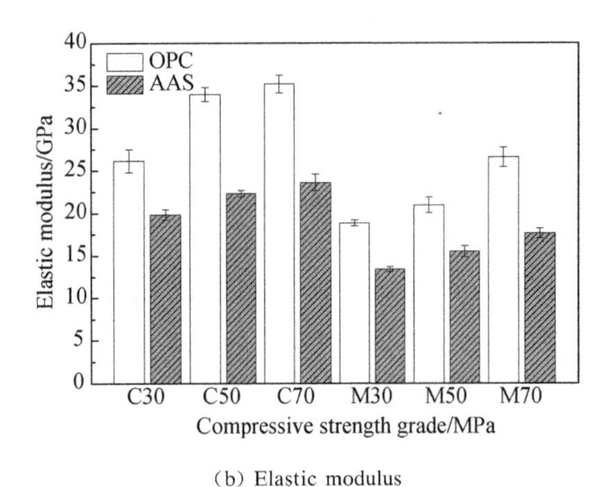

(b) Elastic modulus

Figure 4　Mechanical properties of PC and AAS geopolymer concrete/mortar

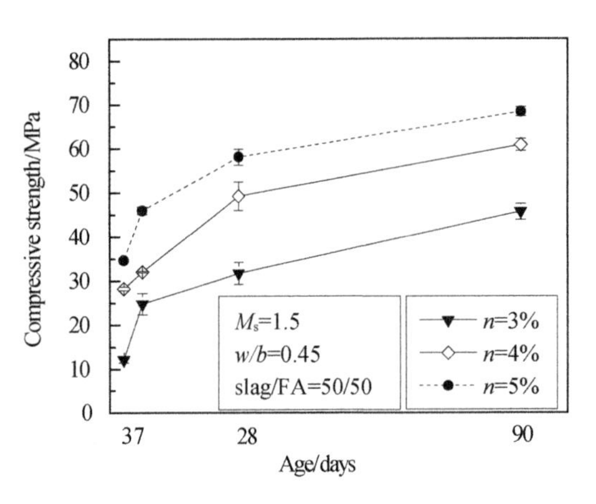

(a) Alkali concentration

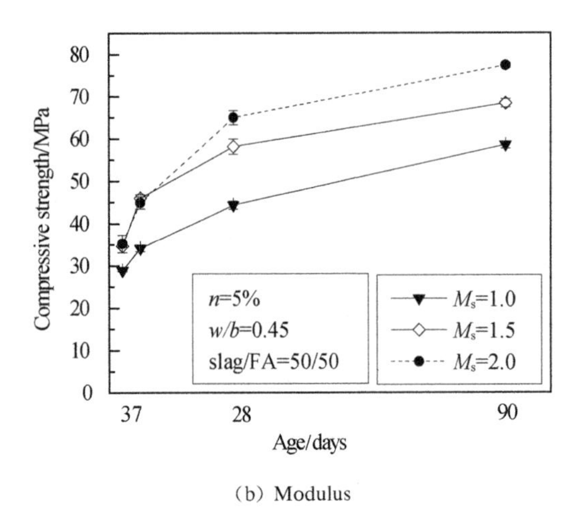

(b) Modulus

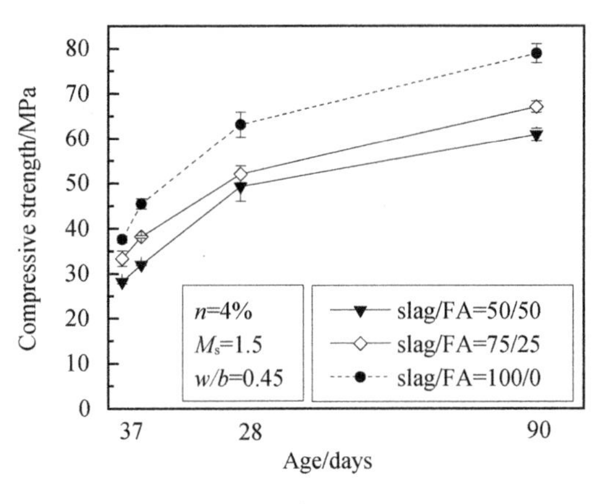

(c) Slag/FA ratio

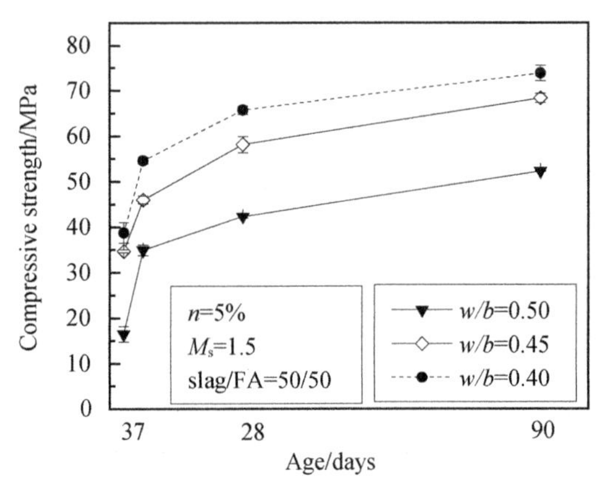

(d) Water/binder ratio

Figure 5　Parametric effects on the compressive strength development of ASFSC geopolymer concrete

The compressive strength developments of AASFC with the alkali concentration, the modulus, the water/binder ratio and the slag/FA ratio are shown in Figure 5. It is seen that the compressive strength of AASFC increased with the curing time and significant strength development could still be observed after 28 days. The compressive strength of AASFC increased with the alkali concentration, the modulus and the slag/FA ratio while decreased with the increase of water/binder ratio. The 3-day compressive strength of

AASFC could achieve almost 40 MPa in many cases. The 28-day compressive strength of AASFC increased from 31.7 MPa to 58.2 MPa (i.e. an 83.6% increase) when the alkali concentration increased from 3% to 5%. The average compressive strengths of AASFC with alkali concentration of 5% were 34.7 MPa, 46.0 MPa, 58.2 MPa and 68.4 MPa, respectively, corresponding to the curing ages of 3, 7, 28 and 90 days. A further 17.5% increase of compressive strength was observed when the curing age increased from 28 days to 90 days.

3.2 Fracture Energy

The fracture energy is an important parameter used to describe the fracture characteristic of concrete material. Three-point bending tests on a notched concrete beam is an indirect method recommended by RILEM TC50-FMC to determine the fracture energy of concrete and mortar. Figure 6(a) shows clearly that the fracture energy of PC concrete and AAS gepolymer concrete increases with the compressive strength. The average fracture energy of PC concrete increases from 127.1 N/m to 177.2 N/m (i.e. a 39.4% increase) when the compressive strength grade increases from C30 to C70. Although the fracture energy improvement of AAS geopolymer concrete is not as significant as that of PC concrete, the value still increases from 177.2 N/m to 207.9 N/m (i.e. a 17.3% increase). Overall, the fracture energies of AAS geopolymer concrete are all higher than those of PC concrete at all three compressive strength grades. The difference of fracture energy between PC concrete and AAS geopolymer concrete appears to reduce as the compressive strength increases, since sodium silicate can reduce the flocculation of slag grains and the wall effect. As a result, the initial porous ITZ around aggregate in AAS geopolymer concrete is smaller than that in PC concrete[10].

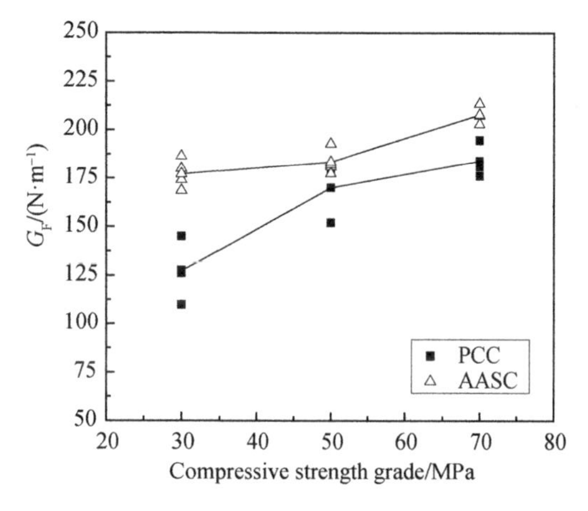

(a) Geopolymer concrete

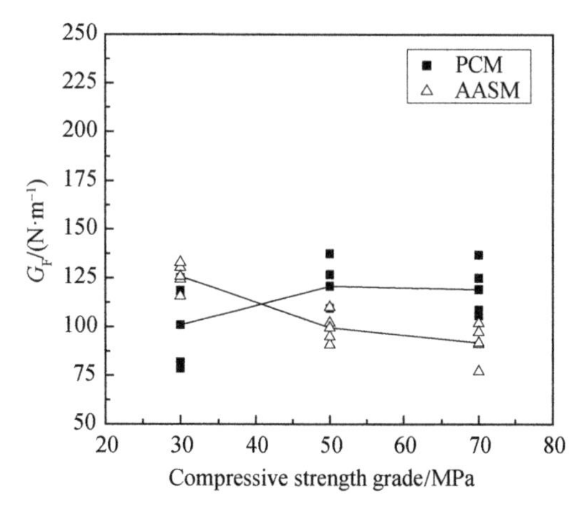

(b) Geopolymer mortar

Figure 6 Fracture energy of PCC/AASC and AASM/PCM with the same compressive grade

The calculated average fracture energy of PCM and AASM is plotted against the compressive strength grade in Figure 6(b). The fracture energy of PCM is found to increase with the compressive strength, but the AASM exhibits an opposite trend. The fracture energy reduction in the case of AASM may be due to the use of a higher alkali concentration activator, which can generate more serious autogenous and drying shrinkage and shrinkage cracks[13]. At M30, the average fracture energy of PCM is 100.9 N/m, which is 24.7% lower than that of AASM. Nevertheless, when the compressive strength increases to M70, the average fracture energy of PCM is 119.1 N/m, which is 23% higher than that of AASM. When the strength is relatively low, it was found that the matrix of AASM is more homogenous than that of PCM and the ITZs in AASM are denser and stronger than in PCM, leading to more energy consumption for crack propagation. When the compressive strength becomes high, it could be found more micro-cracks occurred in AASM than in PCM. Probably due to these existing flaws, AASM exhibits more brittle behavior and the energy consumed for crack propagation during the whole testing procedure is reduced[10].

The comparison of the fracture energy of PC mortar and AAS geopolymer mortar and concrete specimens is shown in Figure 7. Regardless of the matrix type (i.e. PC or AAS), due to the addition of coarse aggregates, the fracture energy of concrete is evidently higher than that of mortar. For both the PC and AAS series, the difference of the fracture energies between the mortar and concrete samples grows with the increase in the compressive strength.

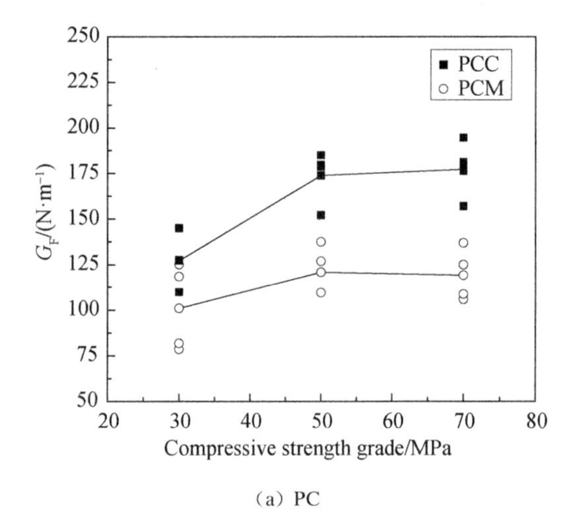

(a) PC

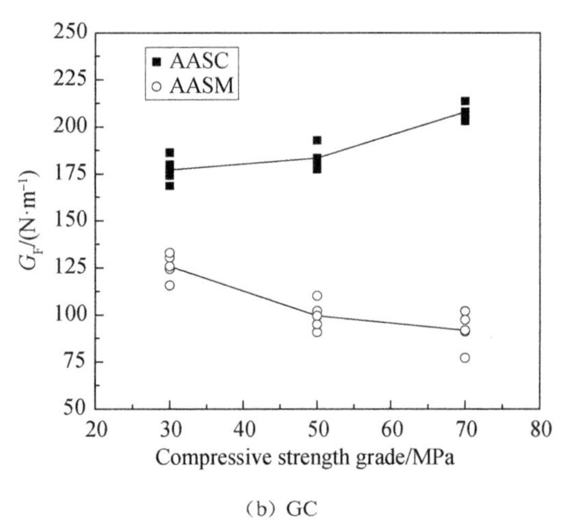

(b) GC

Figure 7 Comparison of fracture energy between mortar and concrete

The fracture energy G_F calculated from the $P-\delta$ curves of AASFC with different parameters are shown in Figure 8. The ultimate loads P_u of the TPB tests are also presented in the Figure 8 for reference. As expected, the ultimate loads P_u and the fracture energy G_F of AASFC beams increased with the alkali concentration, the modulus and the slag/

FA mass ratio while decreased with the increase of water/binder ratio, in other words, increased with the compressive strength. It can be seen from Figure 8(a) that the improvement of the average ultimate load P_u of AASFC beams (i.e. from 2 443 N to 3 214 N) with the increase of alkali concentration from 3% to 5% was 31.5%, and the fracture energy increased from 118.3 N/m to 137.6 N/m with a 16.3% increase. Although the improvement of the ultimate loads P_u of AASFC beams with modulus increase (from 1.0 to 2.0) was not significant (i.e. from 3 202 N to 3 344 N and only a 4.4% increase), the average fracture energy still increased from 128.8 N/m to 157.7 N/m (i.e. a 22.4% increase) Figure 8(b). Similarly, the improvements of the ultimate loads P_u and the fracture energy G_F were 19.2% and 14.1%, respectively (i.e. from 2 993 N to 3 567 N and from 118.7 N/m to 135.4 N/m) with the increase of slag/FA mass ratio from 50/50 to 100/0 (Figure 8(c)). In addition, the increases of the ultimate loads P_u and the fracture energy G_F with the decrease of water/binder ratio from 0.50 to 0.40 were 30.0% and 26.4%, respectively (Figure 8(d)).

3.3 Tension softening behavior

Tension softening is a basic material property which represents the relationship between the cohesive stress and the corresponding crack opening across the fracture process zone (FPZ) of concrete. The softening curve is essential for predicting the fracture behavior of concrete, and can be obtained by direct

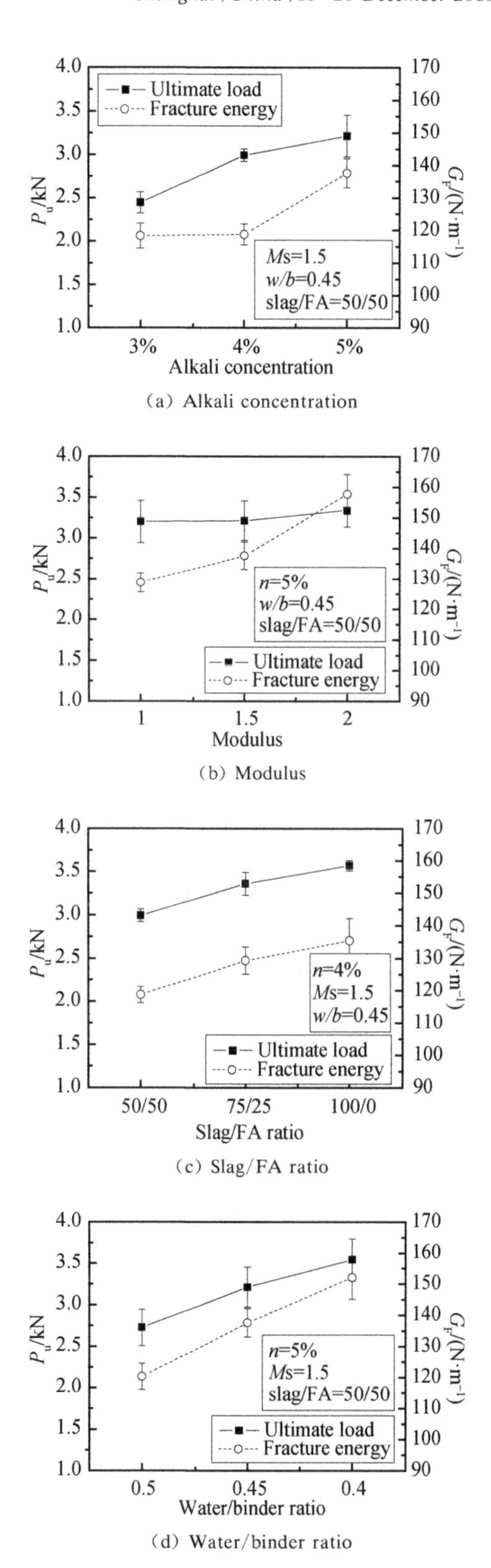

(a) Alkali concentration

(b) Modulus

(c) Slag/FA ratio

(d) Water/binder ratio

Figure 8 Parametric effects on fracture energy of AASFC and ultimate load of the beams

tensile test for practical applications, there are two simplified models for describing the strain-softening relationship of concrete: bilinear[3] and exponential[14]. The bilinear strain-softening diagram is also adopted in the CEB-FIP Model Code.

On the basis of TPB tests, the tension softening curve can be determined indirectly by a backward analysis[15, 16]. The software CONSOFT[17] originally developed by Prof. Volker Slowik and his colleagues at the University of Applied Sciences in Leipzig Germany was utilized to determine the softening curves of the PC and AAS concrete and mortar. For the inverse analyses, the cohesive crack model[15] as well as an evolutionary optimization algorithm[18] is adopted. The applied optimization method is based on a biologically motivated approach. By stepwise updating the assumed softening curve and re-analyzing in several iterations, the numerical results can be fitted to the experimental ones. When a satisfactory fit of the numerical to the experimental results is met, the assumed softening curve is considered to be the one characterizing the behavior of the material[19]. The boundary effect[20] is also taken into account. When the crack tip approaches the specimen boundary, the size of the fracture process zone is reduced and the full amount of the fracture energy can no longer be activated. As a result, the local fracture energy decreases near the end of the crack path.

Figure 9 shows the normalized bilinear tension-softening curves of PC and AAS geopolymer concrete and mortar with compressive strength of 30, 50 and 70 MPa. Figure 9 (a) shows that the normalized bilinear softening curves of PC concrete and AAS geopolymer concrete are generally the same given the same compressive strength although the decrease of the initial descending part of AAS geopolymer concrete is slightly slower than that of PC concrete in the case of C30. Figure 9 (b) clearly shows that PCM usually has gentler first descending slopes than AASM at all the three compressive strength levels. For

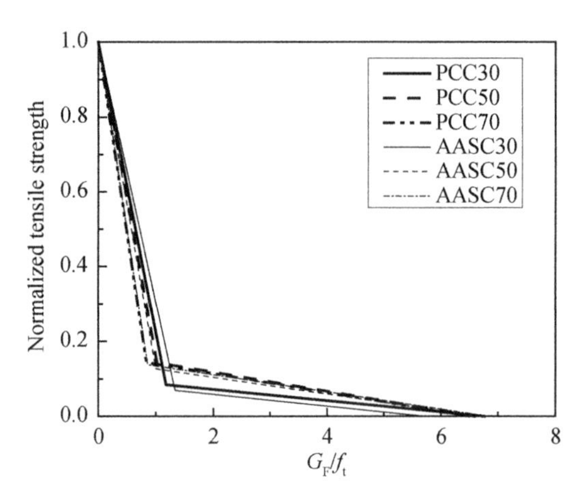

(a) PCC and AASC

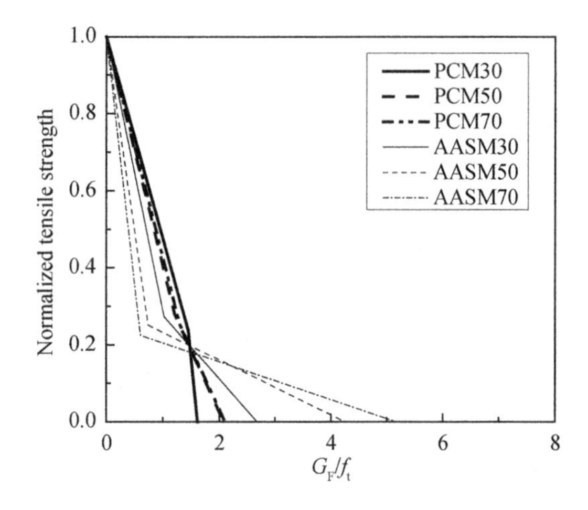

(b) PCM and AASM

Figure 9 Bilinear tension softening curves of PC and AAS geopolymer concrete and mortar

both AAS geopolymer and PC concrete and mortar, the first descending slope becomes sharper with the strength increase.

Figure 10 shows the normalized bilinear tension-softening curves of PC concrete, AAS geopolymer concrete and AASFC gepolymer concrete with different alkali concentration, modulus, slag/FA ratio and water/binder ratio. It is seen for the first descending part all tested parameters had a marginal effect. For the second descending part, the alkali-concentration had the least effect on the normalized softening curve (Figure 10 (a)); the modulus between 1.0 and 1.5 just led to a marginal change of the shape of the softening curve but a further increase of the modulus to 2.0 may increase the free-stress crack width (Figure 10(b)); the effect of slag/fly ash on the softening curves is not deterministic. The slag seems to have an optimal ratio, which may need further research. It is seen that using the slag/FA ratio of 50/50 and using 100% slag almost led to the identical normalized tension-

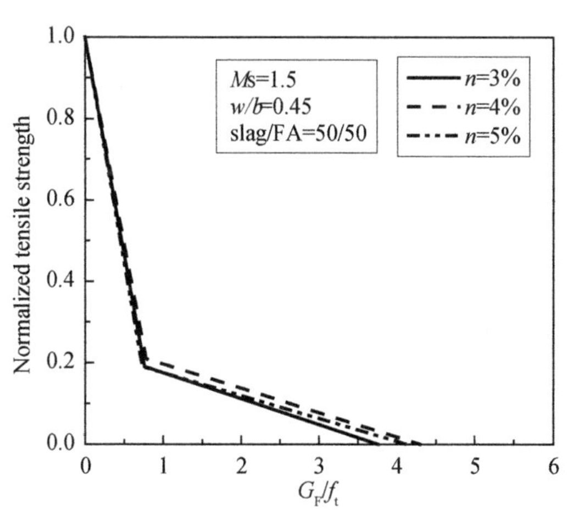

(a) Alkali concentration

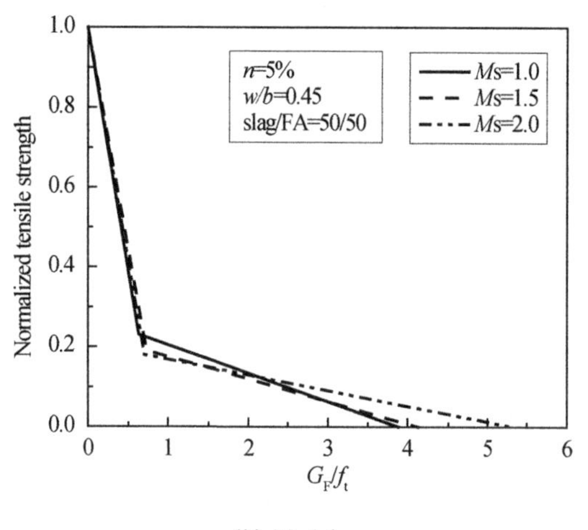

(b) Modulus

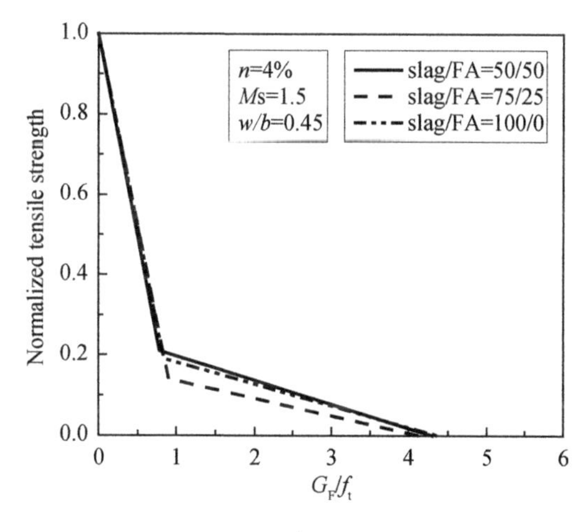

(c) Slag/FA ratio

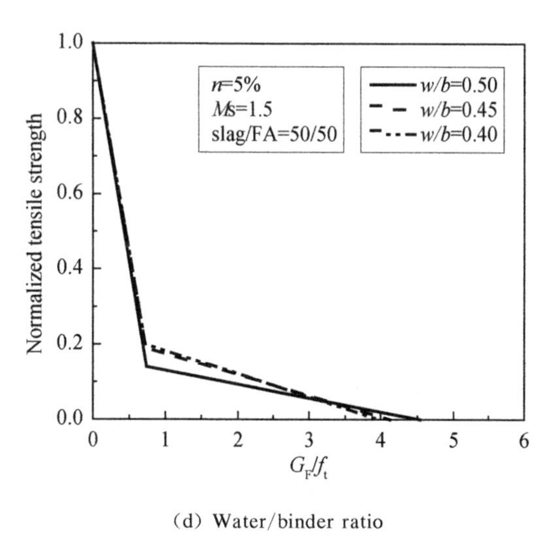

(d) Water/binder ratio

Figure 10 Parametric effects on the bilinear tension softening curves of AASFC

softening curves (Figure 10 (c)); the effect of water/binder ratio is similar to that observed for PC concrete. The higher compressive strength, the more brittle tension softening curves (Figure 10(d)).

4. Conclusions

The fracture properties and tension softening behavior of concrete are important characteristics for structural design and analysis of concrete structures. A series of experimental tests have been conducted to investigate such behaviors of geopolymer mortar and concrete with different strength grades and material compositions. The following conclusions have been reached through the study:

(1) Given the same compressive strength, the elastic modulus of AAS mortar and concrete are generally smaller than their PC counterparts, by around 25%~35%.

(2) The fracture energy G_f of AAS geopolymer concrete increases with the compressive strength, but G_f of AAS geopolymer mortar reduces as the compressive strength increases.

(3) Given the same compressive strength, the G_f of AAS geopolymer concrete is always higher than that of PC concrete due to the denser and stronger ITZs between the AAS pastes and aggregates. However, the G_f of AAS geopolymer mortar at M30 is higher than that of PC mortar due to the more homogenous matrix and denser ITZs, and when the compressive strength further increases, the G_f of AAS geopolymer mortar is lower than its PC counterpart due to more micro-cracks occurring in the former matrix.

(4) The compressive strength of AASFC gepolymer concrete increase with the alkali concentration, the modulus and the slag/FA ratio while decrease with the increase of water/binder ratio. The 3-day compressive strength of AASFC geopolymer concrete can achieve almost 40 MPa and significant further compressive strength development can be observed after 28 days.

(5) The G_f of AASFC geopolymer concrete increases with the alkali concentration, the modulus and the slag/FA mass ratio while decreases with the increase of water/binder ratio, in other words, increases with the compressive strength.

(6) The normalized bilinear softening curves of PC concrete and AASC gepolymer concrete are generally the similar given the same compressive strength although the slope of the initial descending part of AAS geopolymer concrete is slightly smaller than that of PC concrete in the case of C30. PC mortar usually has a smaller first descending slope than AAS gepolymer mortar at all the three compressive strength levels. For both AAS gepolymer and PC concrete and mortar, the first descending slope becomes sharper with the strength increase.

(7) The normalized bilinear tension-softening curves of AASFC gepolymer concrete exhibit the similar first descending part regardless of the varied material parameters. For the second descending part, the alkali-concentration

has the least effect. Changing the modulus from 1.0 and 1.5 leads to a marginal change of the softening curve but a further increase to 2.0 increases the free-stress crack width.

(8) There seems to exist an optimal slag/fly ash ratio in terms of the G_f of AASFC GC. However, a slag/FA ratio of 50/50 and a slag ratio of 100% led to almost the identical normalized tension-softening curves.

Acknowledgements

The authors gratefully acknowledge the financial support provided by the Hong Kong-Guangzhou Technology and Innovation Partnership Programme (Project No. 201807010055), National Science Foundation of China (NSFC) (Project No. 51638008), Innovation Technology Fund (Project code: ITS/009/17) and The Hong Kong PhD Studentship awarded to the second author.

References

[1] SHI C J, QIAN J S. High performance cementing materials from industrial slags—a review [J]. Resources Conservation and Recycling, 2000, 29(3): 195-207.

[2] DUXSON P, PROVIS J L, LUKEY G C, et al. The role of inorganic polymer technology in the development of green concrete [J]. Cement and Concrete Research, 2007, 37(12): 1590-1597.

[3] HILLERBORG A. The theoretical basis of a method to determine the fracture energy GF of concrete [J]. Materials and Structures, 1985, 18(4): 291-296.

[4] DAVIDOVITS J. Geopolymer chemistry and applications [M]. 3rd ed. Geopolymer Institute, 2011.

[5] FERNANDEZ-JIMENEZ A, PALOMO A, CRIADO M. Microstructure development of alkali-activated fly ash cement: a descriptive model [J]. Cement and concrete research, 2005, 35(6): 1204-1209.

[6] AGARWAL S K. Pozzolanic activity of various siliceous materials [J]. Cement and Concrete Research, 2006, 36(9): 1735-1739.

[7] PROVIS J L, BERNAL S A. Geopolymers and Related Alkali-Activated Materials [J]. Annual Review of Materials Research, 2014, 44(1): 299-327.

[8] DING Y, DAI J G, SHI C J. Mechanical properties of alkali-activated concrete: A state-of-the-art review [J]. Construction and Building Materials, 2016, 127: 68-79.

[9] SHANG J, DAI J G, ZHAO T J, et al. Alternation of traditional cement mortars using fly ash-based geopolymer mortars modified by slag [J]. Journal of Cleaner Production, 2018, 203: 746-756.

[10] DING Y. Experimental study on fracture properties of alkali-activated concrete [D]. Hong Kong: The Hong Kong Polytechnic University, 2017.

[11] DING Y, DAI J G, SHI C J. Fracture properties of alkali-activated slag and ordinary Portland cement concrete and mortar [J]. Construction and Building Materials, 2018a, 165: 310-320.

[12] DING Y, DAI J G, SHI C J. Mechanical properties of alkali-activated concrete subjected to impact load [J]. Journal of Materials in Civil Engineering, 2018b, 30(5), 04018068.

[13] YANG L Y, ZHANG Y M, DAI J G, et al. Effects of nano-TiO_2 on strength, shrinkage and microstructure of alkali activated slag pastes [J]. Cement and Concrete Composites, 2015, 57: 1-7.

[14] REINHARDT H W, CORNELISSEN H A W, HORDIJK D A. Tensile tests and failure

analysis of concrete[J]. Journal of Structural Engineering, 1986, 112(11): 2462-2477.

[15] HILLERBORG A, MODÉER M, PETERSSON P E. Analysis of crack formation and crack growth in concrete by means of fracture mechanics and finite elements [J]. Cement and Concrete Research, 1976, 6(6):773-781.

[16] ROELFSTRA P E, WITTMANN F H. Numerical method to link strain softening with failure of concrete [M]//In: Wittmann, F. H., editor. Fracture toughness and fracture energy of concrete, Elsevier Science Publishers, 1986:163-175.

[17] SLOWIK V, VILLMANN B, BRETSCHNEIDER N, et al. Computational aspects of inverse analyses for determining softening curves of concrete[J]. Computer Methods in Applied Mechanics and Engineering, 2006, 195(52): 7223-7236.

[18] VILLMANN B, VILLMANN T, SLOWIK V. Determination of softening curves by backward analyses of experiments and optimization using an evolutionary algorithm [C]// Proceedings of the 5th International Conference on Fracture Mechanics of Concrete and Concrete Structures, USA, 2004: 439-445.

[19] BRETSCHNEIDER N, SLOWIK V, VILLMANN B, et al. Boundary effect on the softening curve of concrete [J]. Engineering Fracture Mechanics, 2011, 78 (17): 2896-2906.

[20] DUAN K, HU X Z, WITTMANN F H. Boundary effect on concrete fracture and non-constant fracture energy distribution[J]. Engineering Fracture Mechanics, 2003, 70 (16): 2257-2268.

Steel-Concrete Composite Immersed Tunnels Structure

J. S. Fan & Y. T. Guo & Y. F. Liu
Key Laboratory of Civil Engineering Safety and Durability of China Education Ministry, Department of Civil Engineering, Tsinghua University, Beijing 100084, China

Abstract

Ocean engineering and construction are promising foundations for the implementation of China's maritime power strategy. Compared to land engineering, ocean engineering requires more complex and demanding construction environments and conditions, poses new challenges to structural engineering. Steel-concrete composite structures have broad application prospects in ocean engineering with significant performance advantages and comprehensive economic benefits as they successfully combine the respective advantages of steel and concrete.

In the field of cross-sea tunnel structural engineering, reinforced concrete structures and steel-shell concrete structures have long been the main structural forms. In the 1980s, with the development of composite structure as well as its application and experience in practice, steel-concrete-steel (SCS) composite structures began to be used in cross-sea tunnels. The steel-concrete-steel composite immersed tube structure is composed of inner and outer steel plates, partition webs, stiffeners and internal concrete, as shown in Figure 1.

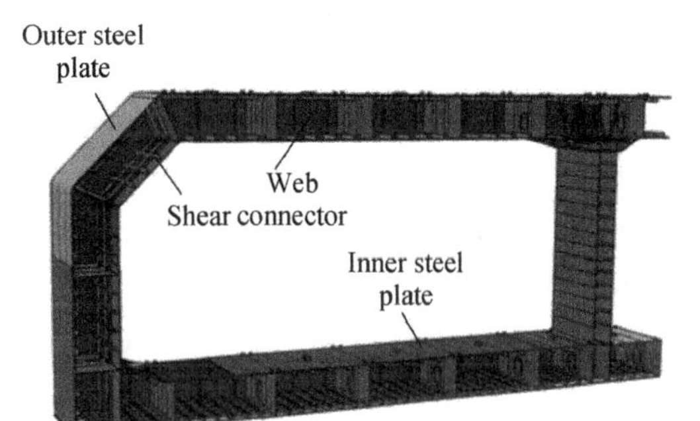

Figure 1 Steel-concrete-steel composite immersed tunnel structure

The steel-concrete-steel composite immersed tunnel structure makes good use of steel properties during construction and operation stage, and has superior bending, shear-resistance, impact-proof, and waterproof performances, while saving costs to a large extent. Besides, it is constructed via the immersed tube method; the immersed tunnel sections are prefabricated in dry docks or large barges, then floated to the construction sites, buried in the design positions, and fixed by the connection measures to finally complete the submarine tunnel. If the steel structure part of the composite tunnel has sufficient out-of-plane stiffness in design to resist the out-of-plane deformation caused by concrete pouring and transportation during the construction process, and the complex temperature effect in the pouring process is accurately analyzed and well controlled, the concrete pouring can be carried out on the water at the construction site, greatly saving the cost associated with the site and transportation.

This study focuses on the mechanical performance of steel-concrete-steel composite immersed tube structure and the complex temperature effect analysis duringits pouring process.The engineering background of this study is the immersed tunnel of the Shenzhen-Zhongshan cross-river tunnel under construction in China, which has already adopted the type of steel-concrete-steel composite immersed tube.

For the part of mechanical performance, a series of studies are conducted that reveal the mechanisms of bending, shear and the interface connection, and propose the corresponding design method.

Firstly, the four-point bending test is used to study the bending performance of seven specimens of the steel-concrete-steel composite structure with a scale ratio of 1 : 2, with the research focused on effects of the factors such as local buckling and concrete casting imperfections[1, 2]. The typical phenomena are shown in Figure 2. The test results show that the main failure mode of the specimen is the yield failure of the tensile steel plate.

Secondly, the three-point bending test is used to study the shear performance

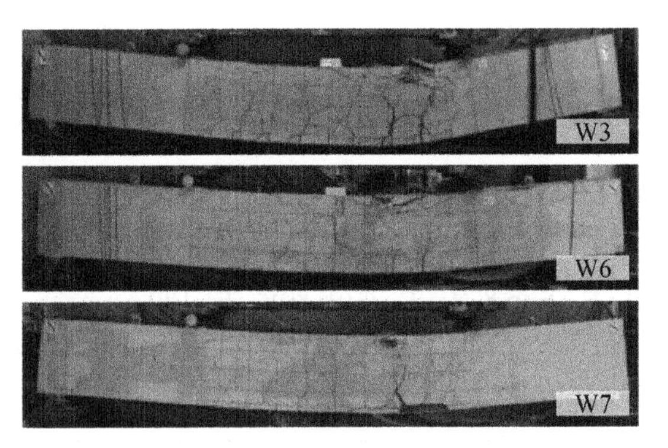

Figure 2 Four-point bending test failure modes

for 16 specimens of compartment steel-concrete-steel composite structure with a scale ratio of 1 : 2[3], with the research focused on the effect of the factors such as shear span-to-depth ratio, concrete width, and bidirectional web layout. The test results show that the structure has good shear performance, high bearing capacity and ductility.

Finally, for the steel shear connectors, push-out test is completed for a total of 78 (three identical pieces per group, 26 groups in total) full-scale models. It is recommended to use T-shape connectors due to their improved force bearing performance.

For the part of temperature effect analysis during the pouring process of steel-concrete-steel composite immersed tunnel structure, a temperature-structural coupling analysis method considering temperature field simulation and structural deformation analysis in the whole process of construction is proposed. The heat source and heat transfer modes considered are shown in Figure 3.

Firstly, the ANSYS finite element is adopted to simulate the construction

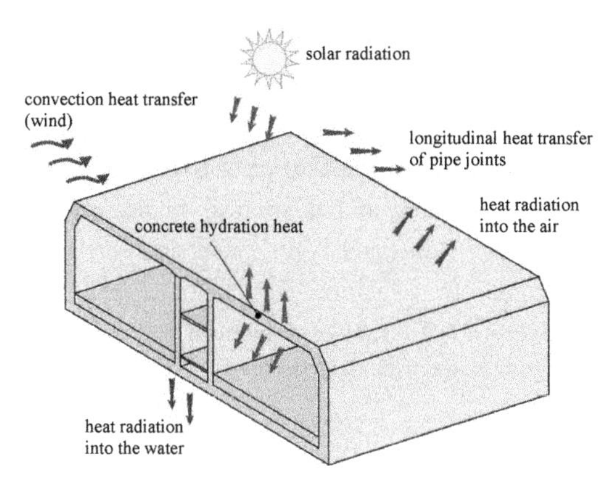

Figure 3 Heat source and heat transfer modes

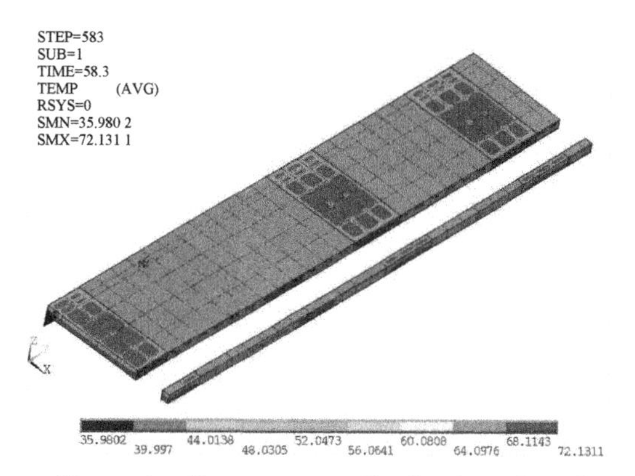

Figure 4 Temperature distribution of roof concrete

process through the birth and death element, taking into account the two temperature loads of hydration heat release and solar radiation during concrete pouring, and the boundary conditions of heat transfer between structure, air and seawater. The simulation of temperature field distribution of time-domain and space-domain during the whole process of tunnel construction is successfully achieved. The simulation results show that the maximum temperature of the roof concrete during the pouring process is about 72 ℃ (as presented in Figure 4), which meets the Chinese codes' requirement.Besides, solar radiation has a limited effect on the peak temperature of the concrete of tunnel, but has a significant effect on the roof and outer surface of the tunnel.

Secondly, the parameter analysis shows that the reasonable selection of construction period with lower environment temperature, and the control of the concrete pouring temperature, can effectively reduce the peak temperature of the concrete.If the outside temperature or pouring temperature reduces by 10 ℃, the peak temperature of concrete can be reduced by 3~4 ℃.

The above research and practice demonstrate that this new type of composite structural system has significant advantages in both mechanical performance and construction performance, thus providing new ideas and choices for ocean engineering construction and effectively promoting the application of steel-concrete composite structures in ocean engineering. Follow-up study will be carried out around the construction monitoring of the full-scale test model of the tunnels.

Acknowledgements

This research is supported by the National Key Research and Development Program of China (Grant No. 2018YFC0809600, 2018YFC0809602) and the National Natural Science Foundation of China (Grant No. 51725803). The authors express their sincere appreciation to their supports.

References

[1] GUO Y T, TAO M X, NIE X, et al. Bending capacity of steel-concrete-steel composite structures considering local buckling and casting imperfection [J]. Journal of Structural Engineering, 2019, 145 (10): 04019102.

[2] GUO Y T, NIE X, TAO M X, et al. Selected series method on buckling design of stiffened steel-concrete composite plates[J]. Journal of Constructional Steel Research, 2019, 161: 296-308.

[3] GUO Y T, TAO M X, NIE X, et al. Experimental and theoretical studies on the shear resistance of steel-concrete-steel composite structures with bidirectional steel webs[J]. Journal of Structural Engineering, 2018, 144(10): 04018172.

General Sessions

Mathematical model for predicting the adiabatic temperature rise of concrete

D. J. Jeong & J. H. Kim

Korea Advanced Institute of Science and Technology, Daejeon, Korea

Abstract

Conventional test method to measure the ATR of concrete needs a sample of larger than 50 L volume. Delicate control for the adiabatic condition requires a heavy insulation and closed-loop circuit to match the chamber temperature with the sample's temperature. Practically, the equipment to measure the ATR has been rarely applied. Semi-adiabatic condition is sometimes used to replace the original equipment. The measurement of ATR of concrete is cumbersome. It is impossible to evaluate the thermal shrinkage of all possible mix proportions in field. The heat of cement hydration can be predicted using a chemical model. The proportion of each compound in Portland cement determines the final heat output achieved by their full hydration. The degree of hydration controls the rate of heat evolution of a given Portland cement compound.

Heat production rate of cement hydration, $Q(t;T)$ is measured using by adiabatic temperature rise test. The rate of heat production is following Eq.(1)[1].

$$\frac{dQ(t;T)}{dt}=f(Q)k(T) \quad (1)$$

where t, T, Q, f, k are hydration time, temperature, heat production, heat development function of chemical reaction, rate constant of cement hydration, respectively. If degree of hydration is defined with the heat production, its development is given by Eq.(2).[2,3]

$$f(Q)=Q_u[1-(Q/Q_u)]^2 \quad (2)$$

where Q_u is the limiting heat production at infinite age. The development function is the same with Bernhardt empirical formula[2], which was previously adopted for the maturity method[4]. Note that a conventional definition for the degree of hydration ranges from 0 to 1, which is $\alpha = Q/Q_n$. Considering the dormant period t_0 and integrating the above equations gives a closed form of heat production, Eq.(3).[4]

$$Q=Q_u k(t-t_0)/[1+k(t-t_0)] \quad (3)$$

Fitting the model function with isothermal microcalorimetry data gives Q_u, $t_0(T)$ and $k(T)$. The total heat production Q_u is constant regardless of the reaction temperature. The dormant period t_0 may be a function of T, but we just need t_0 at initial temperature (20 ℃) of

ATR prediction. The total heat production Q_u is constant at all temperature and it obviously depends on the mix proportion. The reaction rate adopts Arrhenius equation, Eq.(4).[5]

$$k(T) = A e^{\frac{E_a}{RT}} \quad (4)$$

where k, T, A, E_a, R are rate of constant, absolute temperature in Kelvin, pre-exponential factor, activation energy for hydration and universal gas constant, respectively.

Specific heat capacity is a function of a linear or exponential decay with the degree of hydration, where specific heat of portland cement and water are 0.75 J/(g·K) and 4.18 J/(g·K), respectively. Adiabatic temperature rise equation is hard to get a closed form. Considering the thermal expansion of the sample is following Eq.(5).

$$e(t, T) = \frac{1}{2}\kappa\varepsilon_{V1} = \frac{1}{2}\kappa\varepsilon_{V2} \quad (5)$$

where $\varepsilon_V = \alpha\Delta T$, $\kappa(\alpha)$ is thermal coefficient. Some of concrete hydration heat energy is consumed by the volumetric thermal expansion energy.

Adiabatic temperature rise equation considering thermal expansion is Eq.(6). ATR is calculated by integrating the $Q(t, T)$ with increasing T, where $Q(t=0, T) = Q(t, T_0)$. Full hydration cannot be achieved in a sealed curing condition. The achievable degree of hydration is different for cement paste and concrete.

$$ATR(t) = \frac{Q(t, T)B - e(\alpha, T)M/\rho}{c(\alpha)M} \quad (6)$$

where c and α are concrete specific heat function of hydration rate[6-8] and hydration rate.[9] Isothermal microcalorimetry analysis was conducted for cement pasting proportioned by the water-to-cement cement ratio of 0.45 to 0.55 for calculating the ATR using by Eq.(6). The predicted ATR is the verified by adiabatic temperature rise test using a conventional equipment for same mix proportion of isothermal microcalorimetry analysis.

Concrete mix proportions are shown in Table 1. C = 370 kg with W/C = 0.40, 0.45, 0.55; C = 350, 370, 400, 440 kg with W/C = 0.45. ATR test sample consisted of 50 L steel drum with a 400 mm diameter and 400 mm height.

Table 1 shows mix proportion and their hydration temperature values (Q), correction hydration temperature (Q') values at equal origin temperature (20 ℃) and compressive strength. Based on the hydration modelling formula, Q' is corrected to the initial temperature correction for hydration test results was corrected by the weighted average. The effects of the initial temperature are small. The effects are minimal considering the error in hydration experimentation and the error in regression for the exponential function model. Using the results of the experiment, a model formula for the final insulation temperature rise is proposed as $Q = 0.147C$ where Q, C are final hydration temperature and cement content, respectively.

Table 1 Concrete mix proportion and end temp

W/C	C	W	S	G	Q	$T(t=0)$	$Q'(T=20$ ℃$)$
0.40	370	148	937	937	59.8	17.4	59.4
0.45	350	158	933	933	51.8	21	51.9
	370	167	912	912	53.8	22	53.6
	400	180	882	882	59.2	18.1	58.8
	440	198	841	841	65.7	16	64.7
0.55	370	204	863	863	60.0	17.1	59.5

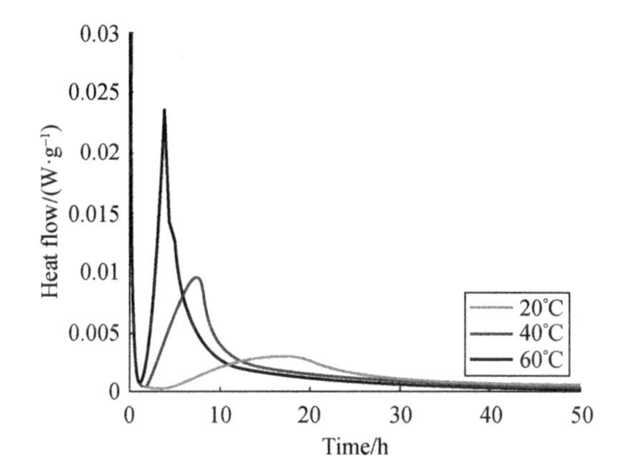

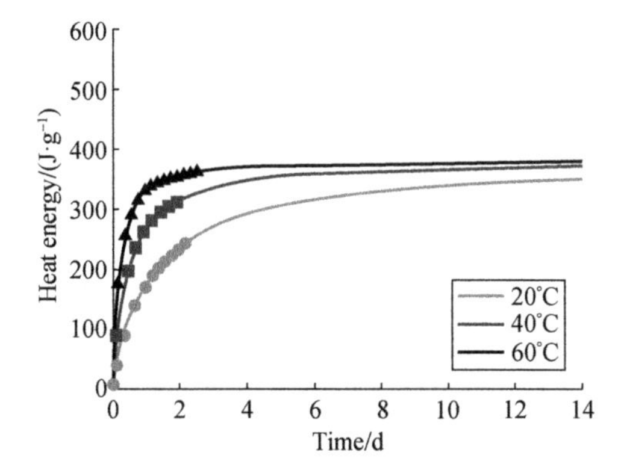

Figure 1 (a) Isothermal microcalorimetry analysis, (b) Integration of Fig. 1(a) for each initial temperature

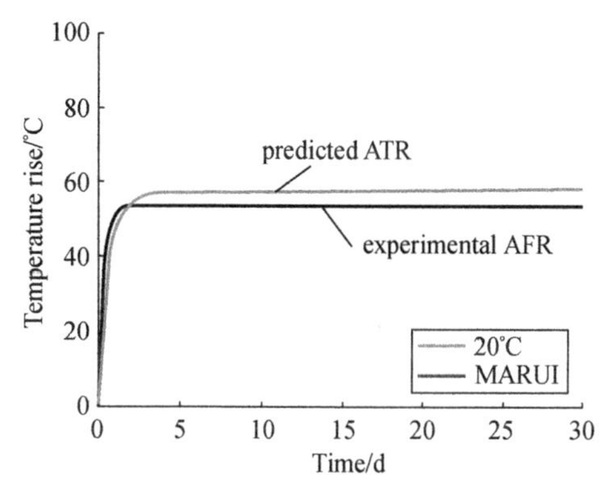

Figure 2 ATR with $C = 370$ kg, W/C = 0.45

The regression lines in Figure 1 are predicted using by Eq.(3). Final hydration energy has no relationship with starting temperature. Final hydration temperature for W/C = 0.45 and $C = 370$ kg is 55.7 ℃. Comparing the predicted ATR with the experimental ATR, the result is shown in Figure 2.

The pre-prediction gives a higher heat production than ATR measurement. Energy dissipation due to the thermal expansion was marginal approximately 0.01% of the total heat production. Sealed conditions for cement paste and concrete was assumably different considering the water content in a unit volume of sample.

In general degree of concrete hydration vary according to the W/C ratio and type of sample[10]. Different between predicted temperature and experiment temperature shown in Figure 3 is the difference of degree of hydration of cement paste and mortar. In Figure 3, reduction quantity is strongly related to hydration content[11].

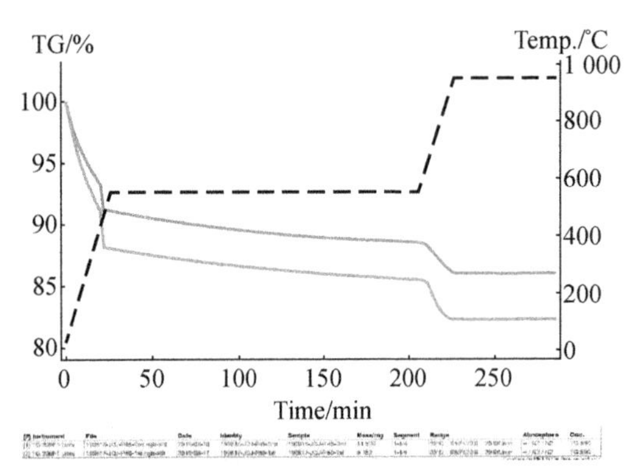

Figure 3 degree of hydration of cement paste and mortar considering proportion of W/C

ATR measurement has high cost, but the isothermal microcalorimetry analysis is very low cost. All the parameters required for predicting the ATR can be obtained through the isothermal microcalorimetry analysis. The predicted hydration temperature is very similar to ATR results.

Acknowledgements

This study was funded by Basic Science Research Program through the National Research Foundation (NRF) of Korea funded by the Ministry of Education (Grant No. NRF-2018R1D1A1B07047321).

References

[1] ASTM C 1074. Practice for Estimating Concrete Strength by the Maturity Method [S]. West Conshohocken: ASTM International, PA, 2000.

[2] BERNHARDT C J. Hardening of concrete at different temperatures[C]// Proceedings of the RILEM Symposium on Winter Concreting. Copenhagen: Danish National Institute for Building Research, 1956.

[3] PLOWMAN J M, OCKLESTON A J, MILLS R H, et al. Maturity and the Strength of Concrete [J]. Magazine of Concrete Research, 1956, 8(22): 13-22.

[4] CARINO N J. The maturity method: theory and application [J]. Cement concrete and aggregates, 1984, 6(2): 61-73.

[5] SIDDIQUI M S. Effect of temperature and curing on the early hydration of cementitious materials [D]. Manhattan: Kansas State University, 2010.

[6] BENTZ D P. Transient plane source measurements of the thermal properties of hydrating cement pastes[J]. Materials and Structures, 2007, 40(10): 1073-1080.

[7] NAIK T R, KRAUS R N, ASCE F, et al. Influence of types of coarse aggregates on the coefficient of thermal expansion of concrete[J]. Journal of Materials in Civil Engineering, 2011, 23(4): 467-472.

[8] PARK J M, KIM H C, LEE Y M, et al. A study on thermal properties of rocks from Gyeonggi-do Gangwon-do, Chungchung-do, Korea [J]. Economic and Environmental Geology, 2007, 40(6): 761-769.

[9] SCHINDLER A K, FOLLIARD K J. Heat of hydration models for cementitious materials[J]. ACI Materials Journal, 2005, 102(1): 24-33.

[10] ZHANG J, SCHERER G W. Comparison of methods for arresting hydration of cement [J]. Cement and Concrete Research, 2011, 41(10): 1024-1036.

[11] XUAN D X, ZHAN B J, POON C S. A maturity approach to estimate compressive strength development of CO_2-cured concrete blocks [J]. Cement and Concrete Composites, 2018, 85: 153-160.

Repair of steel plates by shape memory alloy fiber-reinforced polymer patch: state-of-the-art

V. Kean

Key Laboratory of Performance Evolution and Control for Engineering Structures of Ministry of Education, Tongji University, Shanghai 200092, China & Department of Structural Engineering, Tongji University, Shanghai 200092, China

T. Chen

Key Laboratory of Performance Evolution and Control for Engineering Structures of Ministry of Education, Tongji University, Shanghai 200092, China & Department of Structural Engineering, Tongji University, Shanghai 200092, China & University Corporation for Polar Research, Beijing 100875, China

Abstract

Retrofitting steel plates with shape memory alloy fiber-reinforced polymer (SMA/FRP) patch has become an attractive research direction, which may create reliable retrofitting of existing structures.

The considerations of the most important experimental rest results were obtained from the experimental studies, so the synthesis of research results is divided into the following macro-themes, such as the bond between SMA (wires) and FRP, recovery stress of SMA, influence of materials and surface treatment, and the fatigue life of SMA/FRP/steel plate composite.

SMA/FRP composite has shown its superiority in strengthening cracked components. All of these tests were under pull-out tests, and these patches related to bonding single, three, ten and a sandwich of SMA wires embedded FRP patches[1-3]. To strengthen or retrofit any steel plates, it is very important to understand the common failure modes because debonding was considered to be a major failure mode in SMA/FRP patches, and to improve fatigue life.

The FRP composite patches can bond with cracked metal structures to decrease the stress range near the crack tip and increase the fatigue life. It was proposed to use SMA material to achieve this prestressing effect embedded FRP patch by using the electric current, heating source and water heating. Subsequently, a novel strengthening technique was proposed by El-tahan M et al.[1] using SMA/FRP composites as partially activating prestress by SMA wires which obtained the maximum stress. To make nice performance active hybrid SMA/FRP composites, it is very essential to study the effect of SMA prestress to apply these patches to steel elements.

Properties of materials are very

important for the improvement of bond properties. Among the composites has FRP, SMA, epoxy adhesive and steel plate. Also, the reinforcing effect of FRP with different elastic modulus provided different results of fatigue life studied by many authors. According to previous researches, the fatigue life was the same when the tensile stiffness of FRP was identical[4], however, the higher elastic modulus, the higher fatigue life. Moreover, the number of the SMA wires provided the improvement of fatigue life. A lot of SMA wires can increase the fatigue life ratio that experiments were done with the quantity of SMA wires proposed[3, 5]. Then, the surface treatment used for bond of composites with steel plate that had been conducted by many researchers. On the other, the effect of surface treatment of steel plate through means of grit blasting, needle gun, wire brushing, and rotary wheel, and from the results the advantageous effect on the fatigue life shown.

In general, SMA/FRP patch-steel reinforcement methods include bond the patch to the steel surface with structural epoxy adhesives. Then, there are several failure modes of SMA/FRP patches which was bonded with steel plates under tension fatigue load. Among of these failure modes, the new findings were SMA and adhesive interface debonding, and SMA pull-out[3, 5], as shown in Figure 1. Additionally, epoxy adhesives are a very important connection between steel plate and SMA/FRP patch to grow fatigue life. The study conducted by Li et al.[6] used different adhesives with FRP to investigate the central inclined cracked steel plates that applied a prestressed and non-prestressed system as well as El-tahan M et al.[2] compared both adhesives which were tested to evaluate mechanical properties. The results demonstrated that they powerfully contributed to the failure modes that were the main factor of fatigue performance for applications in any structures. Therefore, epoxy adhesive bonded steel plates to SMA/FRP patch which provided structural strengthening, and analyzed failure modes were essential to improve the cycles of fatigue life in this composite.

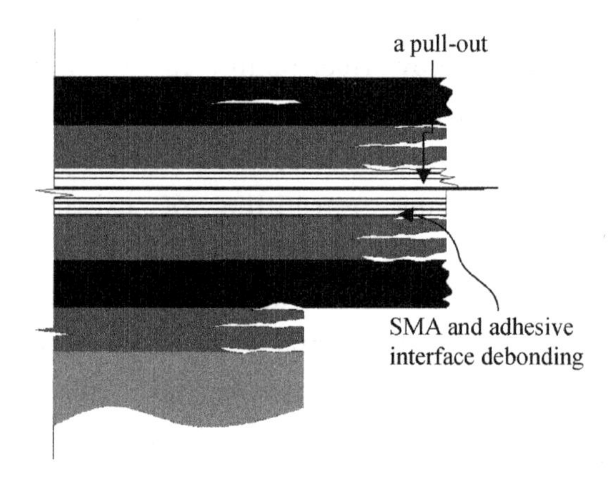

Figure 1 Failure modes of SMA/FRP patch repaired specimens

Finite element method (FEM) was utilized in this study. Moreover, many researchers have studied the fatigue life of all steel plates bonded with SMA/FRP patches analyzed by ABAQUS predicted and tested[3, 4, 7]. The finite element results showed that the SMA/FRP patch had good agreement observed between the predicted and the experimental test which had the longest fatigue life performance. Therefore, all these findings are summarized in Figure 2 and Table 1.

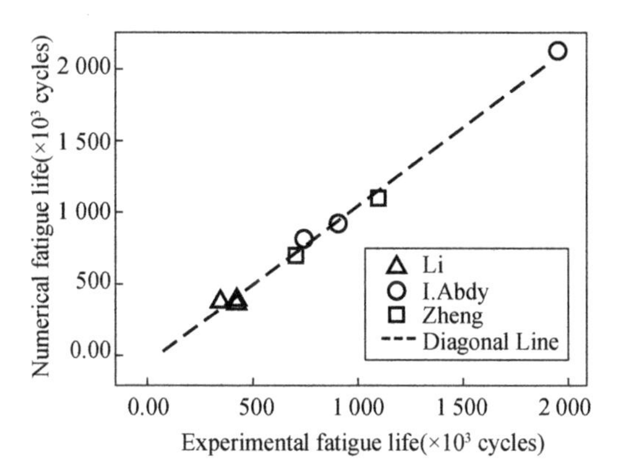

Figure 2 The fatigue life of steel plates

Table 1 Finite element methods

Methodology	Researcher	Detail results
LEFM, EIFS	Abdy, Zheng, Dawood and Li	SIF calculation, parameter analysis

Furthermore, SMA/FRP/steel composite materials need the following contents to study including stress transfer regularity of SMA/FRP composite with steel members after activation, and crack initiation direction and crack propagation law of steel members with Ⅰ/Ⅱ cracks strengthened by SMA/FRP composite under axial tension fatigue load. All of the central point focus on future work.

Based on the review demonstrated shows the SMA/FRP patch can be effectively utilized as a retrofit system. The summaries of SMA/FRP/steel plate composite are followed by experimental test results, numerical methods and future research directions depending on the state-of-the-art.

Acknowledgements

This work was financially supported by the National Natural Science Foundation of China (Grant No. 51978509).

References

[1] EL-TAHAN M, DAWOOD M. Fatigue behavior of a thermally-activated NiTiNb SMA-FRP Patch[J]. Smart Materials and Structures, 2015, 25(1): 15030.

[2] EL-TAHAN M, DAWOOD M, SONG G. Development of a self-stressing NiTiNb shape memory alloy (SMA)/fiber reinforced polymer (FRP) Patch[J]. Smart Materials and Structures, 2015, 24(6): 065035.

[3] ABDY I. Enhancement of fatigue life of steel structures using prestressed CFRP patches with shape memory alloys[D]. Melbourne, Australia: Swinburne University of Technology, 2017.

[4] LI L Z. Fatigue strengthening of central cracked steel plates using SMA/CFRP composite patches[D]. Shanghai, Tongji University, 2019.

[5] ZHENG B T, DAWOOD M. Fatigue strengthening of metallic structures with a thermally activated shape memory alloy fiber-reinforced polymer patch[J]. Journal of Composites for Construction, 2016, 21(4): 04016113.

[6] LI L Z, CHEN T, ZHANG N, et al. Test on fatigue repair of central inclined cracked steel plates using different adhesives and CFRP, prestressed and non-prestressed[J]. Composite Structures, 2019, 216(3): 350-359.

[7] ZHENG B T, DAWOOD M. Fatigue crack growth analysis of steel elements reinforced with shape memory alloy (SMA)/fiber reinforced polymer (FRP) composite patches [J]. Composite Structures, 2017, 164: 158-169.

An analytical model of the concrete cracking induced by the non-uniform corrosion of the steel reinforcement

W. J. Zhu & Y. D. Xu

Shanghai Key Laboratory of Rail Infrastructure Durability and System Safety, Shanghai 201804, China

Abstract

Corrosion of the steel reinforcement is considered as one of the major factors for the durability of the steel reinforced concrete (RC) elements[1,2], which challenges the serviceability and load-bearing capacity of the RC structures seriously[2]. It was reported that the wind could carry the chloride ions into the area as far as 3 km away from the sea[3]. The chloride ions will penetrate into the concrete gradually with the exposure period elapse inevitably. Depassivation happens to the steel reinforcement gradually, and then followed by the corrosion of the steel reinforcement[4]. As the volume of the corrosion products is about 2~6 times larger than that of the original steel[5], the corrosion products would result in the volumetric expansive stress to the concrete cover. Cracking would happen inevitably[6]. So it will very interesting to investigate the cracking propagation of the concrete cover induced by the chloride-induced corrosion of the steel reinforcement.

In this paper, the cracking path of the concrete cover induced by the non-uniform corrosion of the steel reinforcement will be studied. A typical model of non-uniform corrosion will be involved. The Timoshenko[7] theory will be adopted according to the hypothesis of elastic behavior of thick-walled cylinder. A new analytical model to predict the propagation of the cracking pattern of the concrete cover induced by the corrosion will be proposed.

The typical corrosion morphology of Melchers & Val's model was adopted in this investigation. Figure 1 shows the cross-sectional image of the typical corroded steel reinforcement with the corrosion age of 26 years, and the Melchers & Val's model is also included so

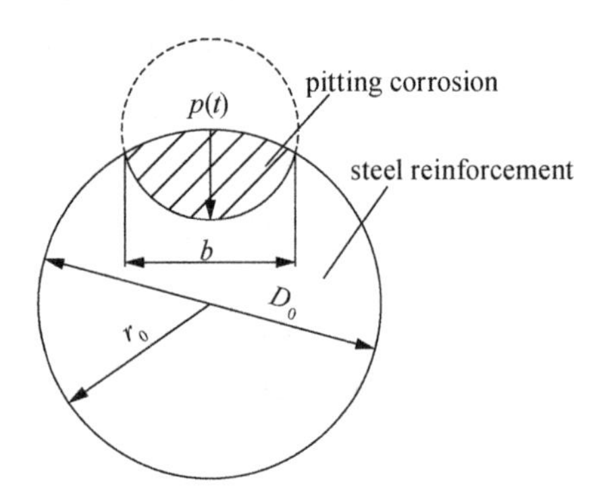

Figure 1 Melchers & Val's model

as to make the comparison. As shown in Figure 1, the pitting corrosion (non-uniform corrosion) is assumed to occur on the surface of the steel reinforcement, which was original from the top edge of the steel reinforcement. The cross-sectional loss of the pitting corrosion could be calculated as follow:

$$A_{rs} = \begin{cases} A_1 + A_2 & p(t) \leqslant \sqrt{2} r_0 \\ \pi r_0^2 - A_1 + A_2 & \sqrt{2} r_0 < p(t) < 2r_0 \\ \pi r_0^2 & p(t) \geqslant 2r_0 \end{cases} \tag{1}$$

where

$$A_1 = r_0^2 \arcsin\left[\frac{p(t)}{r_0}\sqrt{1-\left(\frac{p(t)}{2r_0}\right)^2}\right] - p(t) \cdot \sqrt{1-\left(\frac{p(t)}{2r_0}\right)^2} \cdot \left| r_0 - \frac{p(t)^2}{2r_0} \right| \tag{2}$$

$$A_2 = p(t)^2 \cdot \arcsin\left[\sqrt{1-\left(\frac{p(t)}{2r_0}\right)^2}\right] - \frac{p(t)^3}{2r_0} \cdot \sqrt{1-\left(\frac{p(t)}{2r_0}\right)^2} \tag{3}$$

where r_0 is the radius of the origin steel reinforcement; A_{rs} is the sectional loss of the pitting corrosion; $p(t)$ is the maximum depth of the pitting corrosion.

According to Zhao et al.[8], the layer of concrete filled with the corrosion products developed with the corrosion propagation of the steel reinforcement in a linear way. The depth of concrete filled with corrosion products was marked as shown in Figure 2. The corrosion layer was marked as D_{CL} and the depth of the concrete with the pores filled with the corrosion products was marked as D_{CF}.

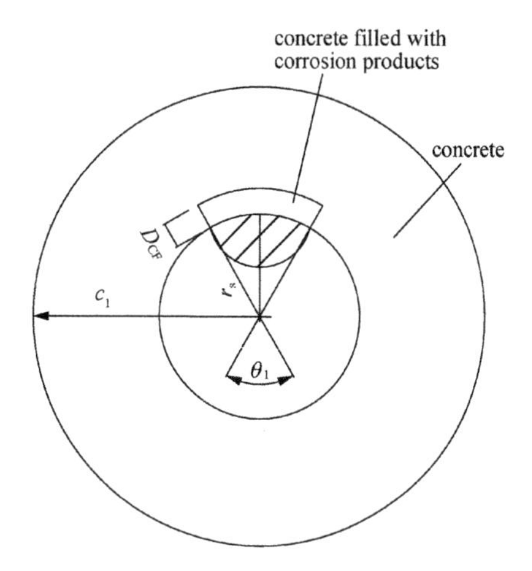

(a) Corrosion products filling with the pores in the concrete without stress

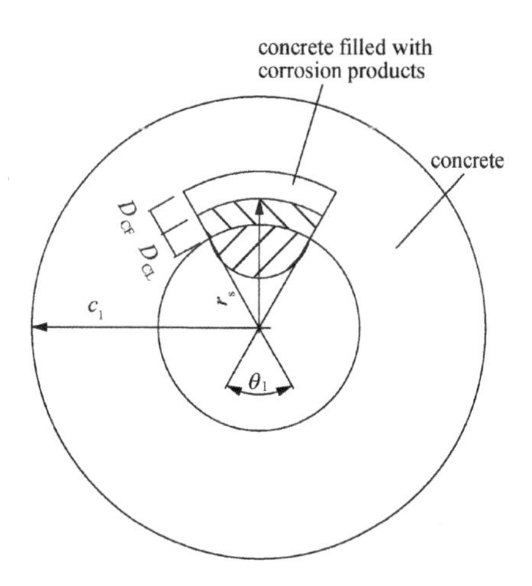

(b) Corrosion products filling with the pores in the concrete with stress

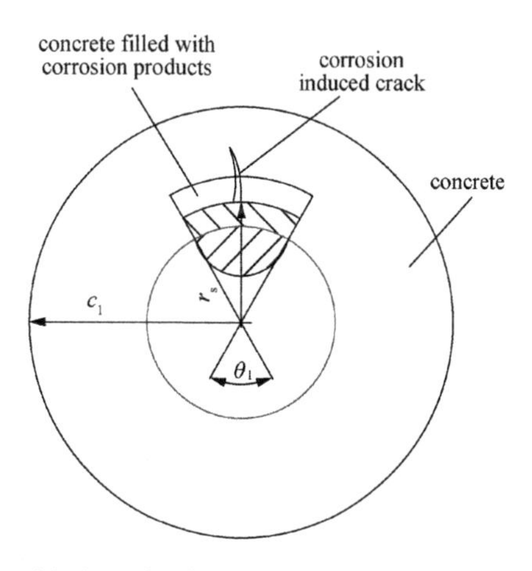

(c) Corrosion-induced crack in the concrete

Figure 2 Three stages of the corrosion-induced cracks

The stress p_{corr}^{crk} contributed by the concrete cover with corrosion-induced crack can be deduced.

$$p_{corr}^{crk} = \frac{f_{ct} \cdot \varepsilon_{cr} \cdot \theta_1 \cdot a_1 \cdot r_s}{2n \cdot w_0}\left(\frac{r_{cr}}{r_s} - 1\right)^2 + f_{ct} \cdot b_1 \cdot \left(\frac{r_{cr}}{r_s} - 1\right) \quad (4)$$

During the different cracking process, the corrosion pressure p_{corr} corresponding to the corrosion zone could be evaluated according to Timoshenko. In the hypothesis of elastic behavior of the cylinder and neglecting the Poisson effects, the pressure of a thick-walled cylinder subjected to an internal radial pressure could be deduced.

Acknowledgements

This work was financially supported by the National Natural Science Foundation of China (Grant No. 51808033).

References

[1] HAY R, OSTERTAG C P. Influence of transverse cracks and interfacial damage on corrosion of steel in concrete with and without fiber hybridization[J]. Corrosion Science, 2019, 153: 213-224.

[2] ANGST U M. Challenges and opportunities in corrosion of steel in concrete [J]. Materials and Structures, 2018, 51(1).

[3] ZHU W J, FRANÇOIS R, FANG Q, et al. Influence of long-term chloride diffusion in concrete and the resulting corrosion of reinforcement on the serviceability of RC beams [J]. Cement and Concrete Composites, 2016, 71: 144-152.

[4] LEE H S, YANG H M, SINGH J K, et al. Corrosion mitigation of steel rebars in chloride contaminated concrete pore solution using inhibitor: An electrochemical investigation[J]. Construction and Building Materials, 2018, 173: 443-451.

[5] MARCOTTE T D, HANSSON C M. Corrosion products that form on steel within cement paste[J]. Materials and Structures, 2007, 40(3): 325-340.

[6] JUNG M S, KIM K B, LEE S A, et al. Risk of chloride-induced corrosion of steel in SF concrete exposed to a chloride-bearing environment[J]. Construction and Building Materials, 2018, 166: 413-422.

[7] TIMOSHENKO S P, GOODIER J N. Theory of elasticity[M]. New York: McGraw Hill Book Co, 2007.

[8] ZHAO Y X, DONG J F, WU Y Y, et al. Corrosion-induced concrete cracking model considering corrosion product-filled paste at the concrete/steel interface[J]. Construction and Building Materials, 2016, 116: 273-280.

Application of BIM in management and maintenance of existing underpass tunnels in Chengdu

W. Y. Zhang
College of Civil and Transportation Engineering, Shenzhen University 518060, China
A. Yuan
School of Architecture and Civil Engineering, Chengdu University 610106, China
H. Z. Cui
College of Civil and Transportation Engineering, Shenzhen University 518060, China

Abstract

In the past 20 years, cities have expanded rapidly with the sharp growth of economy. To avoid the disorderly spreading of the cities to the surrounding suburbs and to meet the sustainable development of cities, the development of urban underground space has been intensified. Underpass tunnel as an important part of urban underground space can alleviate the traffic pressure of cities. Therefore, it is of great significance to improve the visual management level of underpass tunnels, and then timely monitor the damage and repair of underpass tunnels.

This research explored effective methods to combine building information modeling (BIM) with geographic information system (GIS) or three-dimensional laser scanning technology. The methods were applied to municipal infrastructure construction and post-management maintenance, which can effectively provide the solution of the following problems and the construction of smart cities. The contents and results of this study are as follows:

1. **Applying BIM + GIS to the visualized management of existing urban underpass tunnels**

BIM can fully reflect the important information of building design, site, construction and post-management. Nevertheless, it can't reflect the precise site information and environmental information around the building. GIS can fully reflect the geographic information around the building. The data sources of GIS are mainly map data, ground survey data, digital data and text reports. By combining BIM with GIS, this study established the model which can present the macro-geographic information and structural information about the tunnel.

Firstly, in this study the BIM models of the existing underpass tunnels in

Chengdu were established through the completed drawings provided by Chengdu Water Authority. By using the BIM models, the initial state of the just completed tunnel can be reflected accurately.

Secondly, the high-precision topographic data of the area where Tianfu underpass tunnel in Chengdu located by GIS software were obtained. Then using LocaSpaceViewer software to attach the surrounding street scenery to the terrain data generated by InfraWorks, and the street scenery model of Tianfu underpass tunnel was generated. At last, FBX files which collect GIS information and BIM information in InfraWorks software are linked to Navisworks platform for managers to realize cloud roaming (Figure 1).

Figure 1 Underpass tunnel and its surrounding layout

This study explored the methods of parameterization, informatization, visualization and composite design of urban underground space (Figure 2). It can be used to solve the following problems.

(1) The post-designed municipal engineering planning can accurately find the existence underground space from the cloud management platform and then avoid collide the built underground engineering.

(2) Managers can rationally distribute the underground water grid, and effectively manage the complicated water grid with space arrangement in cloud management[1].

(3) Using the information storage function of BIM, the design data of urban underground space can be stored more reasonably, clearly and conveniently, and the real parameterized city and intelligent city can be achieved.

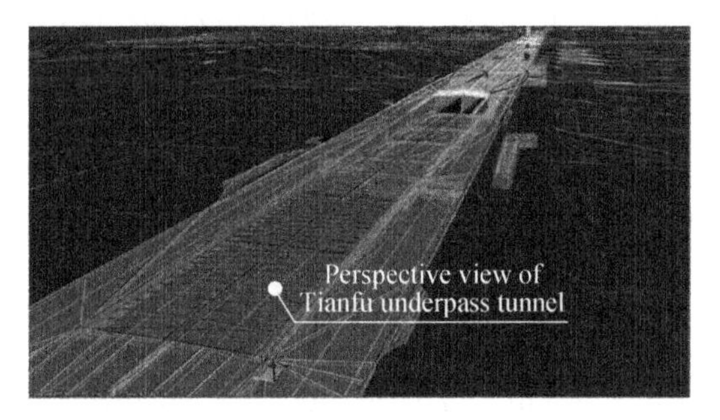

Figure 2 Visualization of urban underground space

2. Applying BIM technology and three-dimensional laser scanning technology to monitor the structural deformation of the completed tunnel

By using a three-dimensional laser scanner to scan the inner surface of the tunnel panoramically, the point cloud data on the inner surface of the tunnel were obtained[2]. By comparing the point cloud data with BIM model which is based on completed drawings, the data of internal structure deformation of the tunnel was obtained. By using the obtained data and referring to the national standard of "Technical Specification of Maintenance for Highway Tunnel" (JTG H12 – 2015), this study formulated the structural safety assessment criteria for existing urban underpass tunnels and compiled maintenance suggestions. It can be mainly used to solve the following problems.

(1) The deformation trend of tunnel structure can be predicted by comparing the point cloud data obtained by scanning the tunnel at intervals of time. However, some tunnels have been built for a long time. When the tunnel was just completed, it was not scanned by a three-dimensional laser scanner, and the initial point cloud data of the tunnel missed, so it is not conducive to deformation prediction.

This study established a BIM model of the initial state of the tunnel when it was completed, and took this model as the initial state of the tunnel, so that the structure without the data of the point cloud at the completion time could have the initial observation value. Then the BIM model was compared with the point cloud model. The data obtained from the comparison presented the deformation of the structure, so as to predict the deformation trend of the structure. This study can ensure that the structural deformation of the tunnel under the tunnel is within the safe range (Figure 3).

Figure 3 Data comparison between BIM model and point cloud model

(2) By inputting structural disease records into the BIM model, managers can easily find and record structural disease data, so as to achieve the purpose of timely recording and eliminating structural safety hazards of urban underpass tunnels (Figure 4).

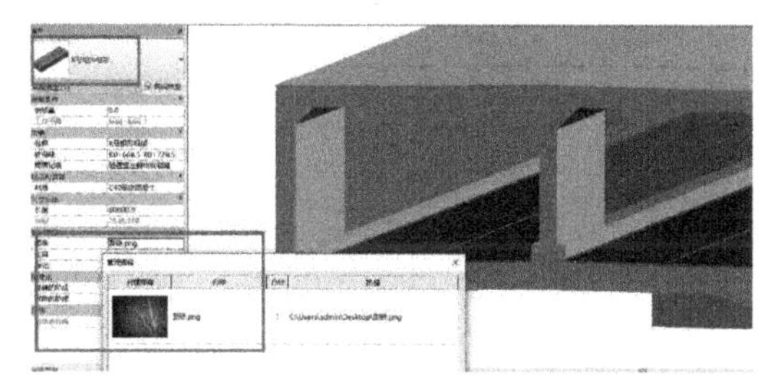

Figure 4 Recording structure detection image and location in BIM

Acknowledgements

This work was supported by the Construction Project of Safety Monitoring and Management System for Underpass Tunnel Structure in Chengdu. The author would like to express her sincere thanks for the research data provided by Chengdu Municipal Government, Chengdu Planning and Management Bureau and Chengdu Water Authority.

References

[1] COSTIN A, ADIBFAR A, HU H J, et al. Building Information Modeling (BIM) for transportation infrastructure-Literature review, applications, challenges, and recommendations [J]. Automation in Construction, 2018, 94: 257-281.

[2] XU Z Y. Research on tunnel sections extraction based on 3d laser scanning point cloud data and its application [D]. Changchun: Jilin Jianzhu University, 2018.

In situ measurements of yield stress for freshly mixed mortar

T. Y. Shin & J. H. Kim
Korea Advanced Institute of Science and Technology, Daejeon 34141, Korea

Abstract

The rheological properties of fresh mortar are directly related to concrete workability. The flowability or filling ability of concrete can be evaluated by measuring rheological properties of mortar with considering size and fraction of coarse aggregates. The previous researches proposed correlation between rheological parameters and flow test results such as mini-slump or channel flow tests. This research aims to establish yield stress equation of freshly mixed mortar based on flow table test results, which is flow test for mortar having relatively poor workability. In this study, a number of flow table tests, numerical simulations and rheological measurements using concentric vane rheometer were conducted, and Sisko & Bingham model can generalize the flow behavior of fresh mortars.

The workability of cement-based materials determines its construction performance in field applications. The workability or flowability can be generalized by rheological properties such as yield stress, viscosities. The relatively high flowable mortar's yield stress can be estimated by simple flow tests using conventional correlation models. Previous researches proposed correlation models to estimate the Bingham parameters through the mini-slump flow test. Tregger et al.[1] established yield stress equations based on theoretical and experimental approaches, and Roussel et al.[2] updated the theoretical yield stress equation with considering surface tension. Kim et al.[3] also proposed correlation equations to obtain yield stress and plastic viscosity of high flowable mortar using channel flow test, which is one of the gravity-induced flow tests for freshly mixed mortar.

This study proposed a model to correlate results of flow table tests with rheological parameters for poor workable cement mortar. The flow table tests and rheological measurements using vane rheometer of 4 test mixes were conducted, and Sisko and Bingham model quantified viscosity curves and flow curves which obtained by rheological measurements. A numerical simulation using volume of fluid technique also gived correlation of rheological parameters and flow table test results. Finally, the yield stress equations with respect to spread diameter after 5 and 25 drops, the two fitting coefficients are also proposed with high determination coefficient (R^2).

The flow table test can evaluate the workability of mortar using an impact-induced disturbance on a sample. The ASTM standards[4, 5] specify the requirements for the flow table and accessory apparatus used in making flow test for consistency of mortar in tests of hydraulic cement, which were shown in Figure 1.

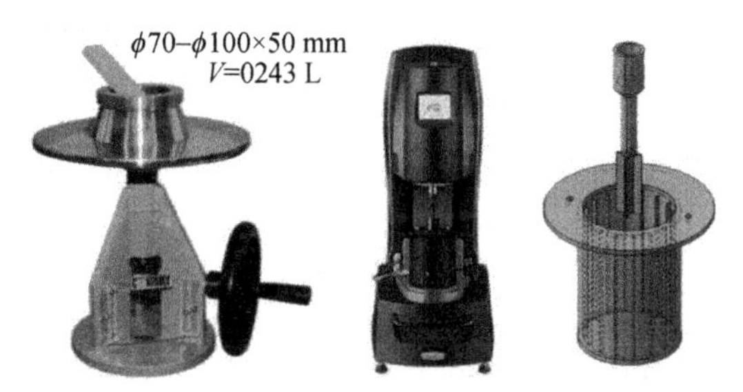

Figure 1 Flow table and rheometer

The mix proportions of a test mortar are regularized by 0.35 : 1 : 1.5 for considering the rheophysical state of SCC, and 0.5 : 1 : 3 for considering standard method of determination of cement strength (ISO679 Methods of testing cements. Determination of strength). The dosage of polycarboxylate ether (PCE) superplasticizer was set as maximum 0.1% with respect to mass of cement. The type of PCE is called methoxy polyethylene glycol (MPEG). The mix proportions are listed on Table 1.

Table 1 Mix proportions of test mortars

Label	W/C	Water /g	Cement /g	Sand /g	PCE /g
MA-PCE00 MA-PCE10	0.35	315	900	1 350	0 1.8
MB-PCE00 MB-PCE05	0.5	225	450	1 350	0 0.45

The procedure for flow table tests is (1) mixing by mechanical mixer for 3 minutes, (2) pouring a sample into a mold and tamping, (3) lifting the mold and measuring spread diameters after 5 drops and 25 drops, and repeating the procedure with interval in an hour. The maximum measurement time is 3 hours. Finally, total 16 flow table tests (4 mixes × 4 intervals) were conducted.

Rheological measurements using commercialized viscometer (Discovery HR - 1, TA Instruments) were also conducted. The concentric vane rotor was used, and its dimensions are ϕ55 mm × 85 mm of container and ϕ15 mm × 38 mm of 4-blades vane. The measurement protocol was flow sweep test which measured viscosity curves ($\dot{\gamma}-\eta$ relations) or flow curves ($\dot{\gamma}-\tau$ relations).

The data measured by rheometer can be generalized by various rheological models[6]. The cement-based material's rheological behavior was commonly generalized by Sisko model for viscosity curves and Bingham model or Herschel-Bulkley model for flow curves.

The correlation between measured yield stress and flow table test database were proposed. Figure 2 showed the experimental and simulated correlation between yield stress (τ_0) and spread diameter (D_5 or D_{25}). The common trend is that increasing rheological parameter is directly related to decreasing spread length as expected. The experimental correlation equation can be functionalized as

$$D_5 \text{ or } D_{25} = k_1 \cdot \tau_0^{k_2} + 100 \tag{1}$$

where k_1 and k_2 are fitting coefficients, and k_1 has the unit of "mm/Pa". The fitting coefficients were determined as $k_1 = 4\ 583$ mm/Pa, $k_2 = -1.216$ for D_5 with high coefficients of determination ($R^2 > 0.85$ for D_5).

The flow table tests of 4 mortar mixes were conducted, and their rheological behavior were also measured by vane rheometer. A numerical simulation using VoF technique can give us the spreadability of mortar accordance with its rheological properties. The measured rheological behavior which is represented by viscosity curves or flow curves could be generalized by fluid models, which are Sisko & Bingham fluid model. The correlations between yield stress and spread diameter obtained by flow table test were presented, and the yield stress equations and fitting coefficients were finally proposed by regression analyses with high coefficients of determination.

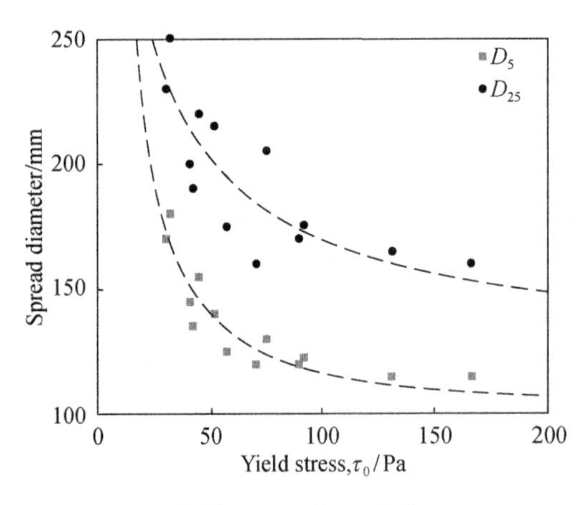

Figure 2 Yield stress-Spread diameter relations

Acknowledgements

This study was funded by Basic Science Research Program through the National Research Foundation (NRF) of Korea funded by the Ministry of Education (Grant No. NRF-2018R1D1A1B07047321).

References

[1] TREGGER N, FERRARA L, SHAH S P. Identifying viscosity of cement paste from mini-slump-flow test [J]. Aci Materials Journal, 2008, 105(6): 558-566.

[2] ROUSSEL N, COUSSOT P. "Fifty-cent rheometer" for yield stress measurements: From slump to spreading flow[J]. Journal of Rheology, 2005, 49(3): 705-718.

[3] KIM J H, LEE J H, SHIN T Y, et al. Rheological Method for Alpha Test Evaluation of Developing Superplasticizers' Performance: Channel Flow Test [J]. Advances in Materials Science and Engineering, 2017(6):1-8.

[4] ASTM. Standard specification for flow table for use in tests of hydraulic cement: C230/C230M-14[S]. ASTM International, 2014.

[5] ASTM. Standard test method for flow of hydraulic cement mortar: C1437 - 15 [S]. ASTM International, 2015.

[6] CHOI B I, KIM J H, SHIN T Y. Rheological model selection and a general model for evaluating the viscosity and microstructure of a highly-concentrated cement suspension[J]. Cement and Concrete Research, 2019, 123, 105775.

Size effect of cementitious materials on CO_2 curing

S. H. Han & Y. Jun & T. Y. Shin & J. H. Kim

Department of Civil and Environmental Engineering, Korea Advanced Institute of Science and Technology, Daejeon 34141, Korea

Abstract

A large amount of CO_2 is emitted in a cement manufacturing process. The CO_2 curing is the one of the ways to use CO_2 gas[1-3]. Reacting early age concrete with CO_2 gas in a chamber increases its strength development. The propose of this study is a quantitative measurement on the efficiency of CO_2 curing.

Table 1 shows the mix proportions of the cementitious materials. The mixtures were fabricated in two methods: (1) Compacting and (2) a general procedure to make mortar samples, following ASTM C109. The compacting method was applied for dry mix samples proportioned by a low water cement ratio (W/C): Paste (W/C = 0.15) and Mortar (W/C = 0.35). The dimension of the paste and mortar compacts was 40 mm cube. The dimensions for the samples, Paste (W/C = 0.4) and Mortar (W/C = 0.5), produced by a general procedure were 25, 40, and 50 mm cube in order to analyze the size effect of CO_2 curing[4-6]. The sealed curing for the premature sample in a mold was proceeded for 24 h at 25 ℃ approximately.

Table 1 Mix proportions of samples

Label	Fabrication method	Mix proportion/g		
		Water	Cement	Sand
Paste (W/C = 0.15)	compacting	150	1 000	0
Mortar (W/C = 0.35)	compacting	157.5	450	1 350
Paste (W/C = 0.4)	general method	400	1 000	0
Mortar (W/C = 0.5)	following ASTM C109	225	450	1 350

We considered two conditions for CO_2 curing. The first consideration for CO_2 curing is following that. The samples in a pressure vessel were subjected to 99.9% purified CO_2 gas for 3 hours. Each sample in Table 1 was subjected to 350 kPa-CO_2 curing for 3 hours, and the successive curing was followed: 21 h-moisture curing for Paste (W/C = 0.15) and Mortar (W/C = 0.35) while 28 day-water curing for Paste (W/C = 0.4) and Mortar (W/C = 0.5). Moisture curing was conducted under relative humidity (RH) of (85 ± 5)% and temperature of 25 ℃ approximately, and water curing was under 23 ℃ approximately. The other CO_2 curing was at 20% concentration of CO_2 where RH was (75 ± 5)% and temperature was 25 ℃. Each sample was placed in a controlled chamber, and the 20% CO_2 curing continued for 28 days. The control samples were also produced by moisture curing for Paste (W/C = 0.15) and Mortar (W/C = 0.35). The moisture curing was at (85 ± 5)% RH and 25 ℃. Those for Paste (W/C = 0.4) and Mortar (W/C = 0.5) took water curing at 23 ℃. The conditions for the control curing were the same with the successive curing after 350 kPa-CO_2 curing.

Internal temperature and pressure in a pressure vessel were monitored. Pressure loss in the pressure vessel (for 350 kPa-CO_2 curing) was monitored using a pressure digital gauge. The compressive strength of the samples was measured in accordance with ASTM C109.

The CO_2 pressure in the pressure vessel decreased over time, where the initial pressure was approximately 350 kPa as injected. Taking the slope at each point evaluated the carbonation rate in a unit of kPa/h. The consumed CO_2 on CO_2 curing was possibly calculated with the ideal gas equation. The carbonation rate in the pressure vessel was then given by $n = PV/RT$ in a unit of mol/h. Normalizing the carbonation rate with the cement mass required for producing the samples in the pressure vessel gave its value per cement mass in a unit of mol/(h · g). The carbonation rate was curve-fitted in a power-law function. Note that the period of 350 kPa-CO_2 curing was assigned by taking the time to get a convergence on the carbonation rate. Integrating the carbonation rate gave CO_2 uptake.

Figure 1 shows the CO_2 uptake of the other samples. Paste (W/C = 0.15) and Mortar (W/C = 0.35) had a higher CO_2 uptake compared to Paste (W/C = 0.4) and Mortar (W/C = 0.5). CO_2 uptake had the effect of specimen size. Cement paste samples show that CO_2 uptake increased with specimen size.

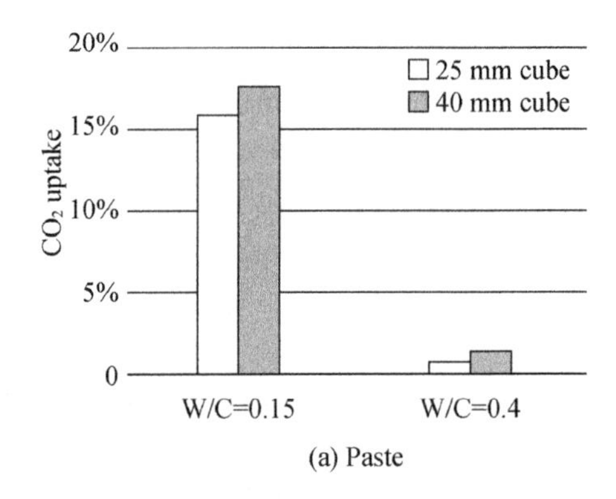

(a) Paste

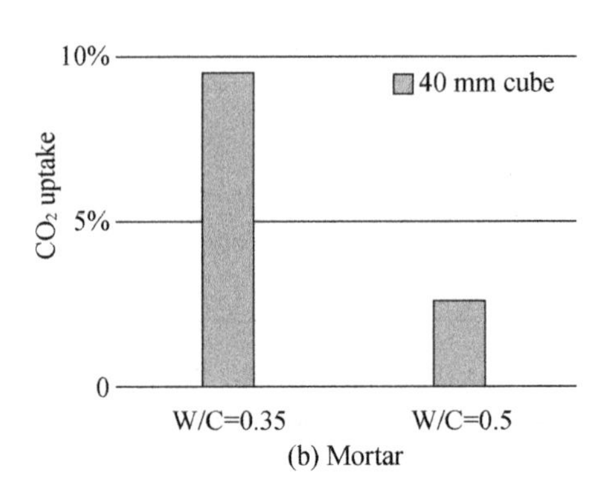

(b) Mortar

Figure 1 CO_2 uptake

Figure 2 shows the strength development of Mortar (W/C = 0.5) by size of specimen, where each trend was fitted in a hyperbolic equation. The 20% CO_2 curing provided a higher strength gains of all samples than other curing conditions.

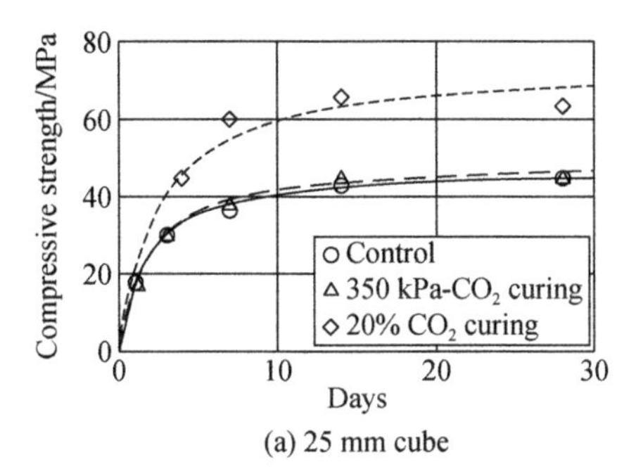

(a) 25 mm cube

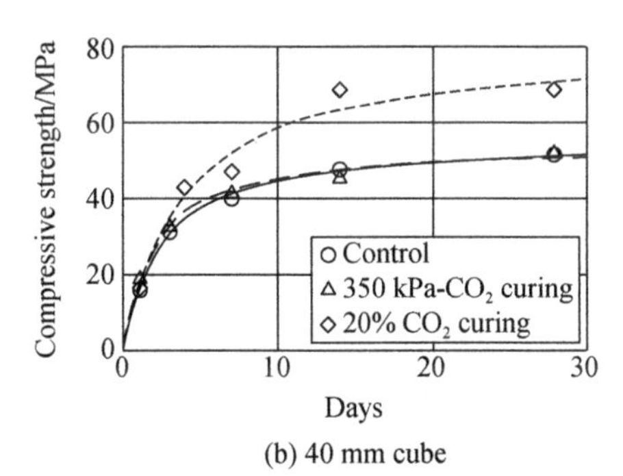

(b) 40 mm cube

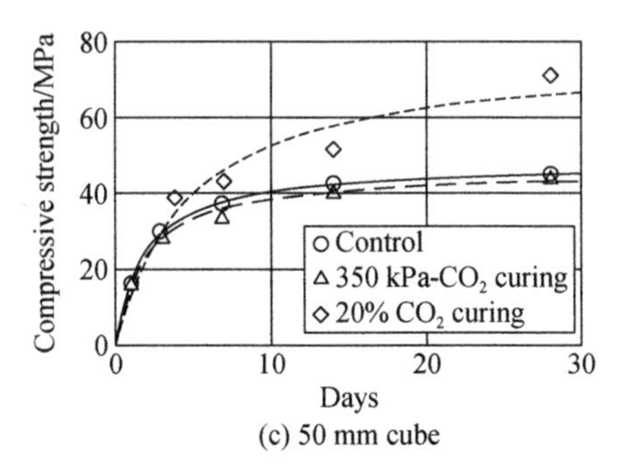

(c) 50 mm cube

Figure 2 Strength-time curve of Mortar (W/C = 0.5)

Acknowledgements

This work was supported by Korea Institute of Energy Technology Evaluation and Planning (KETEP) grant funded by the Korea government (Grant No. 20188550000580).

References

[1] MONKMAN S, SHAO Y X. Assessing the carbonation behavior of cementitious materials[J]. Journal of Materials in Civil Engineering, 2006, 18(6):768-776.

[2] AKCAY B, TASDEMIR M A. Effects of distribution of lightweight aggregates on internal curing of concrete[J]. Cement and Concrete Composites, 2010, 32(8): 611-616.

[3] SHI C J, LIU M, HE P P, et al. Factors affecting kinetics of CO_2 curing of concrete [J]. Journal of Sustainable Cement-Based Materials, 2012 1: 24-33.

[4] KIM J K. Size effect in concrete specimens with dissimilar initial cracks[J]. Magazine of Concrete Research, 1990 42(153): 233-238.

[5] BAŽANT Z P, YAVARI A. Is the cause of size effect on structural strength fractal or energetic-statistical? [J]. Engineering Fracture Mechanics, 2005, 72(1): 1-31.

[6] YI S T, YANG E I, CHOI J C. Effect of specimen sizes, specimen shapes, and placement directions on compressive strength of concrete [J]. Nuclear Engineering and Design, 2006, 236(2): 115-127.

Stress-intensity factors for a circumferential surface crack in a pipe strengthened with CFRP

Z. X. Li

Key Laboratory of Performance Evolution and Control for Engineering Structures of Ministry of Education, Tongji University, Shanghai 200092, China & Department of Structural Engineering, Tongji University, Shanghai 200092, China

T. Chen

Key Laboratory of Performance Evolution and Control for Engineering Structures of Ministry of Education, Tongji University, Shanghai 200092, China & Department of Structural Engineering, Tongji University, Shanghai 200092, China & University Corporation for Polar Research, Beijing 100875, China

Abstract

Pipes are widely utilized in many industries such as petroleum, chemical, nuclear power and pharmaceutical. The flaws of pipes in the process of manufacture and service are inevitable. Crack is a common type of defects in pipe which is also the main cause of fracture. As a result of the occurrence of leakage and fracture accidents caused by crack, the stress intensity factor (SIF) of the cracked pipe is a vital part of the design and evaluation of the pipeline system. To avoid destructive failures of cracked pipes, it is essential to accurately examine the SIF for these cracks. The SIF solutions were used in analysis for a circumferential semielliptical surface crack in a plain pipe.

Raju and Newman made a great contribution on surface cracks. They presented an empirical stress intensity factor equation[1] for a surface crack as a function of parametric angle, crack depth, crack length, plate thickness and plate width for tension and bending. The SIF were obtained from a finite-element analysis of semielliptical surface cracks in finite elastic plates subjected to tension or bending loads. Then they presented stress intensity factors[2] for a wide range of nearly semi-elliptical surface cracks in pipes and rods. The configurations were subjected to either remote tension or bending loads.

Zhao et al.[3] provided an engineering assessment of an oval-shaped clad pipes with a circumferential part-through surface crack subjected to bending moment based upon equivalent stress-strain relationship method (ESSRM) in association with EPRI J-estimation procedure. Plastic limit load equations were developed particularly to identify the equivalent stress-strain relationship of the welded clad pipe.

Hoh et al.[4] developed a methodology to simulate and calculate the stress intensity factors and weld toe magnification factors for semi-elliptical surface cracks in a circumferentially welded pipe. They considered several parameters including depth-to-thickness ($a = t$) ratios and crack shape aspect ratios ($a = c$).

Carpinteri et al.[5] investigated the influence of a circular-arc circumferential notch in a pipe. He identified the stress concentration factor and the stress-intensity factor (SIF) along the surface crack front.

On the other hand, the superior mechanical, fatigue, high strength to density ratio and in-service properties of carbon fibre reinforced polymer (CFRP) composites made them remarkable candidates for strengthening and retrofitting of steel structures. CFRP reinforcement had broad prospects for cracked pipes.

However, little research had been undertaken on SIF for circumferential semielliptical surface cracks in pipes under axial tension repaired by carbon fiber reinforced plastics (CFRP). This article focused on the numerical simulation of circumferential semielliptical surface cracks in a pipe under tension. Due to analytical results available for only limited problems with simple configurations, ABAQUS was used in this article to investigate. A meshing technique with mixed types of tetrahedron and hexahedron elements had been used to achieve simplification in modelling complex geometry of crack region. The finite element model was shown in Figure 1. Circumferential surface crack was shown in Figure 2. By applying the finite element method (FEM), the effects of different relative crack depth, tensile stress, and applying CFRP to reinforce the pipe were considered.

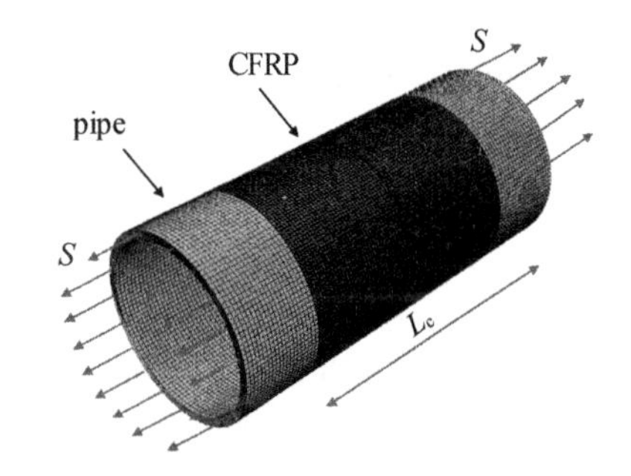

Figure 1 Finite element model

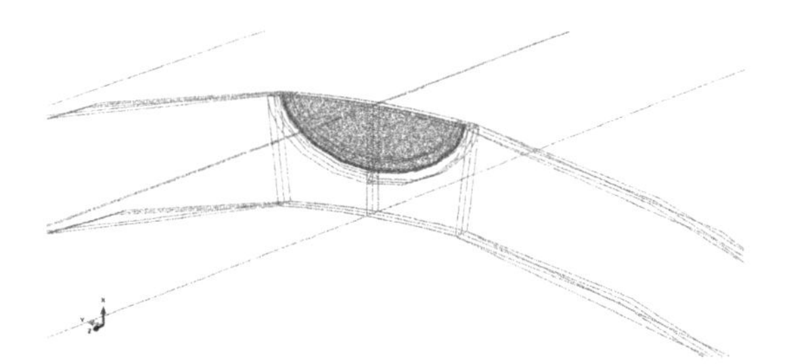

Figure 2 Circumferential surface crack

This research concentrated on the numerical simulation of circumferential surface cracks in a pipe under tension with carbon fiber reinforced plastics (CFRP) reinforcement. Reductions of SIF were surveyed with various parameters including different relative crack depth c/t, tensile stress S, CFRP length L_c and CFRP thickness T_c. The CFRP reinforcement seriously reduced SIF, and the reduction effect became more incredible with increased relative crack depth. Reinforcing the cracked region of the pipe reveals that laminates with larger thickness could be more advantageous to reinforcing a defected structure.

Acknowledgements

This project was supported by National Natural Science Foundation of China (PI) (Grant No. 51678440). This financial support is highly appreciated.

References

[1] NEWMAN Jr J C, RAJU I S. An empirical stress-intensity factor equation for the surface crack [J]. Engineering Fracture Mechanics, 1981, 15(1-2): 185-192.

[2] RAJU I S, NEWMAN Jr J C. Stress-intensity factors for circumferential surface cracks in pipes and rods under tension and bending loads[J]. Fracture Mechanics: Seventeenth Volume, 1986: 789-805.

[3] ZHAO H S, LIE S T, ZHANG Y. Fracture assessment of mismatched girth welds in oval-shaped clad pipes subjected to bending moment[J]. International Journal of Pressure Vessels and Piping, 2017: S0308016117303435.

[4] HOH H J, PANG J H L, TSANG K S. Stress intensity factors for fatigue analysis of weld toe cracks in a girth-welded pipe [J]. International Journal of Fatigue, 2016, 87: 279-287.

[5] CARPINTERI A, BRIGHENTI R, VANTADORI S. Circumferentially notched pipe with an external surface crack under complex loading[J]. International Journal of Mechanical Sciences, 2003, 45(12): 1929-1947.

Finite element analysis of the fatigue performance of central cracked steel plates strengthened by SMA/CFRP composites

M. Ye

Key Laboratory of Performance Evolution and Control for Engineering Structures of Ministry of Education, Tongji University, Shanghai 200092, China & Department of Structural Engineering, Tongji University, Shanghai 200092, China

T. Chen

Key Laboratory of Performance Evolution and Control for Engineering Structures of Ministry of Education, Tongji University, Shanghai 200092, China & Department of Structural Engineering, Tongji University, Shanghai 200092, China & University Corporation for Polar Research, Beijing 100875, China

Abstract

Steel structures are widely used in civil engineering, including steel bridges, steel structure plants and marine platforms, etc. A large number of steel structures are under cyclic loads, for instance, wind, traffic and waves. However, these cyclic loads may induce crack propagation in steel structures, which influence their fatigue performance. Thus, methods should be taken to repair these structures and prolong their fatigue life. Traditional repairing methods, such as welding and stop holes may induce secondary cracks. By contrast, repairing with carbon fiber reinforced polymer (CFRP) may avoid introducing further damage to the structure.

Scholars have studied the fatigue performance of central cracked steel plates repaired with CFRP, using double-sided repairing method. Liu et al.[1] suggested the experimental fatigue life of steel specimens can be increased to 2.2~7.9 times. Emdad et al.[2] considered the effect of prestress in CFRP, experimental results showed that fatigue life extension ratio was 4.0 ~ 35.2, generally higher than that of specimens repaired with non-prestress CFRP, which was 2.9~4.7.

Nevertheless, prestressing CFRP requires adequate operation space, which may limit its application when repairing structures in tiny spaces. When heated in specified temperature range, shape memory alloy (SMA) has a unique property that it would return to its shape before heated. Thus, by attaching prestrained SMA to structural members, prestress can be set when SMA is heated.

SMA can be easily used to prestress structural members, while CFRP has light weight, high tensile strength and high elastic modulus, fitting for structural repairing. Therefore, the author took full advantage of both materials by combining them and then studied the fatigue performance of central cracked steel plates (Figure 1) strengthened by SMA/CFRP composites (Figure 2), using finite element method (FEM).

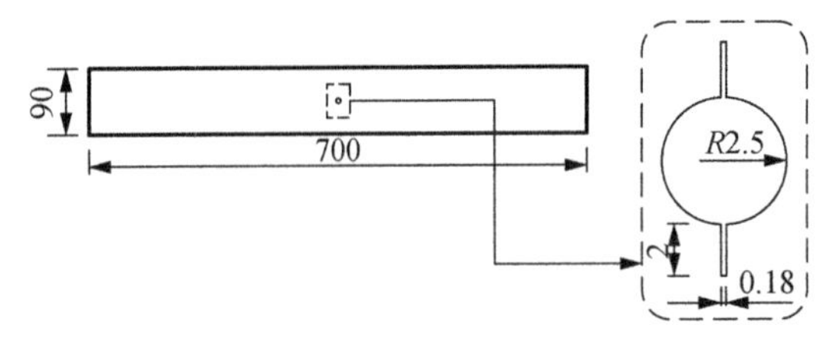

Figure 1 Unstrengthened central cracked steel plate (unit: mm)

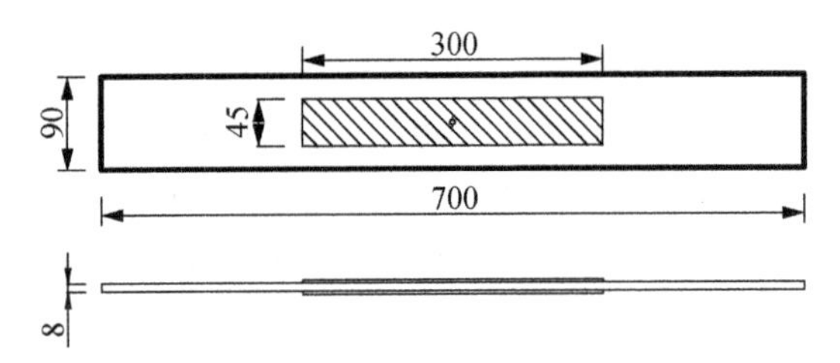

Figure 2 Strengthened central cracked steel plate (unit: mm)

In this research, the author conducted finite element analysis of several specimens strengthened by double-sided composites under tensile fatigue loading. The finite element analysis was carried out by ABAQUS and a typical model mainly comprised four kinds of materials, including steel, adhesive, CFRP and SMA (Figure 3). Three parameters were considered: (1) elastic modulus of CFRP, including normal modulus (230 GPa) and high modulus (640 GPa); (2) whether using SMA or not; (3) if using SMA, considering different prestress levels of SMA. The prestress in SMA was adjusted by cooling method.

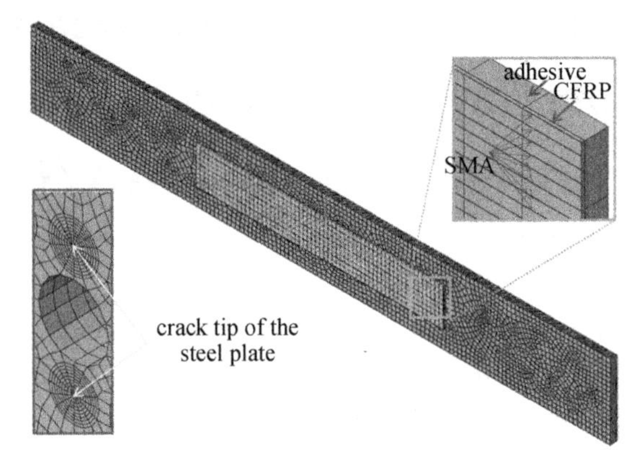

Figure 3 Finite element model

The finite element analysis results showed that using SMA/CFRP composites could reduce the stress intensity factor (SIF) in the surrounding area of the crack tip and thus prolong the fatigue life of central cracked steel plate under tensile fatigue loading. The fatigue life of each specimen was calculated based on the modified Paris model[3]. As for the specimens strengthened only by normal and high modulus CFRP, the fatigue life extension ratios were 1.5 and 2.1, respectively. When using SMA, these indexes increased to 2.9 and 3.8 respectively, and further increased as the prestress level rose (Figure 4).

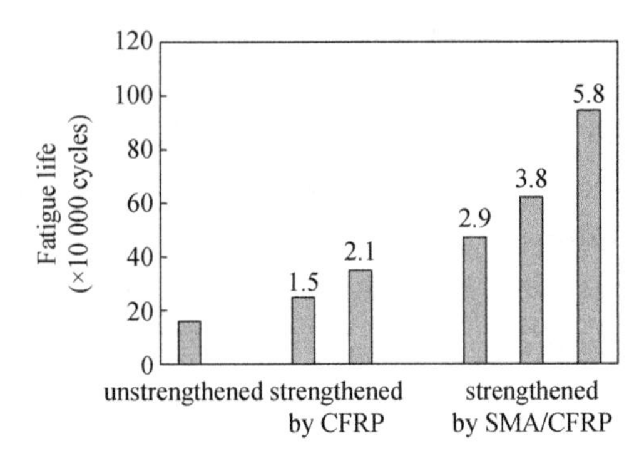

Figure 4 Fatigue life extension ratio

The following conclusions can be drawn: (1) Comparing with CFRP repairing, SMA/CFRP composites repairing can prolong more fatigue life of central cracked steel plates under tensile fatigue loading. (2) Higher prestress level in SMA can lead to longer fatigue life of central cracked steel plates strengthened by SMA/CFRP composites.

Acknowledgements

This work was financially supported by National Natural Science Foundation of China (Grant No. 51978509).

References

[1] LIU H B, AL-MAHAIDI R, ZHAO X L. Experimental study of fatigue crack growth behavior in adhesively reinforced steel structures[J]. Composite Structures, 2009, 90(1):12-20.

[2] EMDAD R, AL-MAHAIDI R. Effect of prestressed CFRP patches on crack growth of center-notched steel plates[J]. Composite Structures, 2015, 123:109-122.

[3] PARIS P, ERDOGAN F. A Critical Analysis of Crack Propagation Laws[J]. Journal of Basic Engineering, 1963, 85(4):528-533.

Porosity based design – an improved design approach for pervious concrete

H. Agrawal & S. Modhe & S. Gupta & S. Chaudhary
Discipline of Civil Engineering, Indian Institute of Technology Indore, Simrol, Indore 453552, India

Abstract

In the last few decades, pervious concrete has found heaps of application in reducing flooding risk, groundwater recharge and is an important sustainable drainage system in various parts of the world[1, 2].

Studies on pervious concrete show that void percent is an important factor for the mechanical and durability properties of pervious concrete[1, 3-5]. Porosity is primarily considered as the basis of mix proportion of pervious concrete[5]. As a common practice trial and error is used for mix proportion of pervious concrete[1]. Many alternate approaches have been suggested for mix proportion of pervious concrete[3, 4, 6]; however, they have been suggested for concrete prepared under similar conditions. While by simply varying compaction effort pore volume was found to be different for same mix proportion[4]; and this can lead to variation in expected results. In this study an alternate design approach has been suggested for mix proportion of pervious concrete.

Within the study different mix proportions have been prepared for pervious concrete and cast to de-sired degree of porosity by controlling compaction efforts. Fresh test for the same are suggested in the study for experimentally verifying the casting; this enables modification or correction in fresh state only without multiple trials.

Porosity of pervious concrete, through theoretical calculations and fresh state tests shows comparable results as compared to experimental observations in hardened state; justifying the use of new design approach (Figure 1) for the field application of pervious concrete.

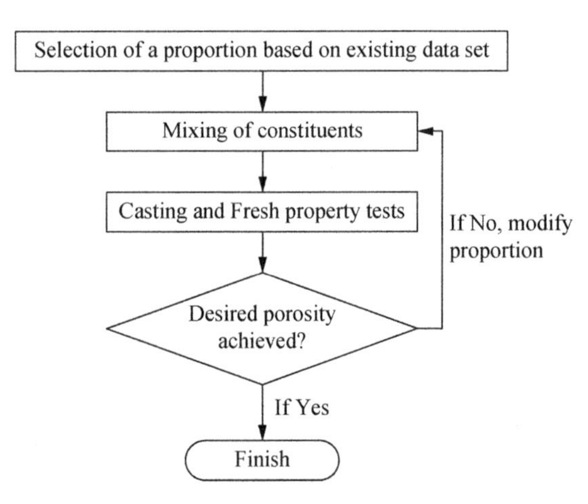

Figure 1 Flow chart of the proposed design approach for pervious concrete

Results of the study indicate a better control over proportion of pervious concrete and can be used for preparing data sets or suitable standards.

References

[1] ZHONG R, LENG Z, POON C S. Research and application of pervious concrete as a sustainable pavement material: A state-of-the-art and state-of-the-practice review[J]. Construction and Building Materials, 2018, 183: 544-553.

[2] SARTIPI M, SARTIPI F. Stormwater retention using pervious concrete pavement: Great Western Sydney case study[J]. Case Studies in Construction Materials, 2019, 11: e00274.

[3] SUMANASOORIYA M, NEITHALATH N. Pore structure features of pervious concretes proportioned for desired porosities and their performance prediction [J]. Cement and Concrete Composites, 2011, 33(8): 778-787.

[4] DEO O, NEITHALATH N. Compressive response of pervious concretes proportioned for desired porosities[J]. Construction and Building Materials, 2011, 25(11): 4181-4189.

[5] ACI COMMITTEE 522. Report on Pervious Concrete [R]. USA, American Concrete Institute: 2010.

[6] YAHIA A, KABAGIRE K D. New approach to proportion pervious concrete [J]. Construction and Building Materials, 2014, 62: 38-46.

Segregation studies on light weight aggregate concretes

S. Gandhi & S. Gupta & S. Chaudhary

Discipline of Civil Engineering, Indian Institute of Technology Indore, Simrol, Indore 453552, India

D. N. Lal

Civil Engineering Department G. B. Pant Institute of Engineering & Technology, New Delhi 110020, India

Abstract

Growing awareness on sustainability has prompted several researchers towards incorporation of waste, like shredded rubber tyres, in concrete as a replacement to aggregate[1, 2]. A major challenge observed while using light-weight aggregate (LWA), like shredded rubber tyre waste, is segregation[2]. Segregation results in a decline of mechanical and durability characteristics of concrete[3]. Segregation limits the application of light-weight aggregates (LWA) in terms of concreting[2, 4].

Segregation in concrete has been attributed to several factors which including rheological properties of the mix, density and shape of aggregate, and energy of compaction[3, 5, 6]. Energy of compaction can be broken down to its two aspects rate of compactional energy and time duration of compaction. Of the two, studies on segregation has been done by varying time duration of compaction for a constant rate of energy[5, 6]. Influence of rate of compactional energy and time duration of compaction was not discussed in the observed literature. For the same, this study experimentally analysed the dependency of segregation on time duration of compaction, rate of compactional energy and total energy of compaction for expanded polystyrene beads (EPS) as LWA.

Fresh mortar of different rheological characteristics, embedded with EPS beads as LWA, have been subjected to varying degree of impact energies for representing different rates of compactional energy and time duration of compaction. Segregation is quantified using distribution of EPS beads in the mortar matrix. Analysis showed that subjected to the rheological properties of the mix, similar total compaction energy delivered at different rate of compactional energy showed different degree of segregation (Figure 1).

Furthermore, segregation is analyzed using two separate approaches, i.e. point counting approach and image processing[5, 7]. These approaches are further explored for determining LWA shift pattern to quantify segregation. Results of this study can be applied for developing a better understanding and control mechanism for segregation.

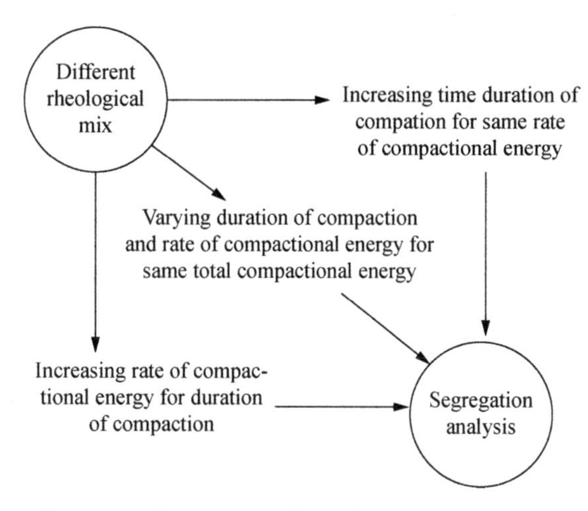

Figure 1 Experimental methodology for the study

References

[1] GUPTA T, SIDDIQUE S, SHARMA R K, et al. Behavior of waste rubber powder and hybrid rubber concrete in aggressive environment[J]. Construction and Building Materials, 2019, 217: 283-291.

[2] KIM Y J, CHOI Y W, LACHEMI M. Characteristics of self-consolidating concrete using two types of lightweight coarse aggregates[J]. Construction and Building Materials, 2010, 24(1): 11-16.

[3] GAO X, ZHANG J Y, SUE Y. Influence of vibration-induced segregation on mechanical property and chloride ion permeability of concrete with variable rheological performance[J]. Construction and Building Materials, 2019, 194:32-41.

[4] VAKHSHOURI B, NEJADI S. Mix design of light-weight self-compacting concrete[J]. Case Studies in Construction Materials, 2016, 4: 1-14.

[5] NAVARRETE I, LOPEZ M. Understanding the relation-ship between the segregation of concrete and coarse aggregate density and size[J]. Construction and Building Materials, 2017, 149: 741-748.

[6] NAVARRETE I, LOPEZ M. Estimating the segregation of concrete based on mixture design and vibratory energy[J]. Construction and Building Materials, 2016, 122: 384-390.

[7] BARBOSA F S, BEAUCOUR A-L, FARAGE M C R, et al. Image processing applied to the analysis of segregation in lightweight aggregate concretes [J]. Construction and Building Materials, 2011, 25(8): 3375-3381.

The effect of nano-silica and micro-silica on the shape stability of mortar against compressive loading

J. Lee & J. Kim

Korea Advanced Institute of Science and Technology, Daejeon 34141, Korea

Abstract

During a digital fabrication of cement based materials, each layer should support the self-weight of the upper layer, where the shape stability, workability, and thixotropy are important factors for successful fabrication[1].

The cement based materials for a digital fabrication mainly consist of water, cement binder, and fine aggregates, which equals mortar, and fine additions such as nano-silica and micro-silica are additionally used to control the shape stability, workability and thixotropy of a mortar. Nano-silica and micro-silica increase the water requirement and decrease workability due to their higher specific surface area[2, 3].

The effect of fine additions on the performance of mortar for a digital fabrication is usually evaluated based on rheological properties, which are measured by shear test using standard rheometer[4, 5]. Compressive method, which is called squeeze flow test, can be used for performance evaluation[6]. Squeeze flow test measures the resistance of samples to the compressive loading. This test shows load-displacement curves, which are generally classified into three stages: elastic, plastic, and hardening stages[7]. If the applied load is larger than yielding point, the sample starts to have cracks on their surface and fails to serve properly as a supporting material. Yielding point exists between the elastic stage and plastic (or hardening) stage, which is, together with stiffness, a critical property to describe the shape stability of mortar.

In this study, the effect of nano-silica and micro-silica on the shape stability of mortar was analyzed through a squeeze flow test. Also, their workability was measured by flow table test[8]. The behavior of mortar over time (9, 30, 60, 90, and 120 minutes after mixing start) were compared to examine the effect of thixotropy.

Type Ⅰ Portland cement was used to produce mortar samples. A standard sand specified in ISO 679[9] was used as fine aggregates. Water to cement ratio is 0.5 and fine aggregates to cement ratio is 3, which is specified mix proportion in ISO 679.

Colloidal silica was used as nano-materials, where its average particle size is 80 - 90 nm and solid content is 0.41. Colloidal silica replaced 1%, 2%, 3% and 4% of cement by weight, respectively. Micro-silica was used to replace 10%, 20% and 30% of fine aggregates by weight, respectively. Its particle size ranges from 0.1 to 0.5 mm.

Mortar samples were mixed during 5 minutes. For squeeze flow test, the same mortar samples were filled into five cylindrical molds, which has 60 mm diameter and 45 mm height. Five molded samples belong to the mortar sample over 9, 30, 60, 90 and 120 minutes after start of mixing, respectively. Each sample is placed between parallel plates and the mold is removed, where bottom plate is fixed and the upper plate moves down applying compressive load to the sample at the rate of 1.5 N/s.

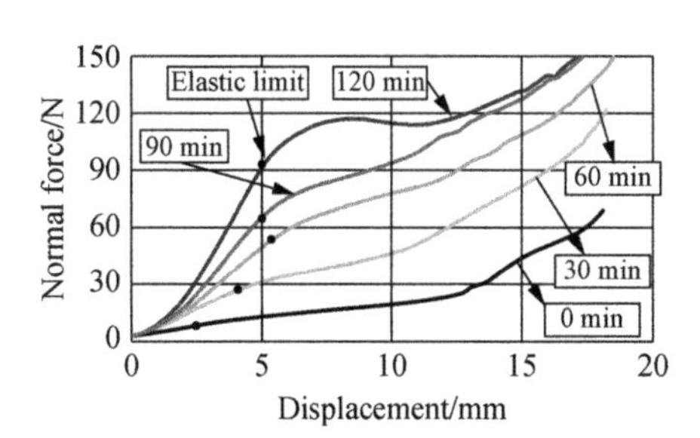

Figure 1 Load-displacement curves for mortar with 10% replacement of micro-silica

Figure 1 shows the compressive load-vertical displacement curves for mortar sample where 10% of fine aggregates are replaced by micro-silica by weight. At low displacement range, load is a linear function of displacement; stiffness is constant. At intermediate displacement range, plastic behavior is observed and hardening occurs at large displacement.

The vertical displacement was set as 0 mm, when the compressive load is 3 N. The tangent point of the load-displacement curves passing through (3 N, 0 mm) is defined as the point of elastic limit. It is assumed elastic limit serves as yielding point and represents the shape stability because the homogeneity of materials starts to break right after elastic limit. Compared to samples that have passed over time, the mortar has a higher stiffness and yield point due to thixotropy.

The workability of mortar according to the replacement of micro-silica was measured through flow table test. 25 drops of compaction were applied to each mortar sample with time. As time passed, the workability of mortar decreased.

Figure 2 and Figure 3 show a trend of yielding point and 25 drops according to the replacement of micro-silica, respectively. As the portion of micro silica increases, both yielding point and workability increases. Since micro-silica induces water requirement, the connection between every particle is enhanced, which results in a higher shape stability. In terms of workability, micro-silica fills the void between relatively larger particles and the continuity between particles and homogeneity are increased, which results in a higher workability.

The replacement of nano-silica shows similar results with the replacement of micro-silica, but the effect on workability is relatively small. This reason seems to relate to the particle size. That is, micro

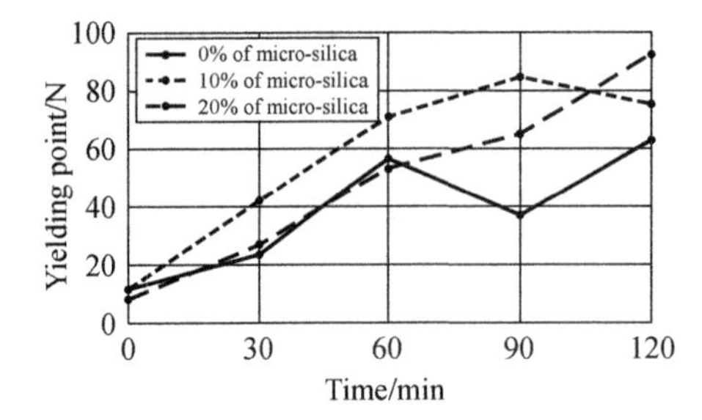

Figure 2 Trend of yielding point with the replacement of micro-silica

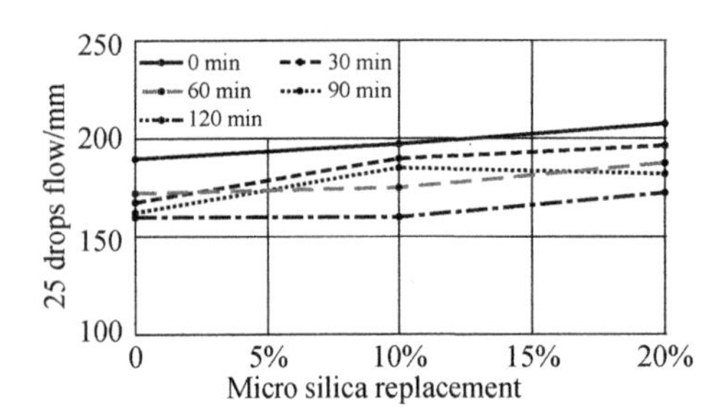

Figure 3 Trend of 25 drops of mortar with the replacement of micro-silica

sized particles contribute more to the increase of homogeneity of mortar.

Acknowledgements

This study was funded by Basic Science Research Program through the National Research Foundation (NRF) of Korea funded by the Ministry of Education, Science and Technology (Grant No. 2018R1D1A1B07047321).

References

[1] SCHUTTER G D, LESAGE K, MECHTCHERINE V, et al. Vision of 3D printing with concrete-technical, economic and environmental potentials [J]. Cement and Concrete Research, 2018, 112C: 25-36.

[2] NOCHAIYA T, WONGKEO W, CHAIPANICH A. Utilization of fly ash with silica fume and properties of Portland cement-fly ash-silica fume concrete[J]. Fuel, 2010, 89(3):768-774.

[3] SENFF L, LABRINCHA J A, FERREIRA V M, et al. Effect of nano-silica on rheology and fresh properties of cement pastes and mortars [J]. Construction & Building Materials, 2009, 23(7):2487-2491.

[4] PAUL S C, TAY Y W D, PANDA B, et al Fresh and hardened properties of 3D printable cementitious materials for building and construction [J]. Archives of civil and mechanical engineering, 2018, 18(1), 311-319.

[5] WOLFS R J M, BOS F P, SALET T A M. Early age mechanical behavior of 3D printed concrete: Numerical modelling and experimental testing [J]. Cement and Concrete Research, 2018, 106:103-116.

[6] PERROT A, RANGEARD D, PIERRE A. Structural built-up of cement-based materials used for 3D-printing extrusion techniques[J]. Materials and Structures, 2016, 49(4):1213-1220.

[7] FÁBIO A, CARDOSO F A, JOHN V M, et al. Rheological behavior of mortars under different squeezing rates [J]. Cement and Concrete Research, 2009, 39(9):748-753.

[8] ASTM Standard C230. Standard specification for flow table for use in tests of hydraulic cement [S]. ASTM International, West Conshohocken, PA, 2014.

[9] ISO 679. Cement-test methods-determination of strength [S]. International Organization for Standardization, 2009.

Analytical model for predicting corrosion-induced concrete spalling

Y. X. Zhao & Y. Z. Wang & J. F. Dong & F. Y. Gong
Department of Bridge Engineering, Zhejiang University, Hangzhou 310058, China

Abstract

Corrosion-induced concrete cover spalling represents severe deterioration of structural durability and has a considerably large impact on structural safety[1]. However, compared to corrosion-induced cracking, fewer researches focus on this field. For predicting the concrete spalling, this paper proposes an analytical model based on mechanical analysis and experimental results.

Six concrete beams (180 mm × 250 mm × 1 200 mm) were casted and the dimensions as well as the reinforcement information are indicated in Figure 1. To accelerate the steel corrosion, the six beams were subjected to sustained loading coupled with wetting-drying cycles simultaneously. During the 4-day wetting cycle, the target corrosion area was wrapped by a sponge saturated with sodium chloride solution of 3.5% concentration and the sponge was unwrapped to let the oxygen into concrete during the 10-days drying cycles, as shown in Figure 2. After test terminated, a spalling test was conducted on the beams. The external force F was applied to the beams and the amplitude of the force F was recorded by the dynameter as shown in Figure 3, then the spalling bending moment was transferred to the spalling concrete by the steel rod and steel block, which can be calculated by Eq. (1) by choosing O_F as pivot point. The crack patterns and the spalling concrete blocks were recorded subsequently as shown in Figure 4.

$$M_{Fa} = F_a L_a = \frac{F_e (L_r - L_m)}{L_m} L_a \qquad (1)$$

where M_{Fa} is the spalling bending moment; F_a is the force applied on the spalling concrete; L_a and L_m is the moment arm; F_e is the external force; L_r is the length of the rod.

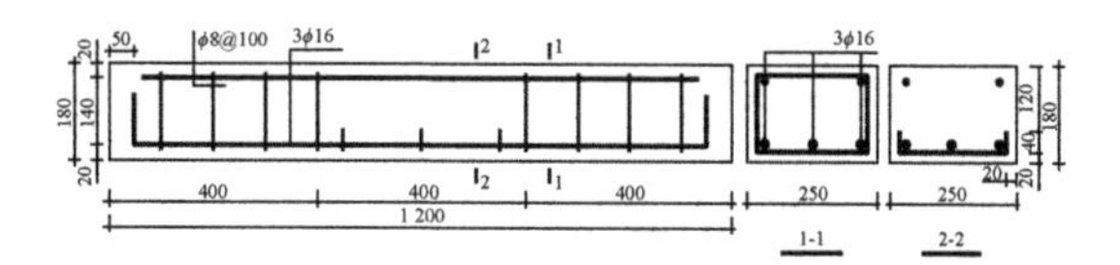

Figure 1 Sample of a figure caption (unit: mm)

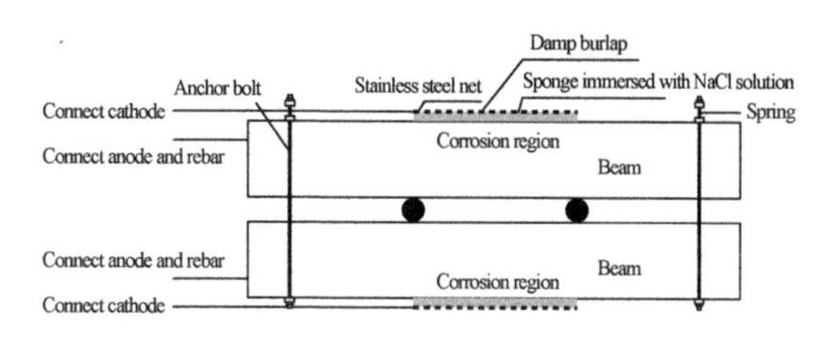

Figure 2 Accelerated corrosion method

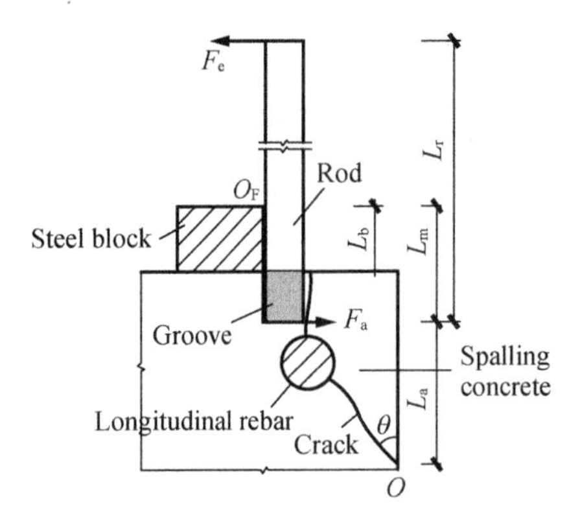

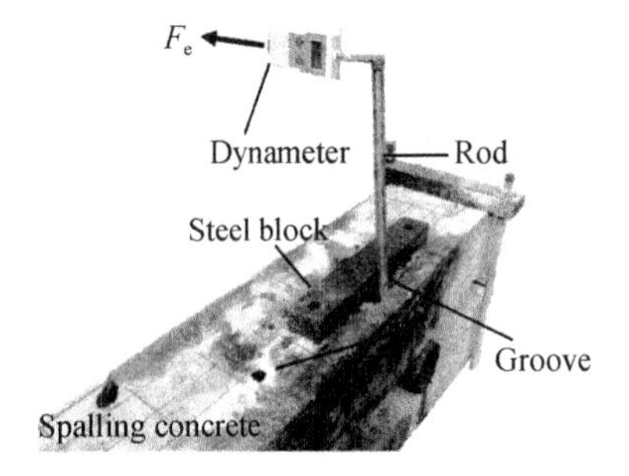

Figure 3 Spalling test setup

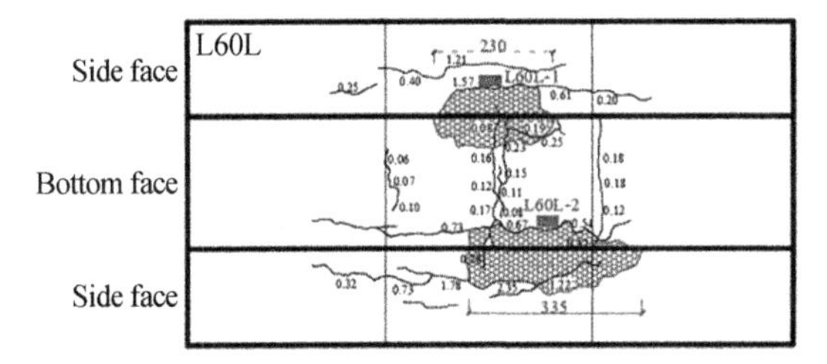

Figure 4 Example for cracking pattern and spalling concrete block

Followed by the spalling test, the geometrical criteria for concrete cover spalling was discussed, the mechanical interaction between the spalling concrete block and concrete beams were analyzed in order to establish the simplified spalling model as shown in Figure 5. The cohesive force in concrete, the cohesive force at the steel-concrete interface and the aggregate interlocking force were considered as the resisting force against spalling. The critical state for concrete spalling is defined as that, when the spalling bending moment M_{Fa} contributed by the external force exceeds the resisting bending moment M_{Con} contributed by the resisting force, the spalling concrete block was separated from the main body of the concrete beams. All the bending moments were calculated by choosing OO′ as the axis and the equation can be described as follows:

$$M_{Fa} = M_{Con} = M(t_{ni}) + M(t_{ns}) + M(t_{nt}) + M(V_{ci}) \quad (2)$$

where $M(t_{ni})$ is the spalling bending moment contributed by the cohesive force at steel-concrete interface t_{ni}, $M(t_{ns})$ is the spalling bending moment contributed by the cohesive force on face CC′OO′ t_{ns}, $M(t_{nt})$ is the spalling bending moment contributed by the cohesive force on face AA′DD′ t_{nt}, and $M(V_{ci})$ is the spalling bending moment contributed by the aggregate interlocking force on the side of the spalling concrete block V_{ci}.

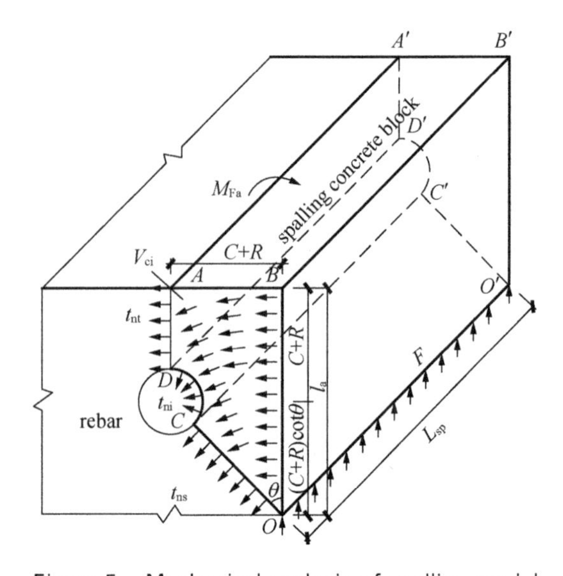

Figure 5 Mechanical analysis of spalling model

The bending moment $M(t_{ni})$ can be calculated based on the corrosion layer thickness[2], $M(t_{ns})$ and $M(t_{nt})$ can be calculated based on the longitudinal crack widths[3] and $M(V_{ci})$ can be calculated based on the transverse crack widths[4]. The predicted value, resisting bending moment M_{Con} and the tested value, spalling bending moment M_{Fa} are shown in Table 1 and a satisfied agreement is reached. It can be concluded that the proposed model predict the corrosion-induced spalling accurately.

Table 1 Comparison of predicted and tested values

Concrete block number	M_{Fa} /Nm	M_{Con} /Nm	M_{Con} / M_{Fa}
1	253.4	201.73	0.80
2	199.6	285.53	1.43
3	288.2	322.63	1.12
4	194.7	204.45	1.05
5	229.7	216.14	0.94
6	246.4	191.51	0.78
7	281.7	197.29	0.70
8	206.5	199.07	0.96
9	182.2	186.78	1.03

Acknowledgements

The financial support received from the National Key Research and Development Program of China (Grant No. 2017YFC0806101 - 02) is gratefully acknowledged.

References

[1] Duracrete. General guidelines for durability design and redesign. Final Technical Report [S]. The European Union-Brite EuRam III, Probabilistic performance based durability design of concrete structures, document B 95-1347/R15, 2000.

[2] CAIRNS J, DU Y, LAW D. Influence of corrosion on the friction characteristics of the steel/concrete interface[J]. Construction & Building Materials, 2007, 21(1):190-197.

[3] LI L D. Application of cohesive traction crack model in crack propagation in concrete [D]. Wuhan: Wuhan University of Science and Technology, 2015.

[4] American Association of State Highway and Transportation Officials. AASHTO LRFD bridge design specifications[S]. 5th Edition. Washington, D.C., 2010.

Parametrical analysis of stress and crack development in concrete due to ASR and DEF based on a discrete model

Y. Wang & P. Jiradilok & K. Nagai
The University of Tokyo, Tokyo 113-8656, Japan

Abstract

Alkali-silica reaction (ASR) and delayed ettringite formation (DEF) are both serious durability problems for concrete structures and surface map cracking is their common feature. The problems have been reported in several countries for the respective damage types and they were also observed occur simultaneously onsite with the effect of interaction. For ASR, the damage mechanism is due to the formation of ASR gel from the reaction of siliceous minerals in aggregates and alkali ions in cement. The damage is usually initiated at the aggregate particles and interfacial transition zone (ITZ). On the other hand, DEF is caused by the elevated temperature curing (normally over 70 ℃) of concrete. Under the high temperature, the formed ettringite will be decomposition and delayed formed after the temperature cooling down. The delayed ettringite formation can cause destructive pressure and cracks after the concrete is hardened. Both types of damage can cause the cracking of concrete inside, which is the reason of significant reduction in concrete strength. Due to the localization of expansion inside, the map cracking pattern is observed at the surface of concrete. However, the stress development with expansion inducing the initiation and propagation of cracks remains unclear since the cracks distribution inside is difficult to detect by experimentation. Also, the quantification of each expansion effect is hard to achieve by setting parameters, as well as the mechanical property evaluation of concrete under a couple of ASR and DEF effects[1].

Since the cracking is of great importance to understanding of the mechanical degradation and hardly to be obtained by experimentation, it is wise to choose a computer-aided tool to simulate the cracking behavior of concrete. In our previous studies, we found that the rigid body spring model (RBSM), a discrete analysis method, has large advantages in simulating the small deformation and cracks. The concept of RBSM was first proposed by Kawai[2]. And then Nagai and his coworkers made their efforts to develop the model from 2D to 3D and to simulate the mechanical performance from concrete materials to complicated reinforced concrete (RC) structures[3]. It has been successfully used to study the frost damage and fatigue damage of concrete, the corrosion of rebar induced bond deterioration and the anchorage effect on beam-column joints and so on. After almost two decades of development, the RBSM can simulate the behavior of RC structures in three-dimensional and visualize the stress and deformation performance. But for the complicated durability problems, the cracking nature has not been paid much attention yet. To achieve a unified model, the RBSM can be further developed to study more comprehensive durability problems and to be a very useful tool for lifetime evaluation.

Therefore, in this study, the stress and crack development due to ASR and DEF are investigated based on a discrete analysis by using the RBSM. The initial strains are applied to the springs of ITZ or interface of mortar elements to introduce the ASR and DEF damage, respectively. Also, regarding the localization expansion in each damage, the percentage of reactive coarse aggregate and intensified expansion area are used for ASR and DEF simulation, respectively. Then the parametrical study on a coupled effect of ASR and DEF are conducted. As shown in Figure 1, the surface cracking behavior of specimens under the combined effects can be obtained. After the ASR/DEF damage, the compressive loading is simulated, the internal stress and cracking

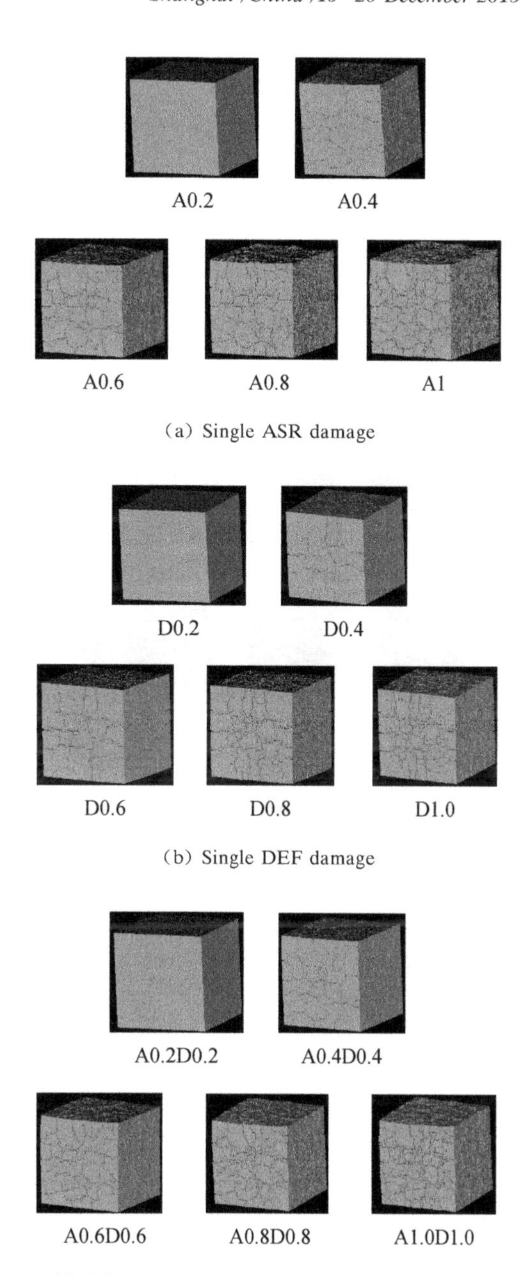

(a) Single ASR damage

(b) Single DEF damage

(c) ASR and DEF damage occurs simultaneously

Figure 1 Surface cracking pattern of concrete damaged by (a) ASR (b) DEF (c) ASR + DEF

Note: A-ASR expansion, D-DEF expansion, Number-the expansion ratio for the certain case.

development of damaged concrete are also recorded, as presented in Figure 2. Besides, the confinement effect of the respective factor can be found from the

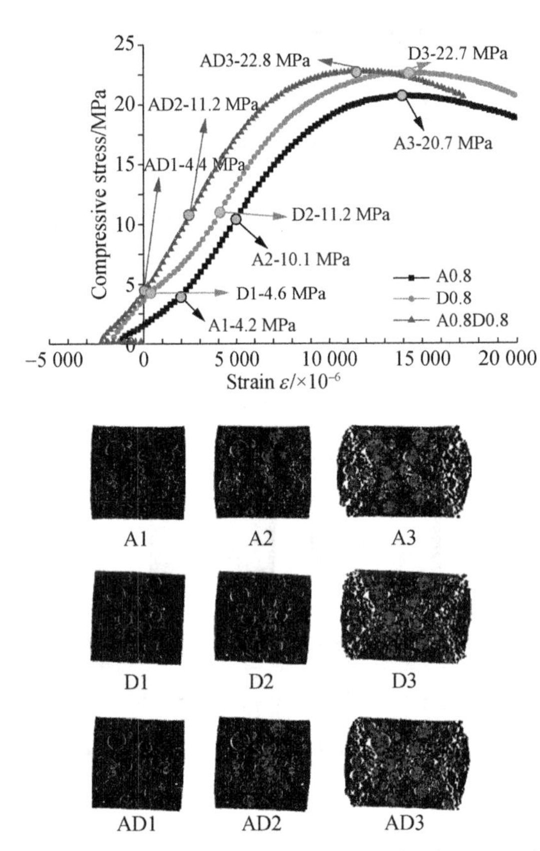

Figure 2 Internal stress and cracking development of ASR, DEF, and ASR + DEF damaged concrete (middle cross-section) under compressive loading (the expansion of A0.8D0.8 is 1%)

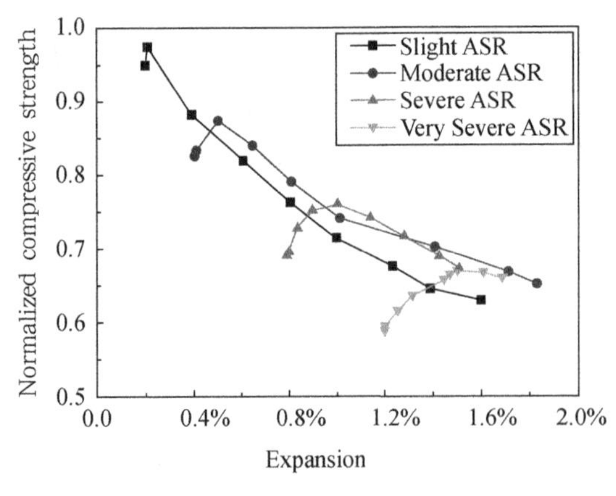

Figure 3 Normalized compressive strength change with expansion for concrete specimen with a coupled effect of ASR and DEF

compressive strength reduction tendency with consideration of different degree of ASR damage, as can be observed in Figure 3. The simulated results are significant to the accurate damage evaluation and clarification of the difference in reduction of mechanical properties of concrete suffered from different durability problems.

Acknowledgements

The first author would like to express his sincere thanks to Japan Society for the Promotion of Science (JSPS) for providing fellowship (P18348) to his postdoctoral study.

References

[1] WANG Y, MENG Y S, JIRADILOK P, et al. Expansive cracking and compressive failure simulations of ASR and DEF damaged concrete using a mesoscale discrete model[J]. Cement and Concrete Composites, 2019, 104: 103404.

[2] KAWAI T. New discrete models and their application to seismic response analysis of structure [J]. Nuclear Engineering and Design, 1978, 48(1): 207-229.

[3] EDDY L, NAGAI K. Numerical simulation of beam-column knee joints with mechanical anchorages by 3D rigid body spring model [J]. Engineering structures, 2016, 126: 547-558.

Study on interfacial moisture transport properties of recycled concrete modified by Nano-materials

X. B. Song & C. Z. Li & D. D. Chen

Department of Structural Engineering, College of Civil Engineering, Tongji University, Shanghai 200092, China

Abstract

In this work, effects of nano-SiO_2 (NS) and carbon nanotubes (CNT) addition on interfacial moisture transport properties of recycled concrete were evaluated. The results indicate that adding NS improved resistance to moisture transport due to its filling effect, of which the transport coefficient decreased by up to 32.0%. While incorporating CNT exerted a negative effect with a significant increase in coefficient could be found at all dosages. This could be explained by the bridge effect, which increased the overall connectivity within the matrix.

Recycled concrete can meet the needs for sustainable and environment-friendly development in construction industry. Interfacial transition zone (ITZ) plays a decisive role in determining the performance of the recycled concrete. Nano-materials are used as a filler in recycled concrete to ameliorate its interfacial weak zone, which proved to be highly beneficial to mechanical properties, durability and microstructure[1-3]. Du et al.[4] demonstrated that adding 0.9 wt% NS could contribute to densification of concrete pore structure by reducing large capillary pores (greater than 50 nm) by about 15% and increasing medium capillary pores (10~50 nm) by about 32% compared with the reference sample. While few literature has been published concerning the interfacial moisture transport property, which is responsible for deterioration of concrete durability. In this work, effects of nano-SiO_2 (NS) and carbon nanotubes (CNT) addition on interfacial moisture transport properties of recycled concrete were evaluated.

Ordinary Portland cement type 42.5R which conforms to Chinese standard GB 175—2007 was employed as the binder material. NS was incorporated into concrete at a dosage ranging from 0.2 wt% to 3 wt% and CNT at a dosage ranging from 0.05 wt% to 1 wt%. A five-phase parallel model of recycled concrete was proposed to measure the transport coefficient of ITZ, as shown in Figure 1. Water transport test was performed according to "Standard Test Methods for

Water Vapor Transmission of Materials"[5]. Average moisture transport coefficient of new ITZ can be determined using Eq. (1)

$$D_{NI} = \frac{D_R - D_{OM}V_{OM} - D_{OI}V_{OI} - D_{NM}V_{NM}}{V_{NI}} \tag{1}$$

where D_{OM}, D_{OI}, D_{NM}, D_{NI} represent transport coefficient of old mortar, old ITZ, new mortar and new ITZ, respectively. V_{OM}, V_{OI}, V_{NM}, V_{NI} stand for volume fraction of old mortar, old ITZ, new mortar and new ITZ, respectively.

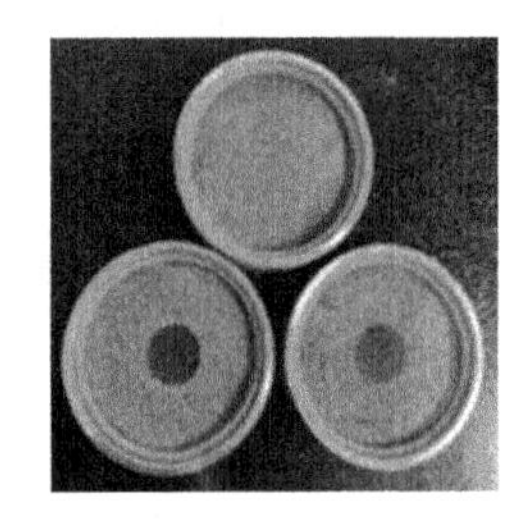

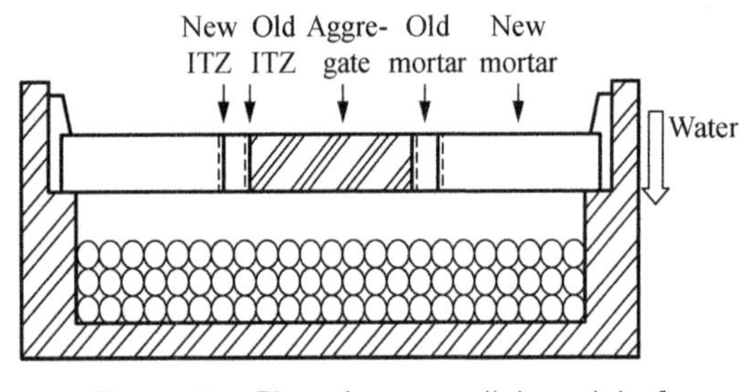

Figure 1 Five-phase parallel model of recycled concrete

Figure 2 shows typical moisture absorption plot of old mortar (OM) as an example. Initially, the mass of the specimens varied nonlinearly with the moisture absorption time. Subsequently, the weight gain curve tended to be linear, indicating that the transport process had reached a steady state at this time. The mass change rate can be characterized by

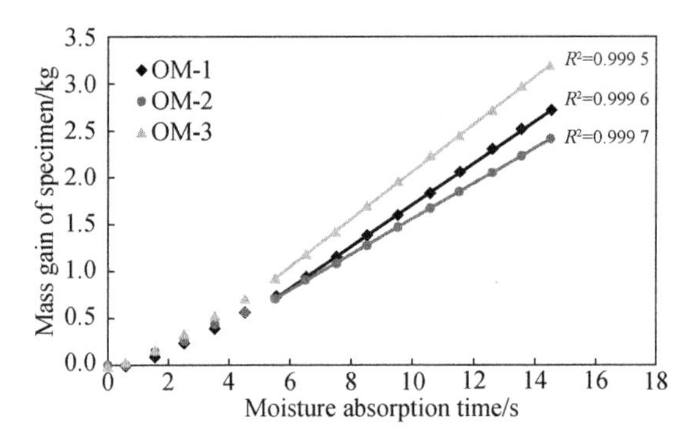

Figure 2 Typical moisture absorption plot with linear regression of old mortar

the slope (k) obtained by linear fitting of 10 data points in this stage.

Moisture transport coefficients (relative value) of new ITZ of recycled concrete modified by NS and CNT are graphically presented in Figure 3. It can be found that samples with NS addition exhibited a lower transport coefficient compared with control sample. Adding 3 wt% NS exhibited an optimum effect regarding water absorption resistance, of which the coefficient decreased by 32.0%. The decreased coefficient probably arise from the filling effect of NS particles, which leads to densification of the microstructure and refinement of pore size distribution, especially in the interface transition zone, as a result, NS-added concrete samples exhibited a better resistance to water migration. The decreased coefficient can also be attributed to the reduced pore volume as well as more disconnected and tortuous transport channels within the matrix. Li et al.[6] indicated that the tortuosity factor may have a more significant effect on concrete permeability than porosity when the porosity is in excess of the percolation

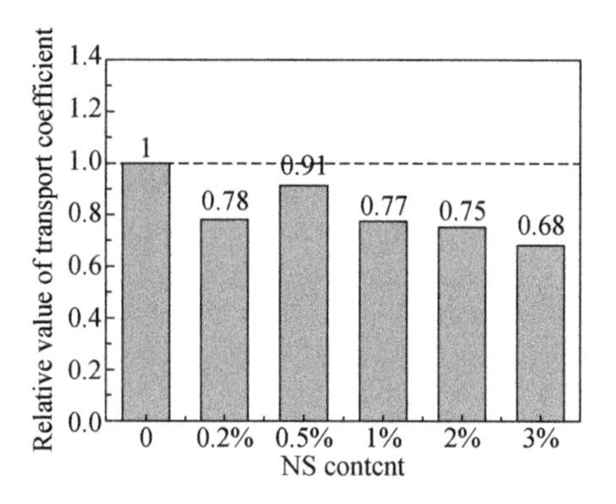

(a) With NS addition

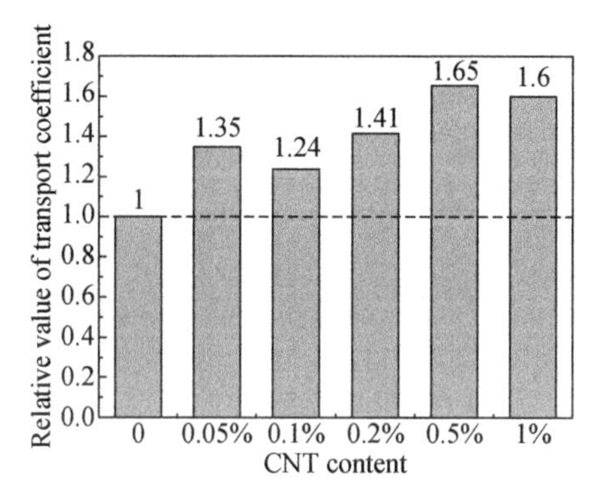

(b) With CNT addition

Figure 3 Moisture transport coefficient of new ITZ of recycled concrete modified by NS and CNT

threshold, above which all capillary pores in the matrix are accessible to the transport process.

It can be also found from Figure 3 that CNT modified recycled concrete exhibited a higher moisture transport coefficient than control sample at all dosages, suggesting a weaker resistance to moisture transport. This could be related to the bridging effect of CNT, which enhanced the overall connectivity of the matrix. Rhee et al.[7] stated that recycled concrete modified by carbon nanotubes presented a more loose interface transition structure, of which the bonding capability between new and old mortars was weak, thereby leading to more larger capillary pores.

Acknowledgements

This study was financially supported by the National Basic Research Program of China (973 Program) (Grant No. 2015CB057703).

References

[1] YING J, ZHOU B, XIAO J. Pore structure and chloride diffusivity of recycled aggregate concrete with nano-SiO_2 and nano-TiO_2 [J]. Construction and Building Materials, 2017, 150: 49-55.

[2] ETXEBERRIA M, VAZQUEZ E, MARI A R, et al. Influence of amount of recycled coarse aggregates and production process on properties of recycled aggregate concrete[J]. Cement and Concrete Research, 2007, 37 (5): 735-742.

[3] XIAO J Z, LI J B, ZHANG C. Mechanical properties of recycled aggregate concrete under uniaxial loading [J]. Cement and Concrete Research, 2005, 35(6): 1187-1194.

[4] DU H J, DU S H, LIU, X M. Durability performances of concrete with nano-silica [J]. Construction and Building Materials, 2014, 73: 705-712.

[5] ASTM. Standard Test Methods for Water Vapor Transmission of Materials: E96/E96M-16[S]. 1996.

[6] LI C Z, JIANG L H, XU N, et al. Pore structure and permeability of concrete with high volume of limestone powder addition [J]. Powder Technology, 2018, 338: 416-424.

[7] RHEE I, ROH Y S. Properties of normal-strength concrete and mortar with multi-walled carbon nanotubes [J]. Magazine of Concrete Research, 2013, 65(16): 951-961.

Behavior and design of cold-formed steel storage rack uprights under localised fires

C. Ren & P. Zhang

Department of Civil Engineering, Shanghai University, Shanghai 200444, China

Abstract

Cold-formed steel has been increasingly used in storage rack upright systems because of its high specific strength and economy. This paper presents a numerical investigation into the axial strength of steel storage rack uprights exposed to a localised fire (the fire source is at the bottom of member) shown as Figure 1. Finite element analyses (FEA), considering perforations and complex cross-sections, were carried out to investigate the heat transfer and buckling behavior. Firstly, the heat translation analysis was carried out to calculate temperature distributions along the length of members. Based on the temperature field, the linear perturbation analysis was performed to investigate eigenvalue buckling analyses in which elastic critical buckling loads and buckling modes were obtained. Finally, the Riks method was employed in nonlinear analysis to predict the ultimate loads and the failure modes of steel storage rack uprights exposed to localised fires. In addition, modified direct strength method (DSM) curves were reported in this paper for predicting the axial strength of cold-formed storage rack uprights under localised fires.

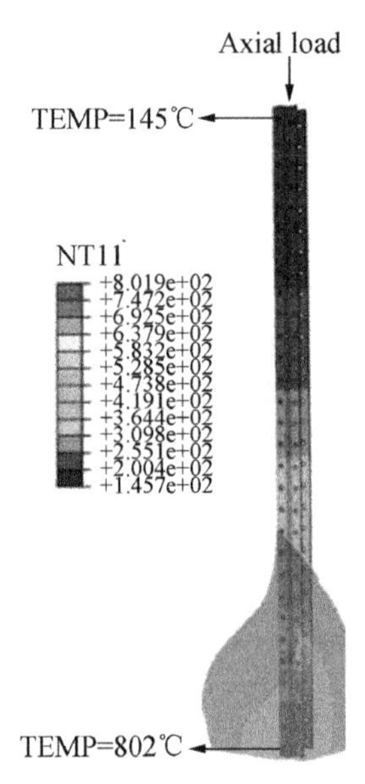

Figure 1 Temperature gradient of uprights under localised fires

Current structural fire design methods are based on international organization for standardization (ISO) 834 fire, where the temperature of gas and steel members exposed in fires are uniform. However, it was highly probable in large compartment that compression members were exposed to non-uniform heat flux in real fires[1]. The temperature gradient of structural components has important negative influences on buckling behaviors, which had been reported by Zhang et al.[2]

Therefore, the temperature gradient along the length of uprights was considered in analytical models to consider a localised fire. According to EN 1991 - 1 - 2[3], the flame length, the diameter of the fire and the rate of heat release were considered in the finite element models. The material properties and thermal performance of cold-formed steel at elevated temperatures were based on EN 1993 - 1 - 2[4]. The results of verification between the numerical results and existing experimental data show that the FEA have good performance for heat transfer[5], elastic critical buckling loads[6] and ultimate loads[7] of cold-formed compression members.

This study employed verified FEA to investigate the buckling behavior of 280 models with types of cross-sections, multiple lengths, different thicknesses and various temperatures (20 ℃, 200 ℃, 400 ℃, 600 ℃ and 800 ℃) aimed at assessing the performance of the current DSM for such uprights under localised fires. Detailed comparisons of ultimate loads and the failure modes between room temperature and elevated temperature were demonstrated in this paper. The results showed that failure modes of uprights were significantly influenced by temperature gradient. Comparison of axial strengths predicted from the current DSM and FEA, this paper reported the predictive equations modified from the current DSM for cold-formed steel uprights exposed to localised fires shown as Figure 2 for 800 ℃.

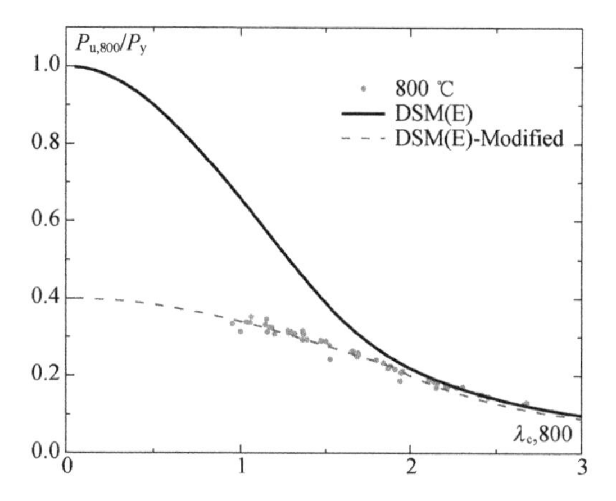

Figure 2 Comparison of the current DSM global curve and modified curve

References

[1] ZHANG C, GROSS J L, MCALLISTER T P, et al. Behavior of Unrestrained and Restrained Bare Steel Columns Subjected to Localized Fire [J]. Journal of Structural Engineering, 2014, 141(10): 04014239.

[2] ZHANG C, Li G Q, USMANI A. Simulating the behavior of restrained steel beams to flame impingement from localized-fires[J]. Journal of Constructional Steel Research, 2013, 83: 156-165.

[3] EN. 1991-1-2: General actions—Actions on structures exposed to fire [S]. Belgium, Brussels: European Committee for Standardization, 2002.

[4] EN. 1993-1-2: General actions—Structural fire design[S]. Belgium, Brussels: European Committee for Standardization, 2002.

[5] CRAVEIRO H D, RODRIGUES J P C, LAÍM L. Cold-formed steel columns made with open cross-sections subjected to fire[J]. Thin-Walled Structures, 2014, 85: 1-14.

[6] KWON Y B, HANCOCK G J. Tests of cold-formed channels with local and distortional buckling [J]. Journal of structural engineering, 1992, 118(7): 1786-1803.

[7] ZHAO X, REN C, QIN R. An experimental investigation into perforated and non-perforated steel storage rack uprights[J]. Thin-Walled Structures, 2017, 112: 159-172.

Numerical study of deformation behavior of FRP-confined reinforced concrete columns under earthquake

P. Gao & T. Y. Wang & Y. H. Zhao & D. Q. Sun

School of Civil Engineering, Hefei University of Technology, Hefei 230009, China

Abstract

In order to realize multi-objective seismic rehabilitation of existing reinforced concrete structures based on displacement performance requirements, this paper studied the seismic behavior of reinforced concrete columns confined with fiber reinforced polymer (FRP). The failure process of reinforced concrete (RC) column confined by FRP under low-cycle repeated lateral forces was simulated by using the finite element software. The concrete damaged plasticity (CDP) material model was adopted in the finite element model. The numerical specimens failure modes and load-displacement curve were verified by the tested results. After that, the influences of the number of FRP layers, the axial compression ratio and the shear-to-span ratio of columns on the failure mechanism and the mechanical behavior of the confined columns were studied. Especially, the relation of the plastic hinge length of the RC columns was carefully analyzed.

Fiber reinforced polymer is widely used in strengthening concrete structure. FRP-confinement significantly improves the seismic strength and ductility of concrete columns[1]. Especially, the deformation ability can be significantly improved. It was reported that the number of reinforcement layers[2], the concrete strength of the component, the axial compression ratio and the loading history in the experiments[3] had significant effects on the deformation capacity of the constrained column. In this paper, the seismic behavior of confined RC columns with different parameters was studied by finite element method.

If selected two typical tested columns from Li's experiments[4] and established model for simulation. One was a control and the other was a strengthened column. The compressive strength of 150 mm × 150 mm × 150mm cubes was 20.1 MPa. The yield strength of the longitudinal bars and transverse bars was 400 MPa and 300 MPa, respectively. The elastic modulus of the CFRP laminate is 210 GPa, and its nominal thickness was 0.111 mm. The three-dimensional finite model setup by ABAQUS[5] was shown in Figure 1. The solid element was used for concrete, the truss element was employed for all the longitudinal and transverse reinforcement bars, and the membrane element was chosen for CFRP. The CDP model was selected for concrete material. The accuracy of the model was verified by the test fact, then the parameter numerical study was done.

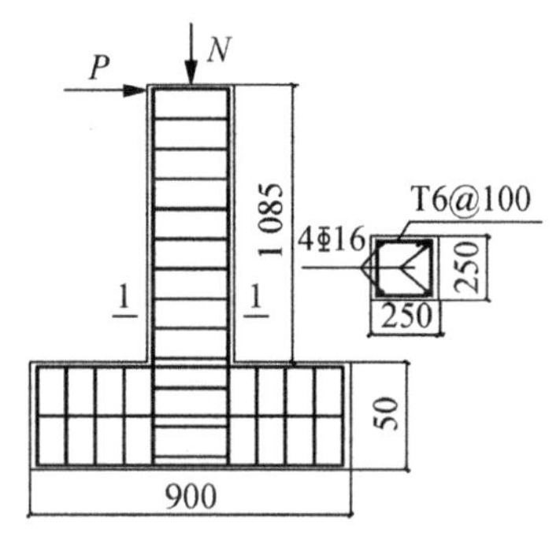

(a) Experimental columns

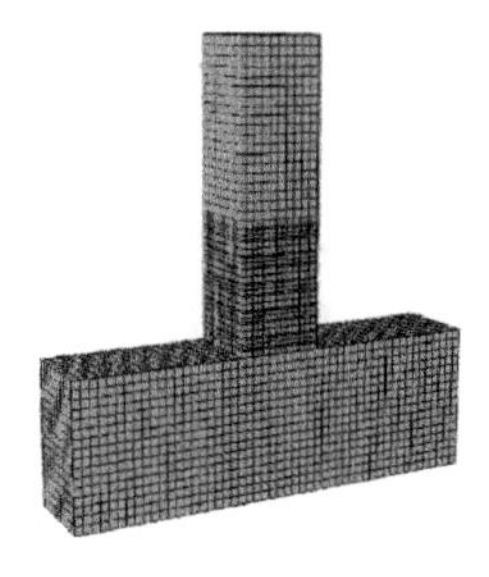

(b) Finite element model

Figure 1 The detail of numerical column (unit: mm)

The fifteen columns of different parameters including axial compression ratio, CFRP layers and shear span ratio were numerical analyzed. The load displacement skeleton curves were all shown in Figure 2.

Then further analysis about the length of column plastic hinge was done, which was defined by two ways including the yielded rebar zone as L_{sy} and the crushed concrete zone as L_{cs} [6] in Figure 3.

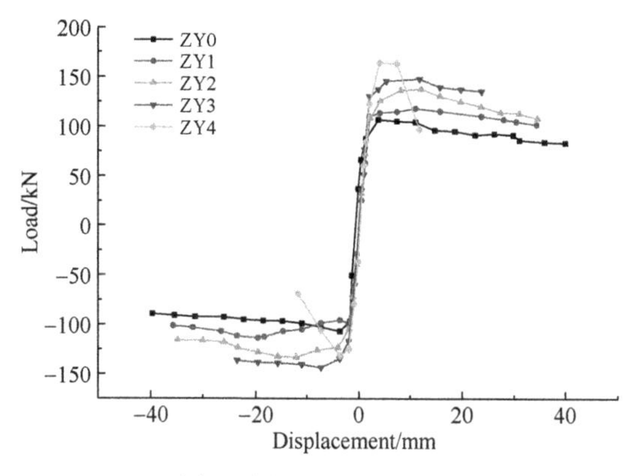

(a) Axial compression ratio

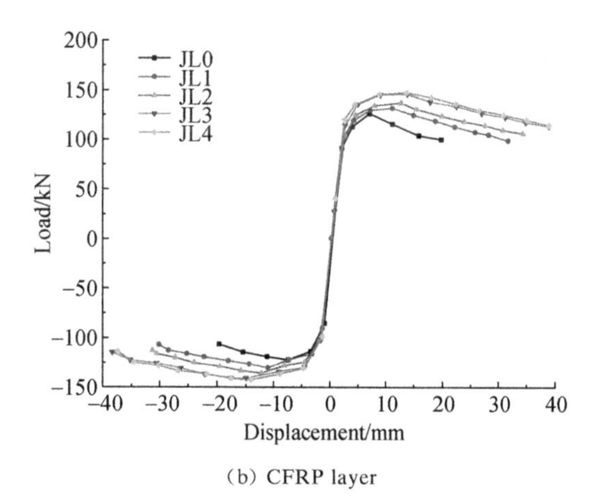

(b) CFRP layer

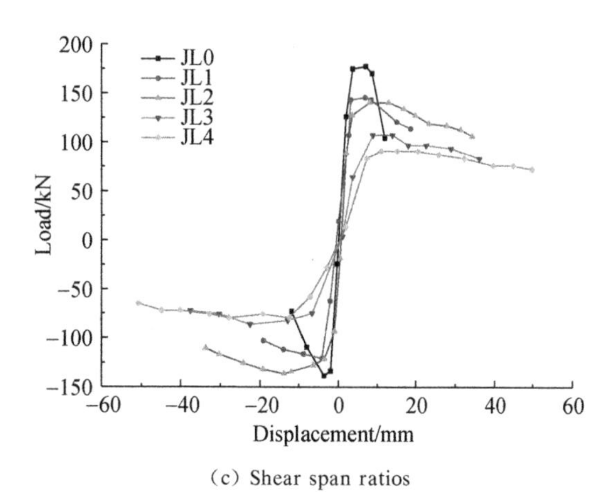

(c) Shear span ratios

Figure 2 Load versus displacement skeleton curves

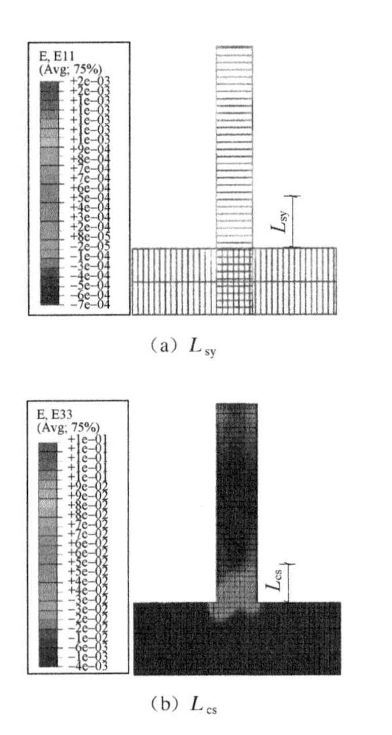

(a) L_{sy}

(b) L_{cs}

Figure 3 The definitions of columns plastic hinge

Conclusions

(1) As the axial compression ratio increases, the failure mode of confined RC column changed from bending to shearing. The ductility increased with the number of CFRP layers, but when the number of layers exceeded 3, the increment was no more obvious. As the shear span ratio increased, the deformation capacity of the confined column increased, but the growth trend gradually disappeared when it reached a certain value.

(2) The results showed that the yielded rebar zone L_{sy} decreased with the increase of axial compression ratio; it also increased first and then decreased with the FRP layers; and it increased with the shear span ratio, while the increment gradually disappeared with high shear span ratio value. The crushed concrete zone L_{cs} increased with the increasing axial compression ratio, but it decreased with the increasing FRP layers; it also increased first with the shear span ratio, and the increment gradually disappeared with high shear span ratio value.

References

[1] OZCAN O, BINICI B, OZCEBE G. Seismic strengthening of rectangular reinforced concrete columns using fiber reinforced polymers[J]. Engineering Structures, 2010, 32(4): 964-973.

[2] GU D S, WU Y F, WU G, et al. Plastic hinge analysis of FRP confined circular concrete columns [J]. Construction and Building Materials, 2012, 27(1): 223-233.

[3] OZCAN O, BINICI B, OZCEBE G. Improving seismic performance of deficient reinforced concrete columns using carbon fiber-reinforced polymers [J]. Engineering Structures, 2008, 30(6): 1632-1646.

[4] LI J, QIAN J R, JIANG J B. Experimental study on deformation capacity of CFS confined concrete columns subjected lateral loading[J]. Concrete, 2004, 8(128): 22-26.

[5] QI W. ABAQUS 6.14 User Manuals[M]. Beijing: Posts and Telecom Press, 2016.

[6] ZHAO X M, WU Y F, LEUNG A Y T. Analyses of plastic hinge regions in reinforced concrete beams under monotonic loading[J]. Engineering Structures, 2012, 34(1): 466-482.

Study on static behavior of steel-concrete composite beams with corroded stud

Y. L. Wang & J. Chen & H. P. Zhang

Structural Engineering Research Institute, Zhejiang University, Hangzhou 310058, China

Abstract

Steel-concrete composite beams both have the advantages of steel structure and concrete structure. And it has been used in lots of engineering in our country, such as Nanbu Bridge and Yangbu Bridge in Shanghai[1]. There are a large number of scholars at home and abroad having studied the mechanic properties[2-5] and durability[6-9] of steel-concrete composite beams. In this paper, the static test and finite element simulation analysis of the positive and negative bending moment of the stud corroded steel-concrete composite beams are carried out.

A constant current was used in the test to accelerate the corrosion, and the studs with different degrees of corrosion were energized at different time. Through the static test (Figure 1 and Figure 2), the overall bearing capacity, the relative slip between the materials, the development of inter-span deflection, and other mechanical properties of composite beams of the positive and negative bending moment were studied before and after the stud corrosion. The finite element analysis software ABAQUS was used to model the composite beams, and the deflection of the composite beams before and after the coring of the studs was analyzed and compared with the experimental values.

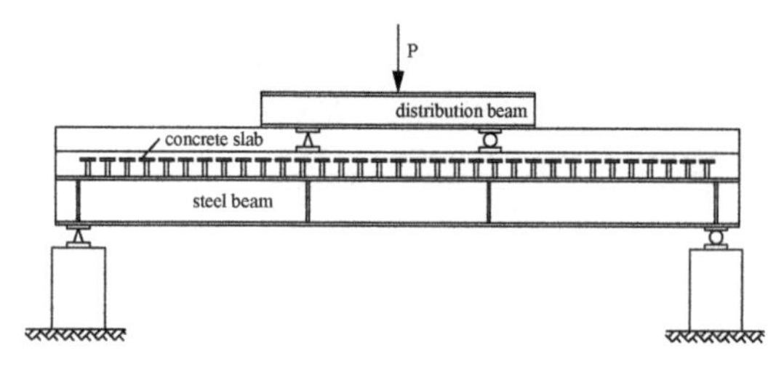

Figure 1 Loading diagram of positive bending moment

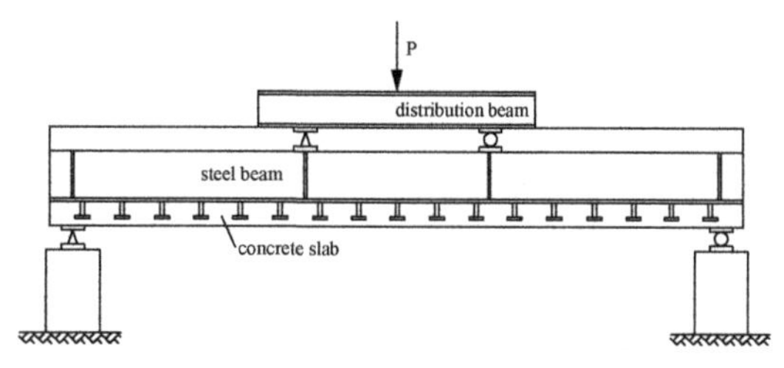

Figure 2 Loading diagram of negative bending moment

During the monotonic loading of the test, the composite beams all failed due to local buckling of the steel beams and cracking of the concrete slab, rather than the fracture of the shear joint (Figure 3 and Figure 4).

It can be seen from the experiment that under the positive and negative bending moment, the bearing capacity of the composite beam decreases with the

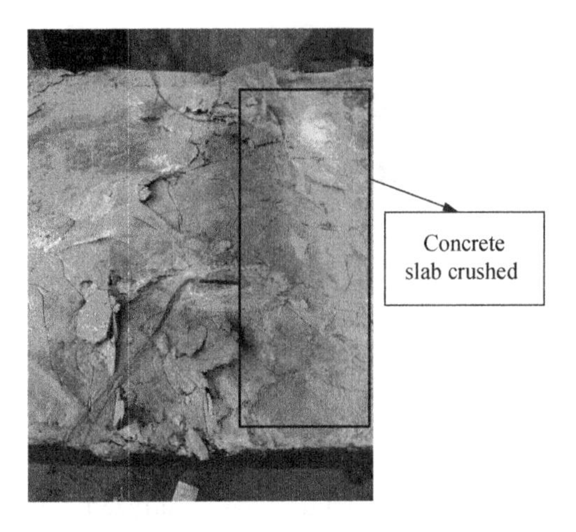

Figure 3 Concrete slab crushed

Figure 4 Steel beam local buckling

increase of the corrosion rate of the studs, and the slip between the concrete slab and the steel beam increases with the increase of the corrosion rate of the studs, with the overall stiffness reduced. And the effect under the action of negative bending moment is more obvious. The reason is that the concrete slab is pulled under the action of negative bending moment, and the studs are subjected to a larger pulling force earlier.

In the simulation of ABAQUS, the concrete constitutive structure selects the concrete plastic damage model, and the steel constitutive structure uses the ideal elastoplastic model. The studs with different diameters are used to simulate the studs with different corrosion rates. T3D2 units are adopted to model steel bars and C3D8R units are adopted to model steel beams, concrete slabs and studs. The simulation of the composite beam support is the same as the test. The support is set at 100 mm from both ends of the beam. One end constrains U1, U2, and U3 to simulate the fixed hinge support, and the other end constrains U1 and U2 to simulate the sliding hinge support. The model is shown in Figure 5.

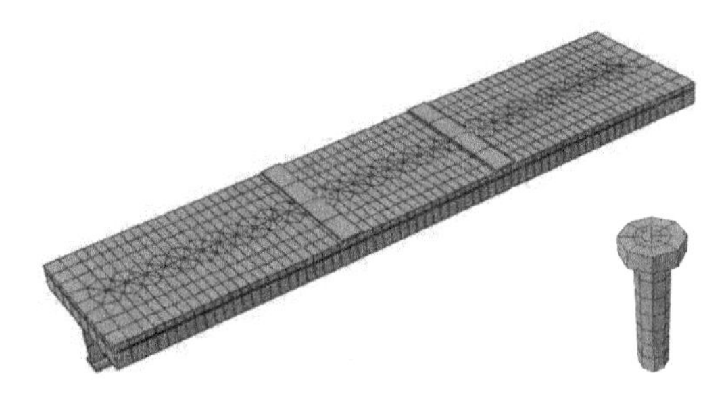

Figure 5 Finite element model of the composite beam

Comparing the simulating results and test results, it can be seen that the stiffness and ultimate bearing capacity of the combined beam test results are different from the finite element values, but it is small. The ultimate bearing capacity of the finite element calculation is slightly smaller than the experimental value. It may be because the constitutive relation of the steel is selected from the ideal elastoplastic model, and the strengthening stage of the steel is not considered; the difference in the initial stiffness of the composite beam may be that there was a gap on the interface between the concrete slabs and steel beams due to the construction errors.

References

[1] LIN Y P. Nanpu Bridge and Yangpu Bridge [J]. Chinese Journal of Civil Engineering, 1995, 28(6): 3-10.

[2] RYU H K, CHANG S P, KIM Y J, et al. Crack control of a steel and concrete composite plate girder with prefabricated slabs under hogging moments [J]. Engineering Structures, 2005, 27(11): 1613-1624.

[3] SHIM C S, CHANG S P. Cracking of continuous composite beams with precast decks[J]. Journal of Constructional Steel Research, 2003, 59(2): 201-214.

[4] NAVARRO M G, LEBET J P. Concrete Cracking in Composite Bridges: Tests, Models and Design Proposals[J]. Structural Engineering International, 2001, 11(3): 184-190.

[5] NIE J G, ZHANG M H. Research on cracks in negative bending moment zone of steel-concrete composite beams[J]. Journal of Tsinghua University (Science and Technology), 1997, 37(6): 95-99.

[6] BERTO L, SIMIONI P, SAETTA A. Numerical modelling of bond behavior in RC structures affected by reinforcement corrosion[J]. Engineering Structures, 2008, 30(5): 1375-1385.

[7] ZHAO C J, QIN X, CHEN W, ZHOU W B. Study on Mechanical Properties of Stud Corroded Steel-Concrete Composite Beam under Negative moment[J]. Technology of Highway and Transport, 2017, 33(6): 60-65.

[8] XUE W, CHEN J, WU L, et al. Experimental investigation of composite steel-concrete beams with corroded studs[J]. Jianzhu Jiegou Xuebao/Journal of Building Structures, 2013, 34: 222-226.

[9] RONG X L, HUANG Q. Experimental study on mechanical properties of corroded stud connectors [J]. Civil Engineering and Environmental Engineering, 2012, 34(2): 15-20.

Durability design of marine concrete structures treated with silane

Y. Zeng & D. W. Zhang & W. L. Jin
Institute of Structural Engineering, Zhejiang University, Hangzhou 310058, China
J. G. Dai
Department of Civil and Environmental Engineering, The Hong Kong Polytechnic University, Hong Kong 999077, China
M. S. Fang
Zhejiang Communication Investment Group Co., Ltd., Hangzhou 310020, China

Abstract

Silane is one of the favorable surface treatment materials in practical engineering due to its superiority in rendering the treated concrete with breathability, durability and aesthetics[1]. In contrast to its wide application, the service life extension effect of silane is not sufficiently considered in most design codes, and the treatment is usually regarded as an additional protective measure, leading to an uneconomic design of cover thickness.

To bridge this gap, the durability design of silane treated concrete structures is carried out in a model-based way. The bilayer physical model proposed by Zhang et al.[2] is used herein to predict the chloride transport process, in which the hydrophobic layer formed after silane impregnation is distinguished from the substrate concrete (Figure 1).

The diffusivity of substrate concrete can be described by the Life-365 model[3], and the coefficient of the treated area is further gained by multiplying an exponential discount function[4]. The water repellent layer thickness can be calculated after incorporating influential factors with the model proposed by Johnasson et al.[5], and the thinning effect of the hydrophobic layer is also considered with an exponential model gained from the regression analysis. The model was built and solved with the COMSOL multiphysics.

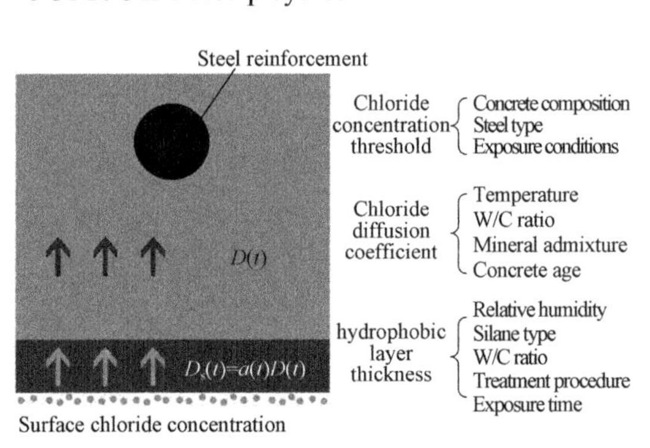

Figure 1 Chloride transport in treated concrete

The inevitable uncertainty of the deterioration process was considered through Monte Carlo simulation, and Latin

Hypercube sampling technique was applied in advance to control the computational workload of solving numerical models. A total number of 1 000 samples were generated according to the distributions of model parameters (Table 1).

Table 1 Statistical properties of parameters

Sym.	Unit	Type	Mean	COV
c_s (Splash zone)	%wc	Gamma	0.36	0.50
c_s (Tidal zone)	%wc	Log-normal	0.70	0.29
c_{cr}	%wc	Beta	0.14	0.41
c_0	%wc	Constant	0.01	—
D_0	m^2/s	Log-normal	7.94E-12	0.20
x	mm	Normal	Variable	0.10
RH	%	Normal	0.75	0.05
T_0	K	Constant	293.15	—
n	—	Constant	0.454	—
t_0	day	Constant	28	—

The model was verified with on-site exposure data gained from Scotland[6], Belgium[7] and Iran[8], and used to conduct durability design.

The design can be conducted in a fully probabilistic way, and the sensitivity analysis reveals that the service life of silane treated structures can be greatly influenced by the cover thickness and water-cement ratio, which is similar to the untreated structures (Figure 2).

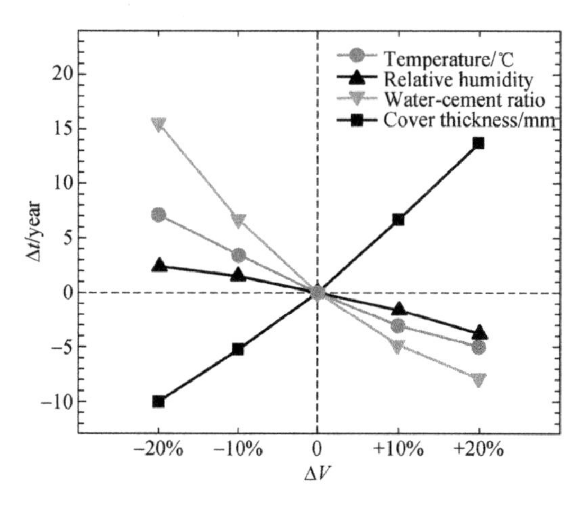

Figure 2 Sensitivity analysis of corrosion-free life

A more practical partial factor method is then proposed for the convenience of engineering practice, and the protection of silane is converted into an additional layer of concrete cover (x_e) in the design equation.

$$\frac{c_{cr,d}-c_0}{\gamma_{Cc}}-\gamma_{Cs}(c_{s,d}-c_0)\cdot\left(1-\mathrm{erf}\left(\frac{x_{d,s}+x_e-\Delta x}{2\sqrt{\gamma_D D_d \eta_t \eta_T t_S}}\right)\right)=0 \tag{1}$$

Further analysis shows that x_e can be influenced by water-cement ratio, surface moisture content and the original cover thickness, but varies little with the temperature and the target reliability index. The exact values for concrete with a water-cement ratio of 0.4 are listed in Table 2.

By choosing the mean values as the characteristic values of random variables, the partial factors can be determined following the first order reliability theory (FROM)[9], and the durability design can be conducted with Eq. (1). The partial factors determined are shown in Table 3.

Table 2 x_e in various conditions (β=1.0, 1.3, 1.5)

Zone	$x_{d,s}$ /mm	x_e /mm			
		RH = 0.6	RH = 0.6	RH = 0.6	RH = 0.6
Splash	40	7/8/9	6/7/8	5/5/6	3/3/3
	50	6/6/7	5/5/6	4/4/5	2/3/3
	60	5/5/6	4/5/5	3/4/4	2/2/3
	70	4/5/5	3/4/4	3/3/3	1/2/2
	80	3/4/4	3/4/4	2/3/3	1/2/2
Tidal	50	8/9/9	7/7/8	5/6/6	3/3/3
	60	7/8/8	6/7/7	5/5/5	3/3/3
	70	6/7/7	5/6/6	4/4/4	2/2/2
	80	5/6/6	5/5/5	4/4/4	2/2/2

Table 3 Partial factors determined with FORM

Zone	β	γ_{Cc}	γ_{Cs}	γ_D
Splash	1.0	1.66	1.17	1.03
	1.3	2.13	1.18	1.04
	1.5	2.69	1.16	1.05
Tidal	1.0	1.73	1.03	1.05
	1.3	2.22	1.03	1.06
	1.5	2.80	1.03	1.07

A representative case (Splash zone, T = 15 ℃, RH = 0.7, W/C = 0.4, β_t = 1.0, t_{SL} = 70a) was chosen to illustrate the process and verify the effectiveness of the proposed method. Through comparisons with the thickness designed by fully probabilistic method, it is clear that the result ($x_{d,s}$ = 69mm) is sufficiently safe and ensures achieve a reliability level of more than 1.20. It is believed that the conservative design is mainly attributed to the consideration of concrete cover thickness tolerance (Δx), which is necessary considering the uncertainty in the construction stage.

Acknowledgements

The financial support from The National Key Research and Development Program of China (Grant No. 2017YFC0806100), the National Science Foundation of China (51820105012), the Fundamental Research Funds for the Central Universities of China (2019 FZA4017) and the research project for traffic engineering construction from transportation department of Zhejiang

(Grant No. 2018035) are gratefully acknowledged.

References

[1] FREITAG S, BRUCE S. The influence of surface treatments on the service life of concrete bridges[R]. New Zealand, New Zealand Transport Agency research report, 2010.

[2] ZHANG J Z, MCLOUGHLIN I M, BUENFELD N. Modelling of chloride diffusion into surface treated concrete[J]. Cement and Concrete Composites, 1998, 20(4): 253-261.

[3] LIFE-365 CONSORTIUM Ⅲ. Life-365 service life prediction model and computer program for predicting the service life and life-cycle cost of reinforced concrete exposed to chlorides[R]. Version 2.3.3, 2018.

[4] PETCHERDCHOO A, CHINDAPRASIRT P. Exponentially aging functions coupled with time-dependent chloride transport model for predicting service life of surface-treated concrete in tidal zone[J]. Cement and Concrete Research, 2019, 120: 1-12.

[5] JOHANSSON A, JANZ M, SILFWERBRAND J, et al. Penetration depth for water repellent agents in concrete as a function of humidity, porosity and time[J]. Restoration of Buildings and Monuments, 2007, 13(1): 3-16.

[6] MCCARTER W J, LINFOOT B T, CHRISP T M, et al. Performance of concrete in XS1, XS2 and XS3 environments[J]. Magazine of Concrete Research, 2008, 60(4): 261-270.

[7] SCHUEREMAS L, VAN GEMERT D A, GIESSLER S. Chloride penetration in RC-structures in marine environments - Long term assessment of a preventive hydrophobic treatment[J]. Construction and Building Materials, 2007, 21(6): 1238-1249.

[8] MORADLLO M K, SHEKARCHI M, HOSEINI M. Time-dependent performance of concrete surface coatings in tidal zone of marine environment[J]. Construction and Building Materials, 2012, 30: 198-205.

[9] LI Q W, LI K F, ZHOU X G, et al. Model-based durability design of concrete structures in Hong Kong-Zhuhai-Macau sea link project[J]. Structural Safety, 2015, 53: 1-12.

Experimental study on fatigue damage detection of prefabricated concrete composite beams based on piezomagnetic effect

Z. Y. Xie & D. W. Zhang & W. L. Jin
Institute of Structural Engineering, College of Civil Engineering and Architecture, Zhejiang University, Hangzhou 310058, China.
J. H. Mao & J. Zhang
School of Civil Engineering & Architecture, Ningbo Institute of Technology, Ningbo 315100, China

Abstract

In the prefabricated concrete structures with wet connection, the interface of the pre-cast layer and the cast-in-place layer is a relative weak part for environmental and load resistance[1]. The fatigue damage evolution of the new-old concrete interface dominates the structure damage under cyclic load. It's significant to use an appropriate damage indicator to quantify the degree of the fatigue damage considering the interfacial degradation. There have been many experimental studies carried out to characterize fatigue damage by using traditional indicators such as stiffness, deflection, energy and strain[2-4]. However, there are still many deficiencies in using the traditional fatigue indicators to characterize the fatigue damage of the composite members. For example, traditional non-destructive testing has difficulty to detect composite member damage in its early stage due to unobvious changes in microscopic and macroscopic physical parameters[5]. The accuracy to quantify damage, especially for interfacial damage and the information available from the curve of macro-level traditional indicators such as stiffness is limited. In this study, a magnetic signal detection method based on piezomagnetic effect is proposed as a new damage indicator to characterize the fatigue damage of prefabricated composite beams. The mechanism of piezomagnetism is that the micro-plasticization process can cause slip dislocation inside the ferromagnetics under cyclic loading, which changes the texture, voids and flaws of the material. The magnetic domain structure of the material can be transformed, which appears as magnetization state changing[6]. Relevant experimental studies have proved that magnetic signal can indicate the fatigue damage process of ferromagnetic materials[7, 8]. However, there is lack of relevant experiments using magnetic signal indicators to measure the development of interfacial damage of composite members.

Static tests and fatigue tests were carried out on 15 T-shaped prefabricated concrete composite beams with 3 different sections designed according to the code of JGJ 1—2014[9]. As shown in Figure 1, PCB indicates the prefabricated composite beam; the following letter is K or S designating the shear key or the pre-cast slab.

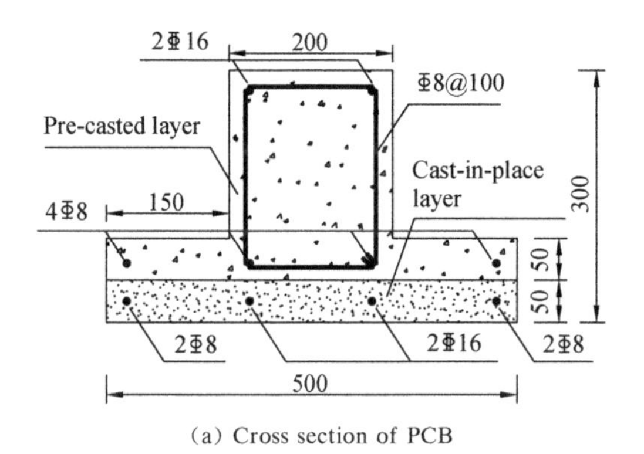

(a) Cross section of PCB

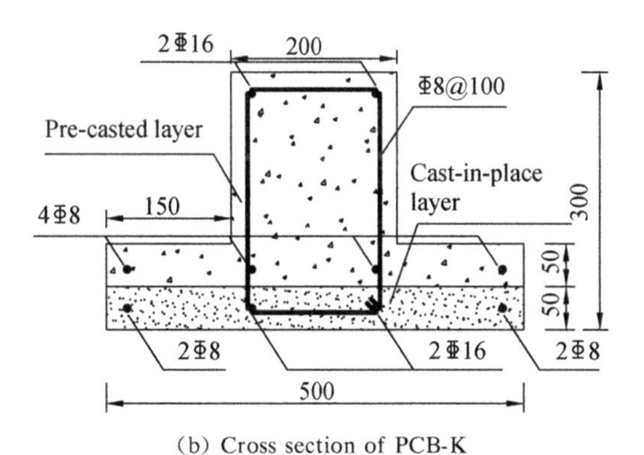

(b) Cross section of PCB-K

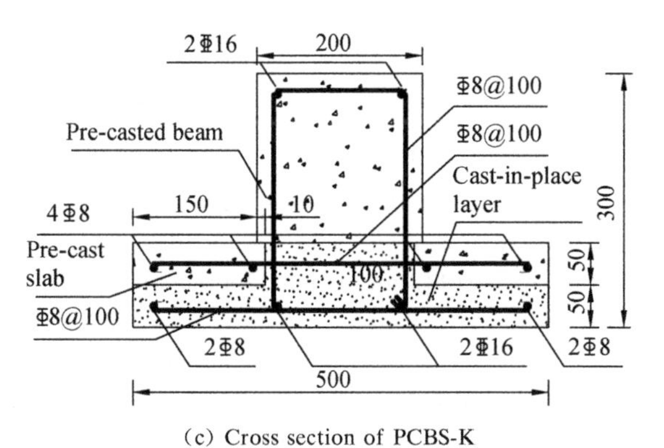

(c) Cross section of PCBS-K

Figure 1 Cross section of composite beams (unit: mm)

The cyclic load level is divided into 50%, 60%, 70%, and 80% of the static ultimate strength.

For failure mode, two kinds of debonding types: full debonding and partial debonding were observed according to the level of the interfacial debonding, which occurred at new-old concrete interface or beam-slab interface. The damage development of the composite members is characterized by the mid-span deflection, stiffness, energy consumption, reinforcement strain and magnetic signal. The traditional damage indicators were compared with magnetic signal as a new indicator.

From time-varying curves it can be found that the shape of the deflection and the strain curve exhibits a sinusoidal waveform, which has little change during the loading process, while as shown in Figure 2, magnetic signals change obviously during cyclic loading and can reveal more information about the fatigue properties of the composite members.

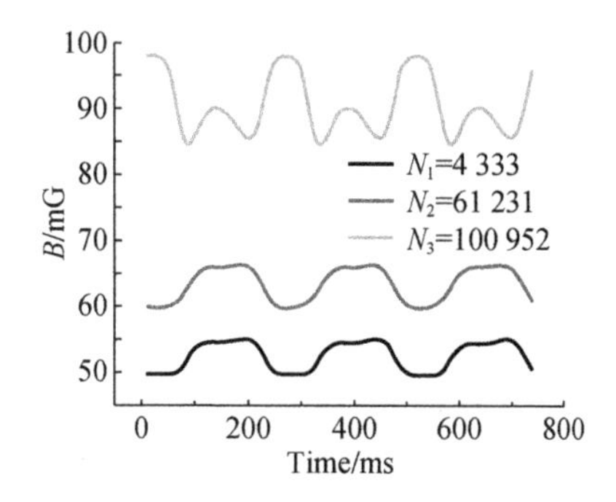

Figure 2 Time-varying curves of magnetic signal in different fatigue stages

The damage degree is quantified by using four indicators of stiffness, deflection, magnetic signal and energy consumption. Based on the Lemaitre damage formula, there has:

$$D = \frac{X_N - X_0}{X_{N_f} - X_0} \tag{1}$$

where X_0 is the initial value of each indicator in the first cycle, X_N is the processing value and X_{N_f} is the value at the designated end of damage evolution. Fatigue damage curves were given using damage indicators of stiffness, deflection, magnetic signal and energy consumption. Figure 3 contains damage curves of PCBS-K-50 using different indicators shown as an example. The differences among several indicators, as well as the sensitivity of parameters such as interfacial damage degree, load level, and setting of shear key were compared.

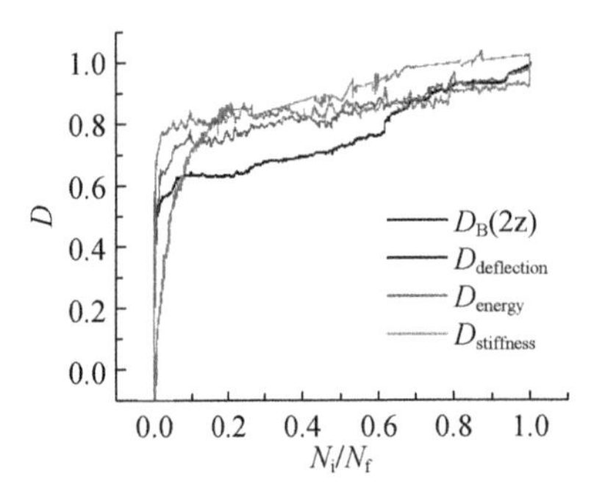

Figure 3 Damage curves of PCBS-K-50 using different indicators

The results show that the magnetic signal can reflect the fatigue damage process of composite members. Setting the shear key at the interface has a great effect on the inhibition of interfacial debonding, of which the performance improvement can be shown more obviously using the magnetic signal. For prefabricated composite members, the interface debonding damage has a great contribution to the overall damage of the composite members, which can influence the whole composite member performance and should be paid more attention. Compared with the traditional indicators, the fatigue damage detection based on the piezomagnetic effect is more sensitive to interface fatigue damage since the piezomagnetic effect and the fatigue damage are at the same mesoscopic level[8, 10]. However, the magnetic signal is greatly affected by the external magnetic field environment and is too sensitive to the magnetic field change[11, 12]. The signal stability problem needs to be solved by subsequent research.

Acknowledgements

The National Science Foundation of China (Grant No. 51878604, 51820105012) and the Fundamental Research Funds for the Central Universities of China (Grant No. 2019FZA4017) are gratefully acknowledged.

References

[1] SANTOS P M D, JÚLIO E N B S. A state-of-the-art review on shear-friction [J]. Engineering Structure, 2012, 45(15): 435-448.

[2] ONESCHKOW N. Fatigue behavior of high-strength concrete with respect to strain and stiffness[J]. International Journal of Fatigue. 2016, 87: 38-49.

[3] YE L. On fatigue damage accumulation and material degradation in composite materials [J]. Composites Science & Technology, 1989, 36(4): 339-350.

[4] PHILIPPIDIS T P, Vassilopoulos A P.

Fatigue design allowables for GRP laminates based on stiffness degradation measurements [J]. Composites Science & Technology, 2000, 60(15): 2819-2828.

[5] ZHANG G Y, ZHANG J, JIN W L, et al. Review of Mechanics and Fatigue Researches on Magnetic Materials Based on Piezomagnetic Effect[J]. Materials Review, 2014, 28(9): 4-10.

[6] ERBER T, GURALNICK S A, Desai R D, et al. Piezomagnetism and fatigue[J]. Journal of Physics D: Applied Physics, 1997, 30 (20): 2818-2836.

[7] BAO S, JIN W L, HUANG M F, et al. Piezomagnetic hysteresis as a non-destructive measure of the metal fatigue process [J]. NDT & E International, 2010, 43(8): 706-712.

[8] JIN W L, ZHANG J, CHEN C S, et al. A new method for fatigue study of reinforced concrete structures based on piezomagnetism [J]. Journal of Building Structures, 2016, 37 (4): 133-142.

[9] Ministry of Housing and Urban-Rural Development of the People's Republic of China. Technical specification for precast concrete structures [S]. Beijing: China Architecture & Building Press, 2014.

[10] ZHANG J. Experimental study on fatigue performance of reinforced concrete structures based on piezomagnetism effect [D]. Hangzhou: Zhejiang University, 2017.

[11] JIN W L, ZHOU Z D, ZHANG J, et al. Experimental research on fatigue properties of corroded steel bars based on dynamic piezomagnetism [J]. Journal of Zhejiang University (Engineering Science Edition), 2017, 51(2): 225-230.

[12] BAO S, JIN W L, HUANG M F. Mechanical and magnetic hysteresis as indicators of the origin and inception of fatigue damage in steel [J]. Journal of Zhejiang University-Science A (Applied Physics & Engineering), 2010, 11(8): 580-586.

Development of an artificial neural network for the deterioration prediction of existing concrete bridges

P. Miao

Graduate School of Engineering, Hokkaido University, Sapporo 060-0808128, Japan

H. Yokota

Faculty of Engineering, Hokkaido University, Sapporo 060-0808128, Japan

Abstract

For existing concrete bridges, maintenance work plays a significant role in keeping the structures safe. However, the maintenance work mainly depends on periodic inspection[1-3], which usually consumes time and human resources. Consequently, there is a growing need for adequate maintenance management to reduce time and human resources. An artificial neural network is extremely efficient with large data sets and would provide a more accurate prediction compared to traditional statistical methods[4]. Therefore, in this paper, we try to apply an artificial neural network to examine its accuracy for deterioration prediction.

An artificial neural network has been applied in the field of civil engineering[5], which is mainly divided into two targets: for laboratory experiments[6, 7] and for practical engineering[8]. In laboratory conditions, many researchers have studied the prediction of performance indicators of concrete, such as compressive strength[6, 7] and chloride diffusivity[9]. Some researchers have also applied an artificial neural network for the mixture design of high-performance concrete[10, 11]. For such applications, many influential factors were investigated, and the effectiveness of an artificial neural network was discussed. Even though the use of an artificial neural network for laboratory experiments has profound significance, it is challenging to consider all influential factors for the application because the accelerated experimental conditions usually used are different from actual conditions on sites; in other words, the experiment is tough to simulate the real situation. Typically, the utilization of neural network in experimental conditions is deviated from the actual terms and only considered a few or even only one influential factor.

Some researchers also have applied an artificial neural network or other methods to inspection[8, 9, 12], evaluation[13, 14], retrofit[15, 16], and scheduling of repair activities[17]. However, most of them neither have long-term practical inspection database nor use traditional expert-score system.

Predictions of deterioration progress have been applied to concrete bridges. However, it is still unclear how the actual conditions influence the overall deterioration of bridges. In this paper, we used an inspection database and an artificial neural network to develop a model to predict deterioration progress of concrete bridges. The database used has been set up by engineers during the long-term maintenance work. To build the model, we considered eight potential environmental factors, such as temperature, traffic volume, and years in service. Moreover, two bridge geometric parameters-length and width are also discussed. The output of the model is the overall deterioration grade of each bridge, which is classified into four grades as per the Inspection Guidelines[1]. Their definitions are Grade a: healthy, Grade b: preventive action required, Grade c: early action required, and Grade d: emergency action required.

Following the flowchart shown in Figure 1, a deterioration prediction model for concrete bridges was developed. At first, we eliminated incomplete data and unreasonable data from the database. Then about 70% of bridge inspection data from all the dataset was used to train the neural network model, and 50% of the remaining dataset was applied to validate the performance of the model. Based on the training and validation sets, the overall performance of the model was adjusted. After the best performance model was got, the remaining 50% data was used to test the model.

Then, the iterated calculation was carried out to finally create the model having the best performance. In the process of this calculation, suitable sets of model parameters were obtained. Finally, by using these model parameters and the best performance model, it can be realized to make prediction of deterioration progress.

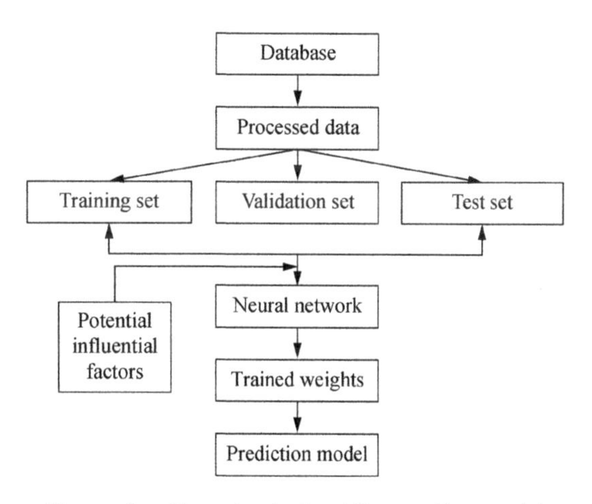

Figure 1 Flowchart of setting up the model

By applying the model to existing bridges, it was found that the prediction results show the accuracy is about 74%, as indicated in Table 1. In the table, the shaded areas represent the prediction numbers in different conditions, and the darker areas indicate the successful prediction numbers. The accuracy was calculated by all the successful predictions (20 + 22 + 30) divided by the total number (96). Even though there are still some works needed to improve the model, the applicability of the preliminary prediction model was verified for practical maintenance work of bridges. This model could be a supplementary method to help us to find appropriate timing of further inspection as well as interventions.

Table 1 The performance of the model

Actual grade	Prediction grade		
	1	2	3
1	20	7	4
2	4	22	2
3	2	1	30
Accuracy			74.04%

References

[1] Guidelines for Regular Inspection of Road Bridges[S]. Japanese: Ministry of Land Infrastructure and Transport, in, 2014.

[2] RYAN T W, HARTLE R A, MANN J E, DANOVICH L J. Bridge Inspector's Reference Manual[R]. Report No. FHWA NHI. 2006: 03-001.

[3] JANDU A S. Inspection and maintenance of highway structures in England [J]. Proceedings of the Institution of Civil Engineers-Bridge Engineering, 2008, 161 (3): 111-114.

[4] VAUGHAN J, SUDJIANTO A, BRAHIMI E, et al. Explainable neural networks based on additive index models[J]. 2018.

[5] REICH I Y. Artificial Intelligence in Bridge Engineering: Towards Matching Practical Needs with Technology[R]. 2007.

[6] OZKAN I A, ALTIN M. Estimating of Compressive Strength of Concrete with Artificial Neural Network According to Concrete Mixture Ratio and Age [C]// International Conference on Advanced Technology & Science, at Antalya, Turkey, 2015.

[7] KIM J, LEE C, PARK S. Artificial neural network-based early-age concrete strength monitoring using dynamic response signals [J]. Sensors, 2017, 17(6): 1319-1330.

[8] TRENT R, GAGARIN N, RHODES J. Estimating pier scour with artificial neural networks[C]//Hydraulic Engineering, 1999.

[9] HODHOD O A, AHMED H I. Developing an artificial neural network model to evaluate chloride diffusivity in high performance concrete [J]. HBRC Journal, 2013, 9(1): 15-21.

[10] YEH I C. Modeling of strength of high-performance concrete using artificial neural networks [J]. Cement and Concrete Research, 1998, 28(12): 1797-1808.

[11] YEH I C. A mix proportioning methodology for fly ash and slag concrete using artificial neural networks[J]. Chung Hua Journal of Science and Engineering, 2003, 1(1): 77-84.

[12] POLLON S, ADAMS L, PALMER R, et al. Expert system for bridge scour and stream stability [C]//Computing in Civil Engineering, American Society of Civil Engineers, 2015.

[13] KUO S S, DAVIDSON T E, FIJI L M. Development of Computer Automated Bridge Inspection Process[C]// Computing in Civil Engineering and Geographic Information Systems Symposium. ASCE, 2010.

[14] KOSTEM C N. Design of an expert system for the rating of highway bridges [C]// Expert Systems in Civil Engineering, 1986.

[15] TEE A B, BOWMAN M D, SINHA K C. A fuzzy mathematical approach for bridge condition evaluation [J]. Civil Engineering Systems, 1988, 5(1): 17-24.

[16] MIKAMI I, TANAKA S, KURACHI A. Expert system with learning ability for retrofitting steel bridges [J]. Journal of Computing in Civil Engineering, 1994, 8 (1): 88-102.

[17] KHAN S I, RITCHIE S G, KAMPE K. Integrated system to develop highway rehabilitation projects [J]. Journal of Transportation Engineering, 1994, 120(1): 1-20.

Knowledge sharing in durability analysis of concrete structures

Y. F. Zhang
Graduate School of Engineering, Hokkaido University, Sapporo 060-0808128, Japan
H. Yokota
Faculty of Engineering, Hokkaido University, Sapporo 060-0808128, Japan

Abstract

Concrete is widely used in infrastructure. Due to its characteristics and environmental impact, concrete structures generally have durability problems[1]. Maintenance management, therefore, must be carried out during the service period of concrete structures. Factors causing the durability problem of concrete structures include carbonation, rebar corrosion, freezing and thawing, alkali silica reaction, chemical erosion, and so on. By analyzing the mechanism and process involved in the durability problem caused by those factors, a maintenance manager can evaluate the deterioration degree, and finally makes decisions on a method of intervention according to the evaluation results.

However, conducting the durability analysis usually involves professional knowledge which is mastered by experts, and a maintenance manager is usually unable to do the analysis alone. The maintenance manager needs to consult experts to get analysis results. The knowledge gap between experts and managers makes knowledge utilization inefficient and leads to inefficiency in management. Moreover, durability needs to be analyzed multiple times throughout the maintenance management process of structures. After using knowledge to build a durability analysis model, it is very efficient to preserve and apply it to a similar durability problem in the future. Therefore, this paper attempts to explore a method to break down knowledge barriers for durability analysis.

Knowledge management[2] is proposed to promote knowledge sharing between experts and non-experts such as managers in terms of framework and process. Knowledge management is mainly composed of the following two steps: knowledge creating and knowledge sharing as shown in Figure 1.

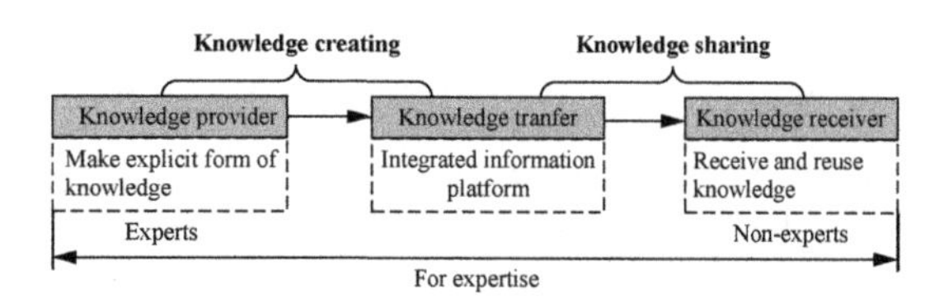

Figure 1 Workflow of the knowledge management

Knowledge is a familiarity, awareness, or understanding of someone or something, which is acquired through experience or education by perceiving, discovering, or learning. Knowledge exists invisibly in the brains of experts in the initial state[3]. In the knowledge creating (which also means the externalization of knowledge), an expert as a knowledge provider should retrieve the corresponding knowledge of durability analysis to build an analytical model and then apply it to a durability problem to be solved with visual tools. Advances in computer software, such as a finite element software and a discrete element software, have made it possible to decode expert's knowledge into a complete visual form such as a simulation model from ideas.

In the information sharing, due to the different needs and purposes of the participants, the levels of knowledge requirements are varied. The knowledge providers focus on the knowledge itself to conceive a model to simulate the reality as completely as possible, so they have high level of knowledge requirements. Since the knowledge receivers (usually as managers without professional knowledge reserve) are mainly engaged in engineering practice, focusing on the application of knowledge. They only need to carry out the durability analysis through a simulation model with simple operation, so they have low level of knowledge requirements. To eliminate the confusion caused by different levels of knowledge requirements in the process of knowledge sharing, the providers need to simplify a knowledge model and customize a user interface to meet the level of knowledge requirements of receivers. Open parameter models would be adopted to realize the simplification of the simulation model by experts, in which various theories applied in the analysis are hidden, while the parameters and one-click calculations are operational. In addition, the results of analysis in open parameter models should be visible. Knowledge receivers can access the knowledge from the user interface and reuse the knowledge through the open parameter models.

The transmission of knowledge requires a carrier that exists in a visual form. A platform which can store and display the knowledge in the form of simulation model is used as a carrier to share the knowledge. Both people of knowledge provider and knowledge receiver can access the knowledge from the platform. The platform is a BIM-based integrated information platform[4], which ensures visualization and persistence in information storage. BIM is designed as a visual tool for management of structures with 3D models, and BIM software can be customized with new features via API (application programming interface) (Figure 2). Cloud technology[5] used for communication via the internet on this platform ensures that managers can access and process information of various forms anytime and can easily share information with others.

A case study on the durability analysis of chloride ion penetration into concrete tunnel was carried out to demonstrate the implementation process of the proposed

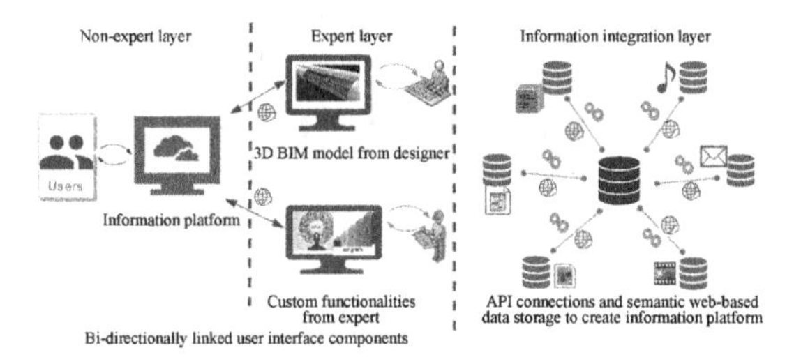

Figure 2 Implementation of BIM-based integrated information platform

approach. An open parameter model was created to simplify the simulation model. Then an information platform was established in BIM software, and eventually the open parameter model was successfully stored and displayed on the information platform. The results showed that knowledge management can be applied to durability analysis with the use of information tools.

The method presented in this paper gives useful instruction in the knowledge sharing between experts and managers breaking the knowledge sharing barriers in knowledge contents and knowledge carrier forms. By using necessary knowledge, a manager can make durability analysis alone to make timely decisions with the analysis results, ultimately improving the efficiency of maintenance management.

Acknowledgements

The authors would like to acknowledge the financial support from China Scholarship Council (CSC) for conducting the present research.

References

[1] JIN W, ZHAO Y. Review and prospect of research on durability of concrete structures [J]. Journal of Zhejiang University (Engineering Edition), 2002, 36(4): 371-380.

[2] BOTHA A, KOURIE D, SNYMAN R. Coping with continuous change in the business environment, knowledge management and knowledge management technology[M]. Chandice Publishing Ltd., London, 2008.

[3] JOHNSON W H A. Mechanisms of tacit knowing: pattern recognition and synthesis [J]. Journal of Knowledge Management, 2007, 11(4): 123-139.

[4] LIU F X, JALLOW A K, SCHOLAR P, et al. Building knowledge modelling: integrating knowledge in BIM [C]// Proceedings of the 30th International Conference of CIB W78, Beijing, China, 2013.

[5] JONES S, IRANI Z, SIVARAJAH U, et al. Risks and rewards of cloud computing in the UK public sector: A reflection on three organizational case studies[J]. Information Systems Frontiers, 2019, 21(2): 359-382.

Three-dimensional vibration control of offshore floating wind turbines using multiple tuned mass dampers

V. Jahangiri & C. Sun

Department of Civil and Environmental Engineering, Louisiana State University, Baton Rouge, Louisiana, 70803, USA

Abstract

Due to severe environmental conditions in the marine area, the FWTs suffer from excessive three-dimensional vibrations which will cause potential instability issues and severe fatigue damage of the structure and mooring cables. To reduce the vibrations caused by wind and wave loading, structural vibration control method is being studied to mitigate the vibrations of offshore wind turbines[1-3]. While most of the current research focuses on reducing the fore-aft vibrations, OWTs suffer from significant side-side vibrations due to vortex induced vibrations or misalignment between wind and wave loading. To address this issue, Lackner and Rotea[4] used dual linear TMDs to control bi-directional vibrations of both fixed and floating wind turbines. The controller was shown to be effective, however the TMD installed in the FWT required a large stroke. To overcome the limitations of dual linear TMDs, Sun and Jahangiri[5] developed a 3D-PTMD to control bi-directional vibrations of a fixed-bottom OWT.

In this paper, a three-dimensional pendulum tuned mass damper (3D-PTMD) and dual linear pounding tuned mass dampers (2PTMDs) are used to mitigate the three-dimensional vibrations of a spar type FWT. An analytical model of a spar-type FWT coupled with the 3D-PTMD and 2PTMDs is established using Euler-Lagrangian equation. The NREL 5MW OC3-Hywind spar buoy wind turbine is used to examine the performance of the 3D-PTMD and 2PTMDs. Results show that the 3D-PTMD with a mass ratio of 0.1% can reduce the FWT tower vibrations by around 32%. It is also found that the 2PTMDs can reduce the spar RMS response by around 18% in roll direction and 50% in pitch direction with an allowable stroke in the spar.

Figure 1 illustrates the schematic model of the FWT with a 3D-PTMD in the nacelle and dual linear pounding TMDs (2PTMDs) installed in the floater. Totally, the coupled system contains 17 degrees of freedom (DOF).

In terms of Euler-Lagrangian equation, the system's EOMs can be established

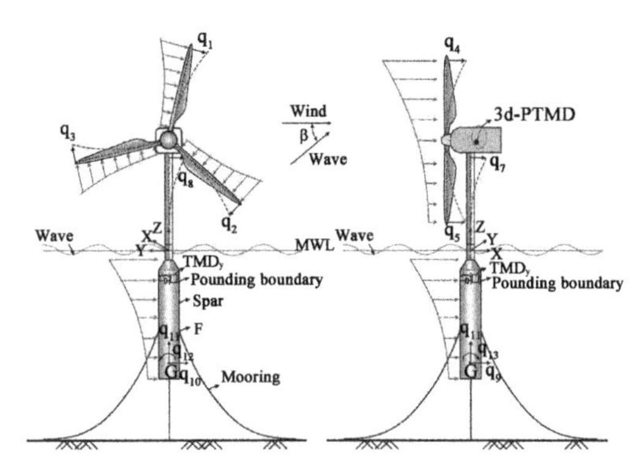

Figure 1 Schematic model of the spar FWT coupled with a 3D-PTMD and 2PTMDs

and written in a matrix form as:

$$\tilde{M}\ddot{\tilde{q}}+\tilde{C}\dot{\tilde{q}}+(\tilde{K}+\tilde{K}_{r})\tilde{q}$$
$$=\tilde{Q}_{wind}+\tilde{Q}_{wv}+\tilde{Q}_{buo}+\tilde{Q}_{moor}+\tilde{F}+\tilde{F}_{p} \tag{1}$$

where $\tilde{M}$, $\tilde{C}$, $\tilde{K}$ are the system mass, damping and stiffness matrices. $\tilde{K}_r$ is the restoring stiffness matrix, $\tilde{F}$ is the generalized force caused by nonlinearity of pendulum, and $\tilde{F}_p$ is the pounding force. Parameters $\tilde{Q}_{wind}$, $\tilde{Q}_{wv}$, $\tilde{Q}_{buo}$, $\tilde{Q}_{moor}$ are the wind, wave, buoyancy and mooring loadings respectively. The responses of the FWT can be determined by solving Eq. (1). Figure 2 portrays the flowchart for solving the response of the FWT system.

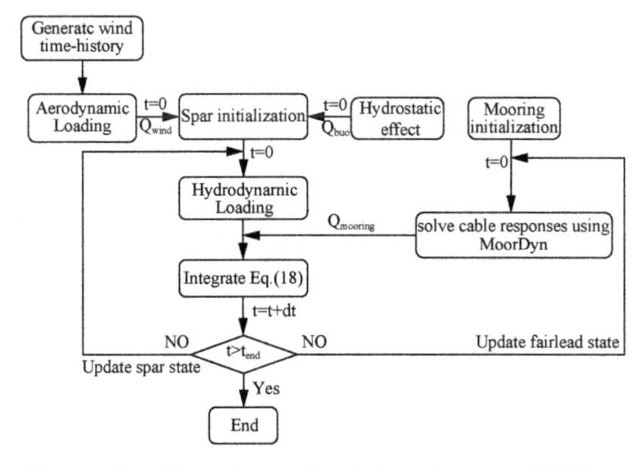

Figure 2 Flowchart of solving the spar FWT system

The established model is validated by comparing the natural frequencies of the established model with that reported in reference [6]. Table 1 shows that the established model is valid.

Table 1 Model verification

DOF	Present model	Reference [6]
Tower FA	0.485 Hz	0.473 Hz
Tower SS	0.493 Hz	0.457 Hz
Spar surge	0.008 Hz	0.008 Hz
Spar sway	0.008 Hz	0.008 Hz
Spar heave	0.032 5 Hz	0.032 Hz
Spar roll	0.033 8 Hz	0.034 Hz
Spar pitch	0.033 8 Hz	0.034 Hz

According to reference [5], a suggested mass ration of 0.1% is chosen with a frequency ratio of 0.97 and a damping ratio of 8%. Figure 3 illustrates the relative displacement of the nacelle with and without the 3D-PTMD under misalignment angles of 30° and 60°. It can be found that the 3D-PTMD can reduce the RMS response by around 32% in fore-aft direction. Also, it can be seen that the 3D-PTMD is ineffective in reducing the structural response of the nacelle in side-side direction. This is due to the reason that the frequency of the 3D-PTMD is tuned to the natural frequency of the tower, while the dominant frequency in side-side direction is around 0.1 Hz caused by wave loading.

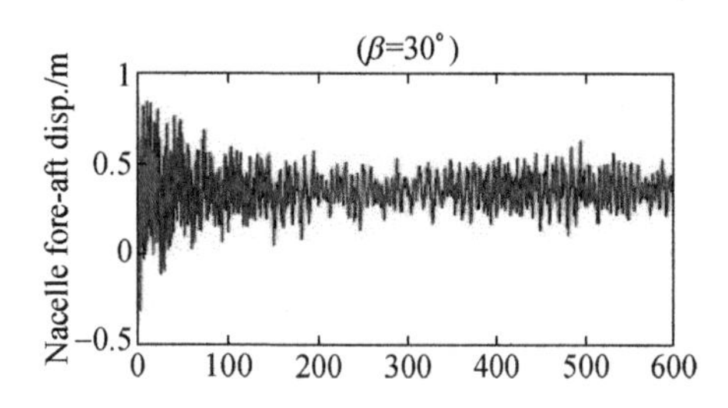

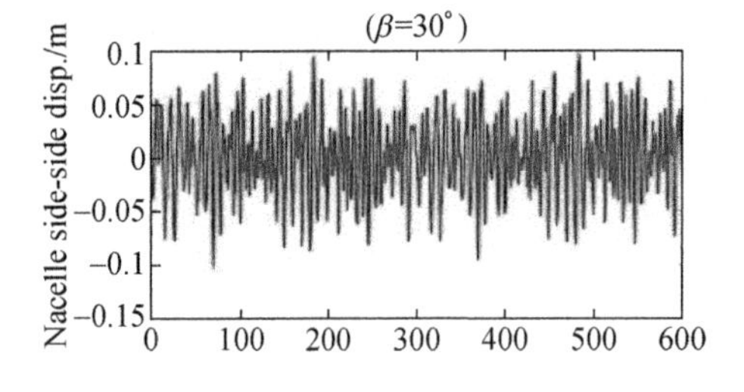

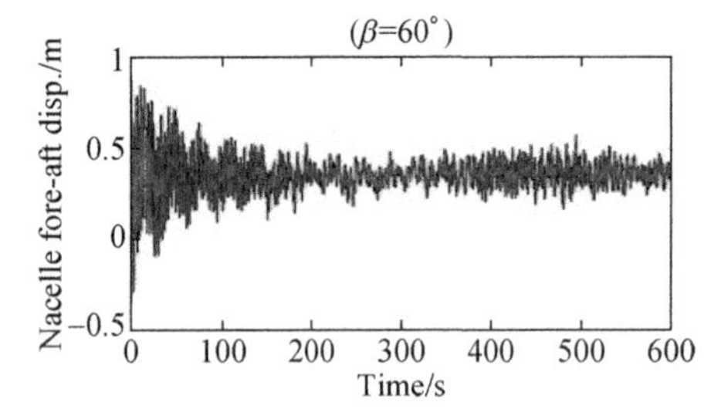

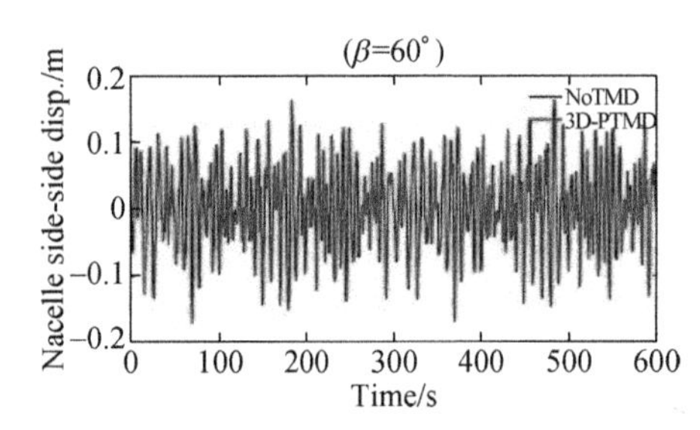

Figure 3 Tower response comparison with and without 3D-PTMD under 30° and 60° of misalignment angle

The 2PTMDs are adopted and located at sea water level to mitigate the response of the spar in roll and pitch directions. A suggested mass ratio of 2% is chosen for the 2PTMDs with a frequency ratio of 0.94 and a damping ratio of 11%. Also, the 2PTMDs properties are optimized and the pounding stiffness is chosen as 10^4 ($N/m^{1.5}$) and coefficient of restitution is 0.3, the gap distance is considered to be 3 m. Figure 4 shows the roll and pitch response of the spar with and without 2PTMDs under 30° and 60° of misalignment angle. It can be found that the 2PTMDs reduce the RMS response by around 18% in roll direction, and 50% in pitch direction. Figure 5 illustrates the TMD stroke in X direction. Although, the TMD stroke is relatively large in X direction, it is much smaller than that of a regular linear TMDs studied in reference [4] where the TMD stroke is up to 20 m.

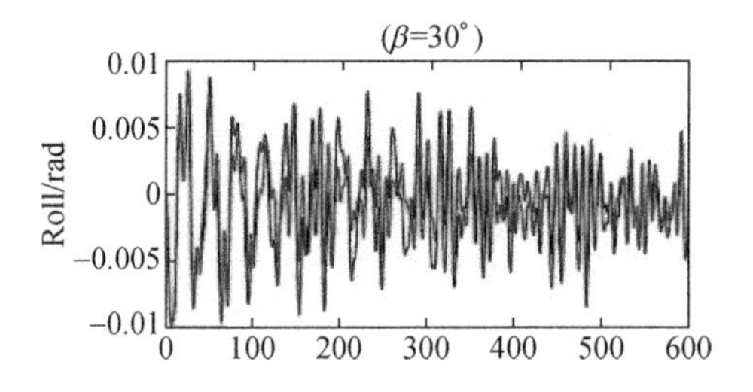

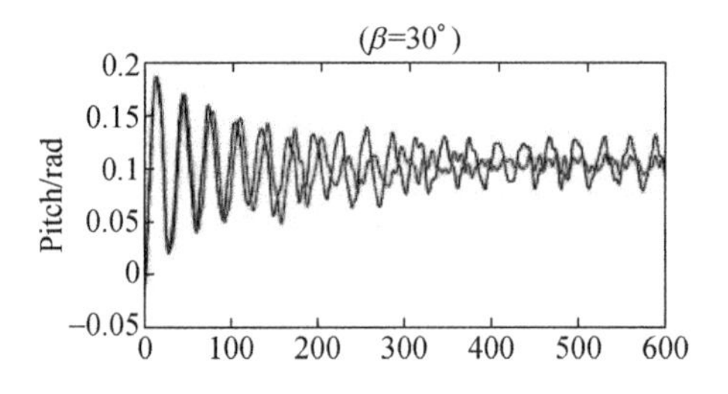

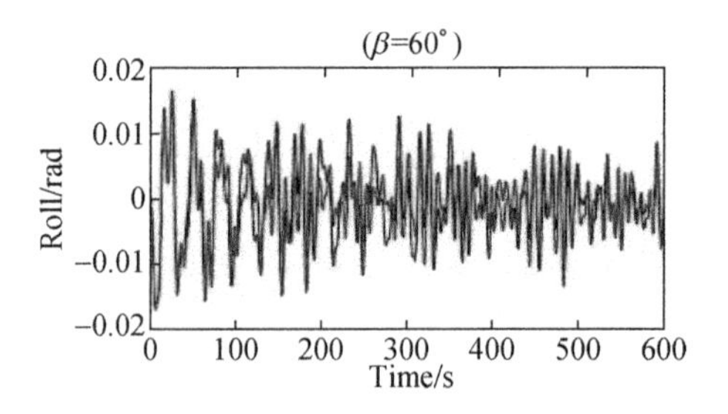

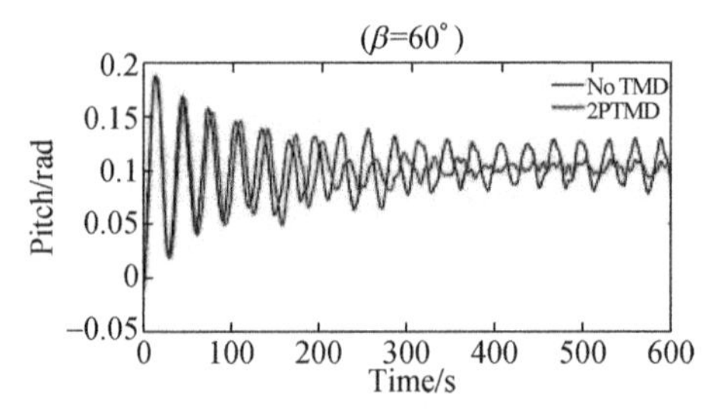

Figure 4 Spar response comparison with and without 2PTMDs under 30° and 60° of misalignment angle

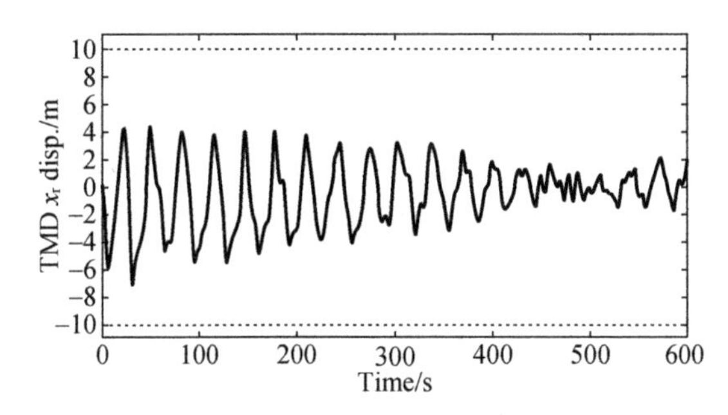

Figure 5 2PTMDs stroke under 30° misalignment angle in x_r direction

References

[1] COLWELL S, BASU B. Tuned liquid column dampers in offshore wind turbines for structural control [J]. Engineering Structures, 2009, 31(2): 358-368.

[2] ZUO H R, BI K M, H H. Using multiple tuned mass dampers to control offshore wind turbine variations under multiple hazards[J]. Engineering Structures, 2017, 141: 303-315.

[3] ZHANG Z L, LI J, NIELSEN S R K, et al. Mitigation of edgewise vibrations in wind turbine blades by means of roller dampers [J]. Journal of Sound and Vibration, 2014, 333(21): 5283-5298.

[4] LACKNER M, ROTEA M. Structural control of floating wind turbines [J]. Mechatronics, 2011, 21(4): 704-719.

[5] SUN C, JAHANGIRI V. Bi-directional vibration control of offshore wind turbines using a 3D pendulum tuned mass damper[J]. Mechanical System and Signal Processing, 2018, 105: 338-360.

[6] MATHA D. Model development and loads analysis of an offshore wind turbine on a tension leg platform, with a comparison to other floating turbine concepts[R]. National Renewable Energy Laboratory, 2009.

Fire endurance tests of insulated reinforced concrete beams shear-strengthened with carbon fiber-reinforced polymer (CFRP) sheets

W. Y. Gao & H. Mahmood

State Key Laboratory of Ocean Engineering, Shanghai Jiao Tong University, Shanghai 200030, China

K. X. Hu & R. Liu

Department of Disaster Mitigation for Structures, Tongji University, Shanghai 200092, China

Abstract

Over the past three decades, externally bonded (EB) fiber-reinforced polymer (FRP) laminates (including wet layup FRP sheets and pultruded FRP plates) are increasingly used for the strengthening and repair of reinforced concrete (RC) structures. The success of the EB FRP strengthening technique is attributed to the superior material properties of FRP composites, including their high strength-to-weight ratio, corrosion resistance, ease of application, tailorable performance characteristics, and minimal alterations to the dimensions of the strengthened member. However, a typical ambient-cure epoxy adhesive for EB FRP strengthening has a low glass transition temperature of approximately 45~80 ℃[1]. When the epoxy adhesive is subjected to elevated temperatures close to this characteristic temperature, it changes from a glassy to a viscous state with severe strength and stiffness degradations[2]. Also, the EB FRP laminates may burn off in fire unless a supplemental insulation layer is provided to separate them from fire. Therefore, fire performance of FRP-strengthened RC members is an important issue that needs to be properly considered during the strengthening design process, especially for the purpose to satisfy the requirements of specified fire-resistant ratings in indoor applications (e. g. in buildings)[3].

Almost all the fire endurance tests in the literature were conducted on flexural FRP-strengthened RC beams[4-6]. There is lacking research on fire performance of RC beams shear-strengthened with FRP laminates. This paper presents the first-ever fire endurance tests on insulated RC beams strengthened in shear with carbon FRP (CFRP) sheets. A total of seven rectangular RC beams were constructed: Three of them were tested at ambient temperature to determine their load-carrying capacity while the remaining four were first exposed to ISO834 standard fire for a duration of 2. 5 h (Table 1 for more details). Figure 1 shows the geometry and reinforcement details of the tested beams. The shear strengthening system consisted of one or two layers of 0.167 mm thick CFRP U-wraps, which were 80 mm wide with a clear spacing of 100 mm between two adjacent U-wraps. For the protected CFRP-strengthened RC beams, the fire insulation was all the same, which was applied to the bottom and two sides of each beam with a 20 mm SJ - 2 layer (Table 1). The fire insulation material (commercial designation of SJ - 2), supplied by a local material company, was a lightweight fire-resistant cementitious plaster that could be manually applied (troweled) onto the surface of the structural member. According to the brochure provided by the manufacturer, it had the thermal properties at ambient temperature as follows: dry density of 500 kg/m^3, specific heat capacity of 1 000 J/(kg · K) and thermal conductivity of 0.12 W/(m · K).

Table 1 Details of the tested beams

Beam	Purpose	Shear strengthening (U-wraps)	Shear span-to-depth ratio	Fire insulation
B1-0	Ambient temp. tests	—	2.4	—
B1-1		One-ply	2.4	
B1-2		One-ply	2.4	
B2-0	Fire endurance tests	—	2.4	20 mm SJ-2 layer
B2-1		One-ply	2.4	
B2-2		One-ply	1.6	
B2-3		Two-ply	2.4	

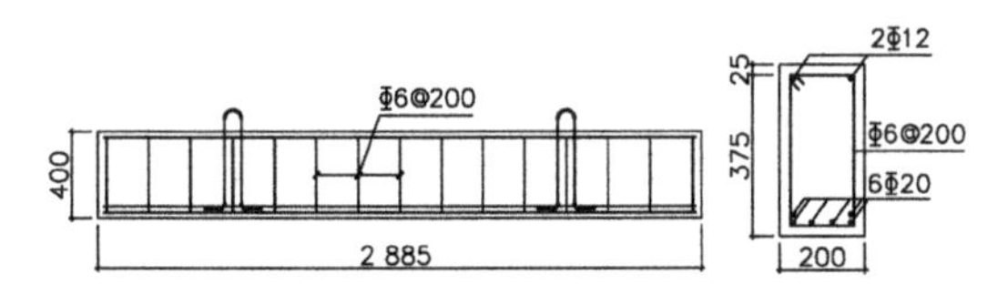

Figure 1 Geometry and reinforcement details (unit: mm)

Figure 2 shows the load versus midspan deflection curves of the three beams tested at ambient temperature. It is seen that the elastic stiffness of these beams are almost the same before the concrete cracking, mainly due to the negligible stiffness contribution of the CFRP U-wraps. The ultimate loads of B1-1 and B1-2, however, were much higher than that of B1 - 0. The gains in ultimate loads were 24% and 30% for B1 - 1 and

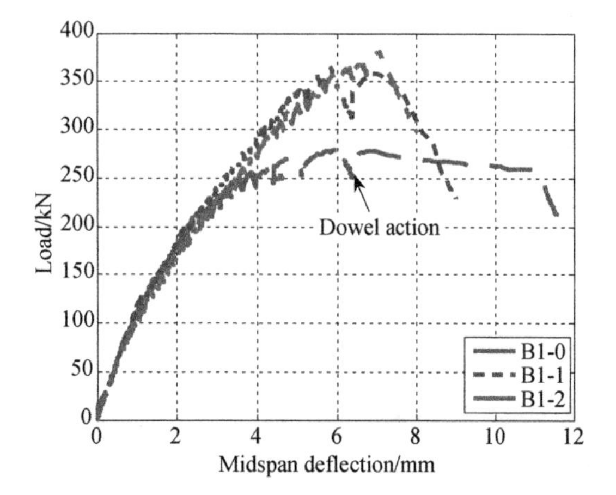

Figure 2 Load-displacement curves (unit: mm)

B1-2, respectively, when compared with B1-0.

Fire endurance tests were conducted on four beams, of which an RC beam without fire protection was tested as reference whereas the remaining FRP-strengthened RC beams were protected with 20 mm SJ-2 layer. Figure 3 shows photographs of the tested beams after fire exposure. The RC beam without fire insulation was characterized by shear failure as a result of the development of diagonal cracks, whereas the insulated CFRP-strengthened beam was well protected and did not fail in fire. Several thermocouples (TCs) were installed over the midspan beam cross-section to record the temperature responses of concrete (TC2 & TC3), steel stirrups (TC6 & TC7) and the CFRP-to-concrete interface (TC1 & TC4 & TC5). Taking the beam B2-3 as an example, Figure 4 depicts the temperatures measured at various locations during fire exposure.

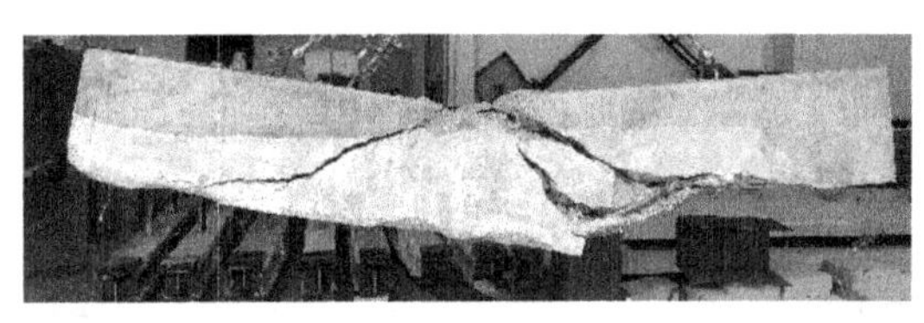

(a) B2-0

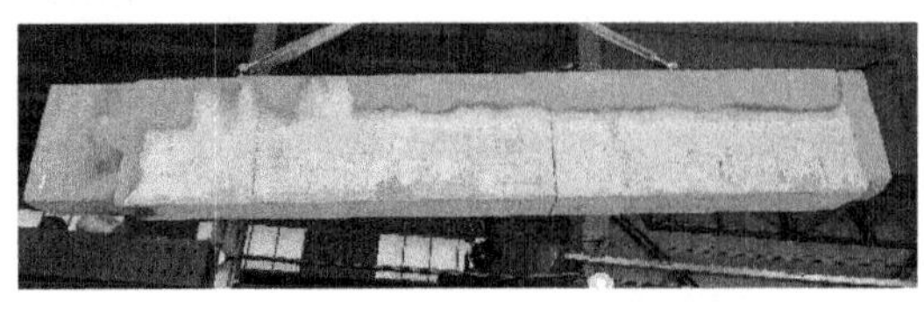

(b) B2-1

Figure 3 Photographs of the tested beams after fire exposure

Figure 5 shows the midspan deflection responses of B2-0 and B2-1 during fire exposure. The un-protected beam B2-0 experienced an abrupt deflection response due to the shear failure, whereas the protected beam B2-1 achieved a satisfactory fire endurance of 2.5 h with the maximum deflection less than 10 mm. The midspan deflection responses of B2-2 and B2-3 were similar to that of B2-1 and thus were not reported herein due to the space limitation.

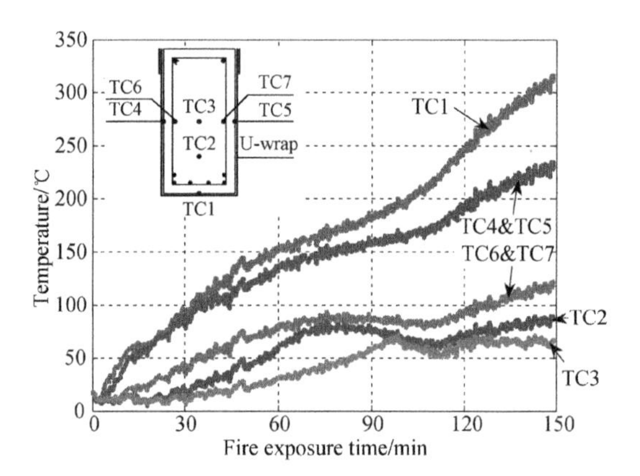

Figure 4 Measured temperatures at various locations

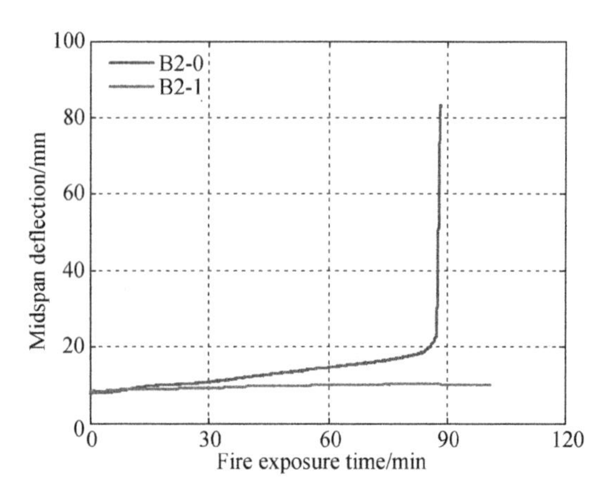

Figure 5 Midspan deflection versus time curves

Acknowledgements

The authors are grateful for the financial support received from the National Natural Science Foundation of China (Grant No. 51978398) and the Natural Science Foundation of Shanghai (Grant No. 19ZR1426200).

References

[1] DAI J G, GAO W Y, TENG J G. Bond-slip model for FRP laminates externally bonded to concrete at elevated temperature [J]. Journal of Composites for Construction, 2013, 17(2): 217-228.

[2] GAO W Y, TENG J G, DAI J G. Effect of temperature variation on the full-range behavior of FRP-to-concrete bonded joints [J]. Journal of Composites for Construction, 2012, 16(6): 671-683.

[3] GAO W Y. Fire resistance of FRP-strengthened RC beams: numerical simulation and performance-based design [D]. Hong Kong: The Hong Kong Polytechnic University, 2013.

[4] AHMED A, KODUR V R. The experimental behavior of FRP-strengthened RC beams subjected to design fire exposure [J]. Engineering Structures, 2011, 33(7): 2201-2211.

[5] FIRMO J P, CORREIA J R, BISBY L A. Fire behavior of FRP-strengthened reinforced concrete structural elements: A state-of-the-art review [J]. Composites Part B: Engineering, 2015, 80: 198-216.

[6] GAO W Y, HU K X, LU Z D. Fire resistance experiments of insulated CFRP strengthened reinforced concrete beams [J]. China Civil Engineering Journal, 2010, 43(3): 15-23.

Experimental study on mechanical properties of hybrid fiber reinforced concrete

S. C. Jiang & H. C. Jiang
College of Civil Engineering, Tongji University, Shanghai 200092, China

Abstract

This study mainly investigated the influence of multi-scale fiber combinations on mechanical properties and strain hardening behavior of fiber reinforced concrete (FRC). After adding to FRC, the fibers of different scales can complement each other and play a full role in different stages of stress and structure, so as to achieve the purpose of crack resistance, toughening and strengthening gradually. Firstly, two types of polymer fibers (PVA fiber and PE fiber) were considered. By comparing the strength and toughness of FRC which incorporated with either polymer fibers, the better polymer fibers were optimized. Then, the optimized polymer fibers were blended with the large size hooked-end (macro-scale) steel fiber in different dosage in order to investigate the strain hardening behavior of FRC prepared with binary combination of polymer-steel fibers. Thus, the best binary combination ratio was found. Macro-scale fibers and meso-scale steel fibers were blended in different dosage as well to determine the best dosage of steel fibers in binary combination. On this basis, the best combination proportion of two groups was applied to the ternary combination which included polymer fiber, meso-scale steel fiber and macro-scale steel fiber. In addition, the micro-scale steel fiber was also selected as the control group in place of the polymer fibers to explore the strain hardening behavior of FRC. Finally, we will add nano-fibers (Figure 1) on top of the ternary combination, so that the FRC was enhanced and toughened at multi-scale and multi-level which included nano-scale, micron and millimeter levels to achieve better strain hardening behavior. Tests showed the mechanical properties in terms of compressive strength, splitting tensile strength, three-point bending fracture energy (Figure 2) and direct tensile performance.

The results indicated that the toughening effect of PE fiber on FRC was better than that of PVA fiber at 0.18 water-binder ratio. With the increase of steel fiber dosage, the compressive strength, splitting tensile strength, three-point bending fracture energy, flexural toughness and tensile strength of FRC with steel-PE and steel fiber combination increased, and both the strength and toughness were improved as well as the strain hardening behavior which was more obviously. Among them, the fiber combination of 1.5% macro-scale steel fiber and 0.5% meso-scale PE fiber showed the best strain hardening behavior as well as the fiber combination of 1.5% macro-scale steel fiber and 0.5% meso-scale steel fiber.

The dosage of macro-scale steel fibers was adjusted on the basis of the binary fiber combination when it comes to ternary combination. The fracture energy obtained from three-point bending test and the flexural toughness were improved on the premise of keeping the peak load almost unchanged. When it came to quaternary combination, nano-fibers were added on top of ternary combination. After adding nano-fibers, the strength and toughness of FRC were improved to a certain extent. The use of carbon nano-fibers (CNF) and carboxyl carbon nanotubes (CNT2) was more conducive to the improvement of fracture energy obtained from three-point bending test, four-point bending and direct tensile performance. After adding 0.1wt% CNF, the three-point bending fracture energy of FRC increased by 37.4% and 4.3%, respectively. The equivalent bending strength of four-point bending increased by 20% and 22.4% respectively. The ultimate tensile strain of direct tension increased by 14.3% and 75% while the direct tensile fracture energy was increased by 26% and 95.4%, respectively. At the same time, the nano-fibers can refine the large pores by filling the pore structure between the hydration products of cement, and improve the pore structure of FRC which makes the microstructure more compact and thus effectively restricts the formation of harmful pores.

The anti-cracking effect of multi-scale fiber combination is multi-step and multi-level, so that FRC has experienced a stable crack generation and expansion process until destroy. Therefore, compared with fiber or single scale, multi-scale fiber combination of FRC has better strength, toughness and better strain hardening behavior.

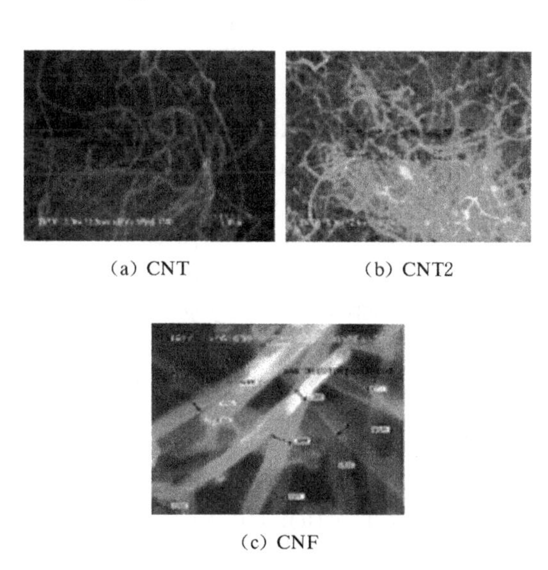

(a) CNT (b) CNT2

(c) CNF

Figure 1 Sample of a nano-fibers

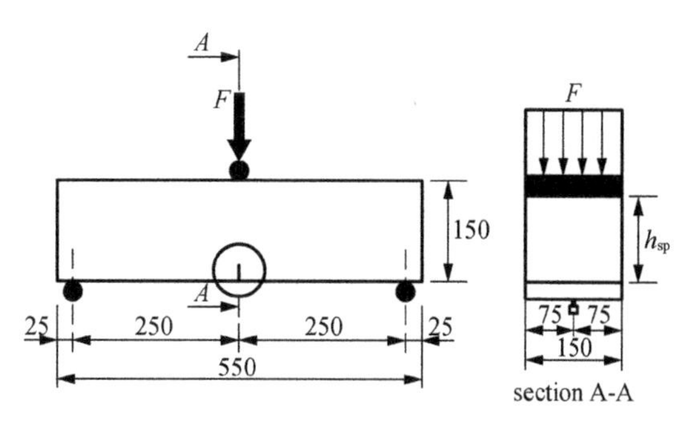

Figure 2 Schematic of three-point bending beam (unit: mm)

References

[1] ALREKABI S, CUNAY A B, LAMPROPOULOS A, et al. Mechanical performance of novel cement-based composites prepared with nano-fibers, and hybrid nano- and micro-fibers[J]. Composite Structures, 2017, 178: 145-156.

[2] YU K Q, WANG Y C, YU J T, et al. A strain-hardening cementitious composites with the tensile capacity up to 8% [J]. Construction and Building Materials, 2017, 137: 410-419.

[3] YOO D Y, YOON Y S. A review on structural behavior, design, and application of ultra-high-performance fiber-reinforced concrete [J]. International Journal of Concrete Structures and Materials, 2016, 10 (2):125-142.

[4] CHANDRASHEKARA A, PALANKAR N, DURGA P, et al. A study on elastic deformation behavior of steel fiber-reinforced concrete for pavements [J]. Journal of The Institution of Engineers (India): Series A, 2019, 100(2): 215-224.

[5] TAI Y S, EL-TAWIL S. High loading-rate pullout behavior of inclined deformed steel fibers embedded in ultra-high performance concrete [J]. Construction and Building Materials, 2017, 148: 204-218.

[6] KANG S T, CHOI J I, KOH K T, et al. Hybrid effects of steel fiber and microfiber on the tensile behavior of ultra-high performance concrete [J]. Composite Structures, 2016, 145: 37-42.

[7] ROSSI P. Influence of fibre geometry and matrix maturity on the mechanical performance of ultra-high-performance cement-based composites [J]. Cement and Concrete Composites, 2013, 37 (1): 246-248.

[8] CUI X, HAN B G, ZHENG Q F, et al. Mechanical properties and reinforcing mechanisms of cementitious composites with different types of multiwalled carbon nanotubes[J]. Composites Part A: Applied Science and Manufacturing, 2017, 103: 131-147.

[9] YOO D Y, BANTHIA N, YOON Y S. Impact resistance of ultra-high-performance fiber-reinforced concrete with different steel fibers [C]// Proceedings of 9th Rilem International Symposium on Fiber Reinforced Concrete, Vancouver, Canada, 2016.

[10] YU R, SPIESZ P, BROUWERS H J H. Static properties and impact resistance of a green Ultra-High Performance Hybrid Fiber Reinforced Concrete (UHPHFRC): Experiments and modeling[J]. Construction and Building Materials, 2014, 68: 158-171.

Self-healing behavior of mortar under different levels of pre-damage

Z. Y. Chen & Q. Q. Yu
Key Laboratory of Performance Evolution and Control for Engineering Structures of Ministry of Education, Tongji University, Shanghai 200092, China & Department of Structural Engineering, Tongji University, Shanghai 200092, China
X. Wang & W. T. Li
Key Laboratory of Advanced Civil Engineering Materials of Ministry of Education, Tongji University, Shanghai 200092, China

Abstract

This paper studied the self-healing behavior of cementitious materials with fly ash (FA). Various parameters were considered, such as FA replacement ratio, preloading level and times of preloading. Mechanical tests were conducted to evaluate the self-healing performance. The results revealed that the impact of self-healing was declined with the increase of preloading level and preloading times. Besides, a moderate FA replacement ratio was beneficial to activate self-healing behavior, where its residual strength after twice preloading of 80% was increased by 11% in comparison with the undamaged ones. This study extended the understanding of self-healing technique with addition of FA.

1. Introduction

In recent years, autogenous self-healing has shown great promise in the improvement of concrete durability[1-4]. However, a limited number of reports were published on its performance when subjected to repeated loading. Sahmaran et al.[5, 6] repeatedly preloaded engineered cementitious concrete (ECC) with FA or slag to 70% of their splitting tensile deformation capacities. Sneock et al.[7] preloaded the strain-hardening cementitious composites twice to 1% of their maximum strain. The pre-damage level of both studies was restricted. Also, the tested specimens were limited on advanced concrete instead of the common mortar. In this paper, influences of various degrees of pre-damage on self-healing behavior of mortars were investigated based on mechanical tests.

2. Experimental program

Three mixture proportions of the mortars were prepared with different FA replacement ratios of 0%, 35% and 55% by mass of binder. Prisms with a cross section of 40 mm×40 mm and a length of 160 mm were tested under a four-point bending test as shown in Figure 1. Pre-damage was defined as the ratio of preloading to the ultimate strength of the specimen, i.e. 30%, 60% and 80%. The specimens were first applied by different levels of preloading, and then stored in lime-saturated water at (23 ± 2) ℃ for 30 days. Afterwards, one batch of the specimens were loaded till failure while the other were again preloaded and immersed to examine the impact of repeated loading.

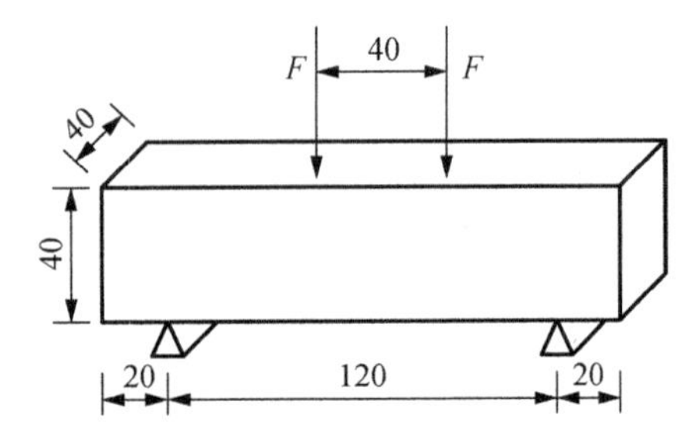

Figure 1 Four-point bending test (unit: mm)

3. Results and discussions

The failure load of specimens after the healing process was shown in Figure 2. They were normalized by the strength of the ordinary mortar without preloading. To get rid of the effect of strength variation with time, properties of undamaged specimens cured for the same days were considered.

Figure 2(a) illustrated the results of specimens after the first healing. Most specimens recovered very well in strength when compared with the undamaged ones showed by blank columns. Among them, specimens with FA replacement ratio of 35% performed best, whose flexural strength improved by 22%, 17% and 16%, when preloading levels of 30%, 60% and 80% were applied, respectively. In Figure 2(b), different from those in Figure 2(a), the remaining strength was lower than the reference (white), demonstrating the decline of self-healing effect under repeated loadings. However, specimens with FA replacement ratio of 35% under 80% f_{max} did not comply with the findings, whose strength went up by 11%. The interesting phenomenon could be resulted from the pozzolanic reaction of FA[3, 8, 9], where healing products like C-S-H gels were generated to fill the crack. Adequate space was required to activate this reaction. Under the high preloading level such as 80%, the key requirement was satisfied.

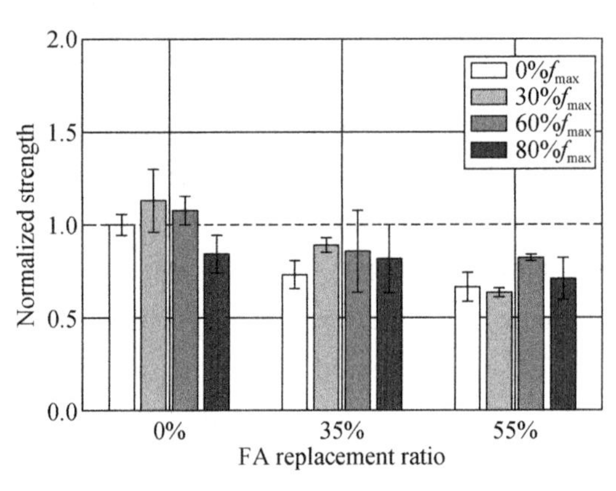

(a) Under monotonic load

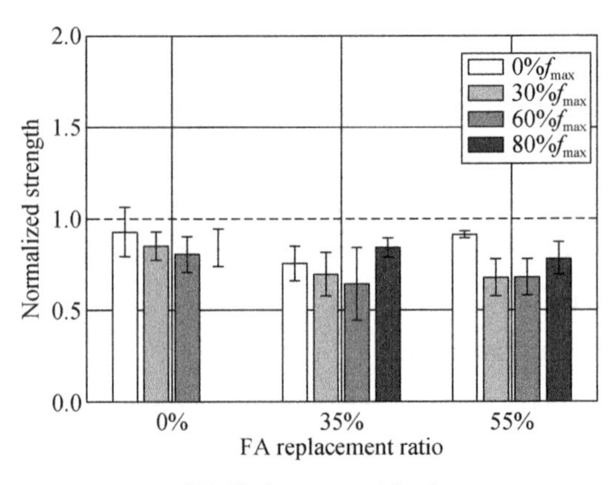

(b) Under repeated load

Figure 2 Strength restoration

4. Conclusions

Self-healing behavior of mortar under different levels of pre-damage was investigated in this study. Based on the results in mechanical property, some conclusions were drawn.

(1) A proper replacement ratio of FA was beneficial to activate the self-healing performance. Even under repeated 80% f_{max}, remarkable self-healing behavior was observed in the specimen with FA replacement ratio of 35%.

(2) For most specimens, the self-healing performance would be weakened by the increase of preloading level and times.

Acknowledgements

The authors gratefully acknowledged the financial supports provided by Top Interdisciplinary Funds of Civil Engineering for Tongji University.

References

[1] ALTOUBAT S, LANGE D A. Creep, shrinkage and cracking of restrained concrete at early age [J]. ACI Materials Journal, 2001, 98(4): 323-331.

[2] ŞAHMARAN M, KESKIN S B, OZERKAN G, et al. Self-healing of mechanically-loaded self consolidating concretes with high volumes of fly ash[J]. Cement and Concrete Composites, 2008, 30(10): 872-879.

[3] ZHANG Z G, QIAN S Z, Ma Hui. Investigating mechanical properties and self-healing behavior of micro-cracked ECC with different volume of fly ash[J]. Construction and Building Materials, 2014, 52: 17-23.

[4] TITTELBOOM K V, GRUVAERT E, RAHIER H, et al. Influence of mix composition on the extent of autogenous crack healing by continued hydration or calcium carbonate formation [J]. Construction and Building Materials, 2012, 37: 349-359.

[5] YILDIRIM G, SAHMARAN M, AHMED H U. Influence of hydrated lime addition on the self-healing capability of high-volume fly ash incorporated cementitious composites[J]. Journal of Materials in Civil Engineering, 2014, 27(6): 04014187.

[6] SAHMARAN M, YILDIRIM G, NOORI R, et al. Repeatability and pervasiveness of self-healing in engineered cementitious composites[J]. ACI Materials Journal, 2015, 112(4): 513-522.

[7] SNOECK D, DE BELIE N. Repeated autogenous healing in strain-hardening cementitious composites by using superabsorbent polymers [J]. Journal of Materials in Civil Engineering, 2015, 28(1): 04015086.

[8] TERMKHAJORNKLT P, NAWA T, YAMASHIRO Y, et al. Self-healing ability of fly ash-cement systems[J]. Cement and Concrete Composites, 2009, 31 (3): 195-203.

[9] SAHMARAN M, YILDIRIM G, ERDEM T K. Self-healing capability of cementitious composites incorporating different supplementary cementitious materials [J]. Cement and Concrete Composites, 2013, 35 (1): 89-101.

Coupled effects of moisture and chloride transport in cement mortar during wetting-drying cycles

X. X. Zhao & Z. L. Jiang & W. W. Li

Guangdong Provincial Key Laboratory of Durability for Marine Civil Engineering, Shenzhen University, Shenzhen 518060, China

Abstract

In the marine environment, the concrete in the tidal zone is subject to the periodic drying and wetting, where chloride-induced steel corrosion is the most severe. During the wetting period, the chloride penetrates into concrete quickly along with capillary absorption. During the drying period, the water moves towards the exposed surface and the diffusion process of chloride ions became dominant. The wetting and drying cycles accelerate the chloride penetration into concrete. In this study, the coupled effects of moisture and chloride transport in cement mortar (including two mixing ratios, the water-cement ratio is 0.5, and one group is mixed with 15% fly ash) during wetting-drying cycles were investigated. Three different tests, i.e. the water vapor sorption, capillary absorption and cyclic drying-wetting tests, were conducted for cement mortar specimens. The water retention of cement mortar with different chloride concentrations was measured by water vapor adsorption tests[1]. The hysteretic behavior of the adsorption isotherms was also studied[2]. Under an ambient environment with constant temperature and relative humidity, the drying rates of the mortar containing different amount of chlorides were measured. In the capillary absorption tests, the water penetration depths were measured for the mortars exposed to 0%, 1%, 3% and 5% NaCl solution. The chloride penetration depths were measured by $AgNO_3$ colorimetry, potentiometric titration and Electron Microprobe (EPMA), respectively[3, 4]. The comparison of the measured depths by different methods indicated that the $AgNO_3$ colorimetry was reliable for the chloride penetration depth. The water penetration depths were also analyzed by Photo colorimetry and Nuclear magnetic resonance and the two methods were compared. The results also showed that the water penetration depth of mortar specimens was significantly greater than the chloride penetration depth the same exposure time[4]. Finally, the chloride transport tests were conducted under cyclic drying-wetting conditions[5]. The effects of dry-wet ratio on chloride ion and water transport in mortar were analyzed. The interaction between the chloride and water transport during the cycles was discussed based on the measured results.

Acknowledgements

This work was supported by the National Natural Science Foundation (Grant No. 51808346) and the New Faculty Start-up Research Project of Shenzhen University (Grant No. 2018024).

References

[1] BAROGHEL-BOUNY V, THIÉRY M, WANG X. Modelling of isothermal coupled moisture-ion transport in cementitious materials [J]. Cement and Concrete Research, 2011, 41(8): 828-841.

[2] Lou M Y, Zeng Q, Pang X Y, et al. Characterization of pore structure of cement-based materials by water vapor sorption isotherms [J]. Journal of the Chinese Ceramic Society, 2013, 41: 1401-1408.

[3] Zhang P, Hou D S, Liu Q, et al, Water and chloride ions migration in porous cementitious materials: An experimental and molecular dynamics investigation[J]. Cement and Concrete Research, 2017: 161-174.

[4] Liu Q. Study on transport mechanism of moisture and chloride cement mortar [D]. Shandong: Qingdao Technological University, 2016.

[5] Zhang Q Z. Similarity study on accelerated test of chloride erosion of concrete in Tidal Zone [D]. Shanghai: Tongji University, 2012.

Comparison of different experimental methods for pore characteristics of cement paste

J. F. Lu & Z. L Jiang & W. W. Li

Guangdong Provincial Key Laboratory of Durability for Marine Civil Engineering, Shenzhen University, Shenzhen 518060, China

Abstract

Pore structure characteristics, including porosity, pore size distribution, pore connectivity and tortuosity, play an important role in the durability of cement-based materials. The gradual improvement of computational capabilities has enabled nondestructive imaging methods such as X-ray computed microtomography (X-ray μCT) to yield high-resolution, three-dimensional (3D) representations of pore spaces. In this paper, X-ray μCT was used to derive the 3D images of hardened cement paste specimens with water-to-cement (W/C) ratios of 0.3, 0.4 and 0.45 at curing ages of 1, 3, 7, 28 days at the resolution of 1 μm/voxel. By defining the upper threshold value in the basis of the inflection point of the cumulative volume fraction curve, the pores were distinguished from solid phases (i.e. hydration products and unhydrated cement grains) in the microtomography images of each specimen[1-3]. The 3D pore structures of cement-based materials were obtained and the pore structure characteristics were analyzed in detail by the application of the segmentation and 3D reconstruction software Avizo. The pore characteristics by X-ray μCT were compared with the results from other experiment methods, including mercury intrusion porosimetry (MIP), backscattered electron (BSE), N_2 adsorption, vapor adsorption, vacuum water saturation[4], nuclear magnetic resonance (NMR)[5] and small-angle X-ray scattering (SAXS)[6, 7]. Furthermore, chloride ion diffusion test was conducted to investigate the relationship between the pore structure and chloride diffusivity of cement-based materials. Based on the pore structure characteristics obtained by experiment methods, chloride ion transmission coefficient yielded from chloride diffusion was analyzed qualitatively and quantitatively, by which we identified the most critical factors affecting chloride ion transmission coefficient in pore structure characteristics. In the meanwhile, by fitting the experiment data on a theoretical basis, a more optimized model for the relationship between pore structure characteristics and chloride ion transmission coefficient was proposed.

Acknowledgements

This work was supported by the National Natural Science Foundation (Grant No. 51808346) and the New Faculty Start-up Research Project of Shenzhen University (Grant No. 2018024).

References

[1] WONG H S, HEAD M K, BUENFELD N R. Pore segmentation of cement-based materials from backscattered electron images [J]. Cement and Concrete Research, 2006, 36(6): 1083-1090.

[2] GALLUCCI E, SCRIVENER K, GROSO A, et al. 3D experimental investigation of the microstructure of cement pastes using synchrotron X-ray microtomography [J]. Cement and Concrete Research, 2007, 37 (3): 360-368.

[3] IASSONOV P, GEBRENEGUS T, TULLER M. Segmentation of X-ray computed tomography images of porous materials: A crucial step for characterization and quantitative analysis of pore structures [J]. Water Resources Research, 2009, 45 (9).

[4] ČÁCHOVÁ M, KOŇÁKOVÁ D, VEJMELKOVÁ E. Pore Distribution and Water Vapor Diffusion Parameters of Lime Plasters with Waste Brick Powder [J]. Advanced Materials Research, 2014, 1054: 205-208.

[5] ZHAO H T, QIN X, LIU J P, et al. Pore structure characterization of early-age cement pastes blended with high-volume fly ash[J]. Construction and Building Materials, 2018, 189: 934-946.

[6] METWALLI E, PLANK J, MÜLLER-BUSCHBAUM P, et al. Occurrence of intercalation of PCE superplasticizers in calcium aluminate cement under actual application conditions, as evidenced by SAXS analysis [J]. Cement and Concrete Research, 2013, 54: 191-198.

[7] YANG R H, LIU B Y, WU Z W. Study on the pore structure of hardened cement paste by SAXS [J]. Cement and Concrete Research, 1990, 20(3): 385-393.

Introduction to performance-based wind design for high-rise buildings

S. Y. Jeong & T. Kang

Department of Architecture and Architectural Engineering, Seoul National University, Seoul, Korea

Abstract

US building design code, ASCE 7-16[1] defines the same return periods concept for seismic and wind loads based on the probability of collapse of buildings. The return periods of risk targeted maximum considered earthquake (MCE_R) in ASCE 7-16 are from 1 000 to 3 000 years (National Institute Building Sciences Building Seismic Safety Council, 2015). The design basis earthquake (DBE) is 2/3 of MCE_R; however, considering importance factor of 1.0 to 1.5, converted return period of DBE is up to 3 000 years. Wind loads in ASCE 7-16 use return periods from 300 to 3 000 years. Although the return periods of seismic and wind loads are similar, there is a large difference in target performance of structure. An elastic design is used for wind loads, while inelastic behavior is permitted under seismic loads. An elastic design is carried out using seismic loads reduced by response modification coefficient (*R* factor) considering ductility of system. The difference of target performance of two loads causes problems in the design of high-rise buildings.

The design seismic load can be reduced by *R* factor, because buildings are assumed to have sufficient ductility with a desirable yield mechanism. The design wind load can be larger than reduced seismic load in the high-rise buildings with aspect ratio larger than 3[2], and the elastic design is required. To satisfy elastic design for wind load, the size of horizontal members is increased. It results in yielding of vertical members and joints rather than horizontal members like beams and braces under strong earthquakes. The yield of vertical members and joints prior to horizontal members leads to a brittle system as shown in Figure 1. The initial stiffness and strength of system increase, but the ductility decreases. Thus, the *R* factor with assumed sufficient ductility in the seismic design becomes invalid.

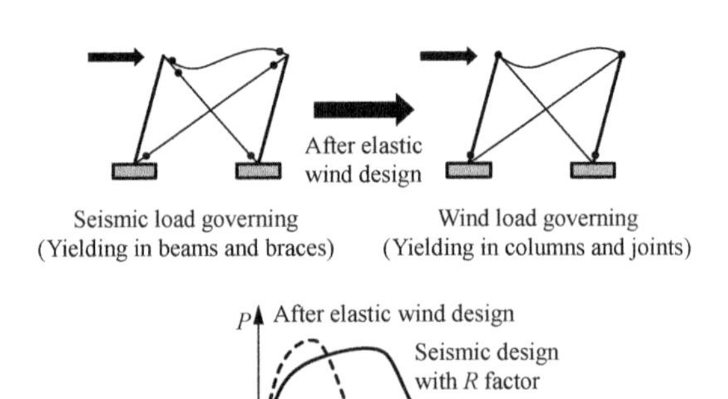

Figure 1 Simplified yield mechanism issue of a frame in high-rise buildings caused by elastic wind design

Because the real seismic load without R factor is still larger than wind load, the safety of structure cannot be guaranteed. Due to this issue, the necessity of inelastic wind design, so-called performance-based wind design (PBWD) is recognized.

Although no detailed implementation method has been proposed so far, ASCE recently published Prestandard for Performance-Based Wind Design[3]. It proposed limited inelastic performance objectives for wind load with return periods from 700 to 3 000 years.

To carry out PBWD, verification of performance by nonlinear analysis is necessary, and before that an initial design is required. The R factor is used in initial seismic design (prior to performance-based seismic design), but the whole wind load cannot be reduced like seismic design. Wind load is composed of mean, background, and resonant components. The mean and background components are essentially quasi-static loads, so it makes sense that a structure cannot be designed to be inelastic under these components. It appears more reasonable that the response modification factor for wind load (R_W) would be applied to the resonant component only[4]. Because the duration of wind load is much longer than that of seismic load, the R_W factor should be less than R factor considering fatigue (i. e. 1.5 to 2.0). Because the portion of resonant component in the along-wind load increases for increased building height and the across and torsional wind loads are mostly composed of resonant component[5], the design wind load can be effectively reduced by an R_W factor of 1.5 or 2.0. By reducing initial design wind load, overdesign of horizontal members can be precluded. After the initial design, exact dynamic performance under original wind loads should be verified by nonlinear time history analysis. Time histories of wind load are obtained from wind tunnel tests or generated from power spectral density functions.

As buildings become higher, PBWD is expected to be necessary to secure the structural performance and efficiency.

Acknowledgements

This work was supported by the Korea Agency for Infrastructure Technology Advancement (KAIA) grant funded by the Ministry of Land, Infrastructure and Transport (Grant No. 19CTAP-C151831-01).

References

[1] ASCE. Minimum Design Loads and Associated Criteria for Buildings and Other Structures [S]. Reston, VA: American Society of Civil Engineers, 2017.

[2] KANG, T H K, JEONG S Y, Alinejad H. Understanding of wind load determination according to KBC 2016 and its application to high-rise buildings[J]. Journal of the Wind Engineering Institute of Korea, 2019, 23: 83-89.

[3] ASCE. Prestandard for Performance-Based Wind Design [S]. Reston VA: American Society of Civil Engineers, 2019.

[4] El Damatty A A, Elezaby F Y. The integration of wind and structural

engineering[C]// The 2018 World Congress on Advances in Civil, Environmental, & Materials Research, Incheon, Korea, 2018.

[5] KANG, T H K, JEONG S Y, ALINEJAD H. Comparative study on standards of along-wind loads for the design of high-rise buildings [J]. Journal of the Wind Engineering Institute of Korea, 2019, 23: 135-142.

Analysis of short fibers distribution in glass fiber reinforced concrete using X-ray Nano-CT

L. Hong

Department of Structural Engineering, Hefei University of Technology, Hefei 230009, China & Department of Civil and Environmental Engineering, University of California, Irvine, CA 92697, USA

T. D. Li

Department of Structural Engineering, Hefei University of Technology, Hefei 230009, China

L. Z. Sun

Department of Civil and Environmental Engineering, University of California, Irvine, CA 92697, USA

Abstract

Fibers have been successfully adopted as reinforcing phase to improve the properties of materials and structures due to their excellent anti-cracking performance[1]. The reinforced efficiency is significantly influenced by the distribution of fiber[2]. Many researchers[3-6] have analyzed the distribution or orientation of steel fibers in steel fiber reinforced concrete (SFRC), however, there is a lack of study focusing on the short flexible fiber distribution in fiber reinforced concrete (FRC). In this paper, Nano-CT analyses were adopted firstly to reveal a detailed microstructure of glass fiber reinforced concrete, the typical "slice" of the sample and the 3D distribution of glass fiber in FRC were shown in Figure 1, respectively, in which the physical extent was 2.99 mm×2.99 mm×2.99 mm.

Additionally, the distribution of glass fiber in FRC was analyzed using statistical theory. Here the inclination angle of glass fiber, θ and φ were illustrated in Figure

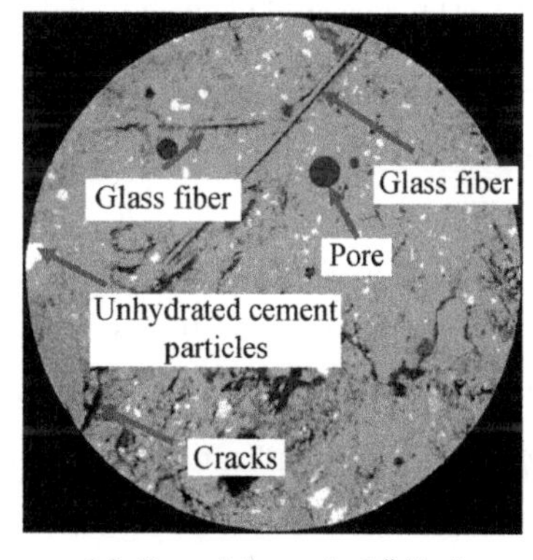

(a) One of the typical "slice"

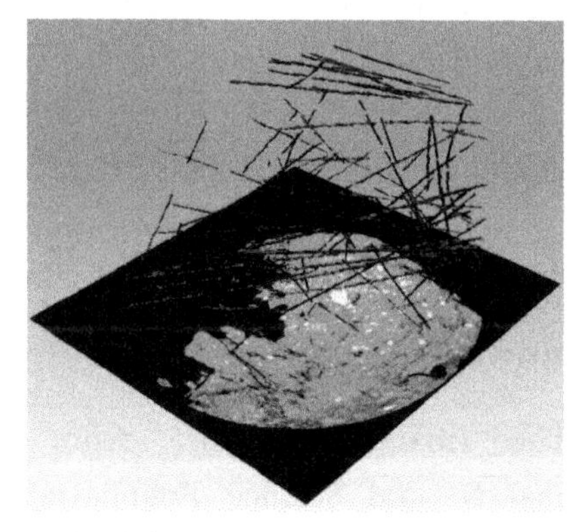

(b) The distribution of glass fiber

Figure 1 Image of glass fiber reinforced concrete from Nano-CT

2, and the probability density histogram of θ and φ were described in Figure 3. It can be seen in Figure 3(a) that most of the θ was between in 80° and 90°, while φ was mainly in the range from 0° to 30°.

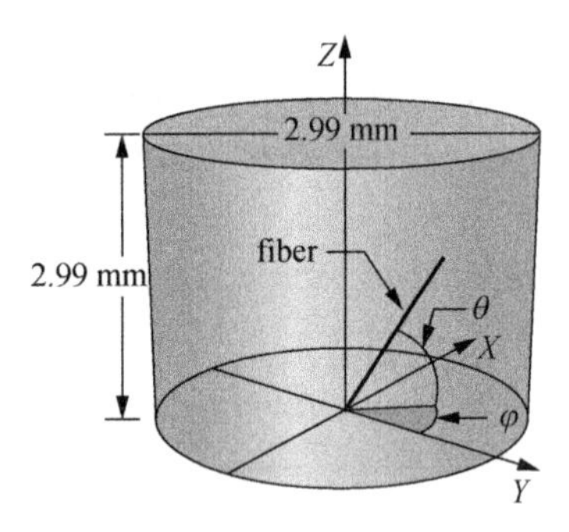

Figure 2 The inclination angles of glass fiber

According to previous researches in references [4 – 6], the parameters used to describe fiber orientation η_1, η_2, η_3, S_{n_1}, S_{n_2}, S_{n_3} and S were summarized in Table 1. It can be obtained from Table 1 that η_1 equaled to 0.97, which was larger than either η_2 or η_3. This indicated the glass fibers mostly tended to *XOY* plane. Moreover, S_{n_1} equaled to 10.81, which was smaller than either S_{n_2} or S_{n_3}. This also means the glass fibers more tended to *XOY* plane than either *YOZ* plane or *XOZ* plane.

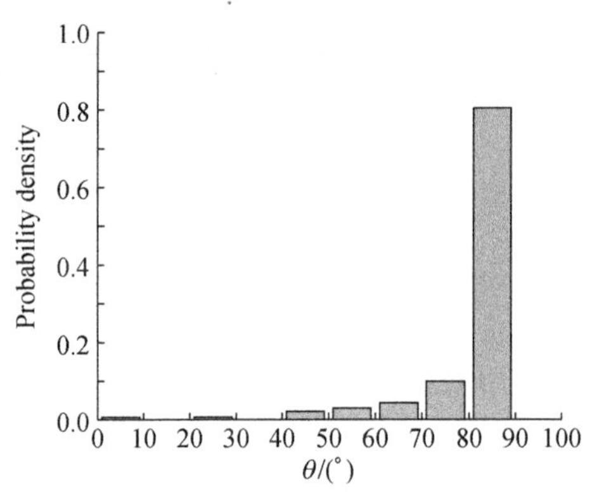

(a) Probability density-θ

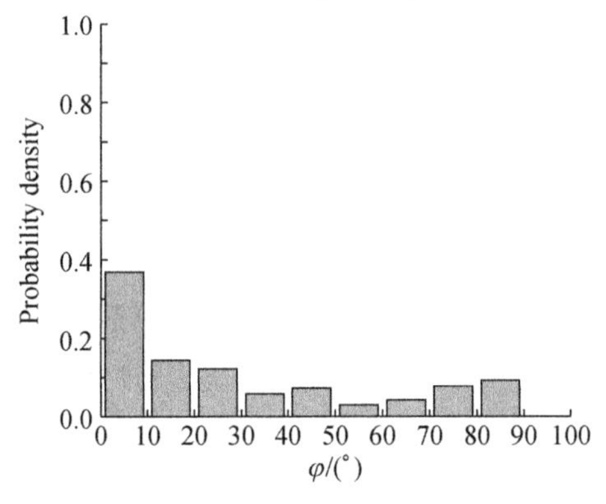

(b) Probability density-φ

Figure 3 Probability density histogram of inclination angle

Table 1 Summary of the fiber distribution parameters of the samples

Number of the fibers	η_1	η_2	η_3	S_{n1}	S_{n2}	S_{n3}	S
215	0.97	0.81	0.46	10.81	25.15	27.63	5.22

Acknowledgements

The financial support from the National Natural Science Foundation of China (Grant No. 51508146) and the China Scholarship Council (Grant No. 201766195021) are acknowledged and sincerely appreciated.

References

[1] AKKAYA Y, SHAH S P, ANKENMAN B. Effect of fiber dispersion on multiple cracking of cement composites[J]. Journal of Engineering Mechanics, 2001, 127(4): 311–316.

[2] KANG S T, LEE B Y, KIM J K, et al. The effect of fibre distribution characteristics on the flexural strength of steel fibre-reinforced ultra high strength concrete[J]. Construc-

tion & Building Materials, 2011, 25 (5): 2450-2457.

[3] ZHOU B, UCHIDA Y C. Relationship between fiber orientation/distribution and post-cracking behavior in ultra-high-performance fiber-reinforced concrete (UHPFRC) [J]. Cement and Concrete Composites, 2017, 83: 66-75.

[4] HERRMANN H, PASTORELLI E, KALLONEN A, et al. Methods for fibre orientation analysis of X-ray tomography images of steel fibre reinforced concrete (SFRC) [J]. Journal of Materials Science, 2016, 51(8):3772-3783.

[5] SUURONEN J P, EIK M, PUTTONEN J, et al. Analysis of short fibres orientation in steel fibre-reinforced concrete (SFRC) by X-ray tomography [J]. Journal of Materials Science, 2013, 48(3): 1358-1367.

[6] SOROUSHIAN P, LEE C D. Distribution and orientation of fibers in steel fiber reinforced concrete [J]. ACI Materials Journal, 1990, 87(5):433-439.

Probabilistic assessment of scaling for sub-sea wells

Y. Guan & J. J. Qin & J. D. Sørensen & M. H. Faber
Department of Civil Engineering, Aalborg University, DK - 9220 Aalborg Ø, Denmark & Danish Hydrocarbon Research Technology Center, Aalborg, Denmark

Abstract

Scaling is a common and severe phenomenon of degradation for sub-sea wells of offshore oil and gas production facilities. Scaling itself, as well as intervention activities aiming to remove scaling may lead to significant production losses over their service lives and this calls for a careful consideration of how the management of scaling formation may best be optimized. However, since scale formation is associated with substantial uncertainty, its management comprises a rather non-trivial challenge. To achieve this challenge, it is of crucial importance to understand the processes leading to and governing scaling, and to represent the best available knowledge on these processes consistently.

Based on scaling models proposed by Zhen-Wu et al.[1] and Dawe and Zhang[2] together with ongoing research undertaken at the Danish Hydrocarbon Research and Technology Center, the present paper introduces novel probabilistic models for the representation of the available knowledge and dominating uncertainties associated with calcium carbonate ($CaCO_3$) and barium sulphate ($BaSO_4$) scaling processes[3]. The proposed probabilistic models are formulated as functions of the main parameters governing scaling processes, including the temporal variability of the chemical composition of the production, temperature and pressure. The dependency of scale propagation, on temporal and spatial variations in production temperature and pressure is represented through a Poisson square wave process. On this basis, it is possible to model the scale growth probabilistically over the service lives of sub-sea wells - and this in turn forms the basis for optimal sub-sea well integrity management. The proposed probabilistic models are illustrated through a principal example, assessing the significance of model assumptions and the sensitivities of the probabilistic characteristics of scale formation with respect to the considered uncertainties.

Acknowledgements

The authors would like to appreciate the Danish Hydrocarbon Research and Technology Center (DHRTC) for providing the financial support and technical engagement for this project.

References

[1] ZHEN-WU B Y, DIDERIKSEN K, OLSSON J, et al. Experimental determination of barite dissolution and precipitation rates as a function of temperature and aqueous fluid composition [J]. Geochimica et Cosmochimica Acta, 2016, 194: 193-210.

[2] DAWE R A, ZHANG Y P. Kinetics of calcium carbonate scaling using observations from glass micromodels [J]. Journal of Petroleum Science and Engineering, 1997, 18(3-4): 179-187.

[3] ODDO J E, TOMSON M B. Why scale forms in the oil field and methods to predict it[J]. SPE Production & Facilities, 1994, 9(1): 47-54.

Corrosion behavior of steel bars in simulated concrete pore solution under tensile stress

Z. H. Jin & X. L. Gu
Department of Structural Engineering, Tongji University, Shanghai 200092, China

Abstract

Corrosion of steel reinforcement bars in concrete could lead to critical damage to concrete structure durability, possibly even lead to structure failure[1]. In practice, reinforced concrete structures in coastal areas, such as bridges and ports, inevitably experience variable mechanical loads and chloride penetration. Different kinds of loads may have an impact on the depassivation and corrosion processes in metals. Some researchers have conducted studies with respect to the influence of stress on the corrosion behavior of different metals. It is found that stress could affect the corrosion speed of metals regardless of the environment[2-5]. However, only a few studies were done in a concrete environment[6, 7], and these studies merely presented qualitative conclusions. The influence mechanism of stress on steel corrosion has not been studied well. Besides, reinforcing steel bars embedded in concrete are always intersected with each other to form rebar mesh. The macro-cell corrosion between dissimilar crossed steel bars has not been well studied before.

Therefore, the present study aims to investigate the influence of tensile stress on corrosion behavior of single steel bar and macro-cell corrosion between dissimilar crossed steel bars.

In this study, the specimens were divided into two groups. In Group 1, each specimen was pre-stressed and immersed into simulated concrete pore solution within chloride ions alone, as shown in Figure 1.

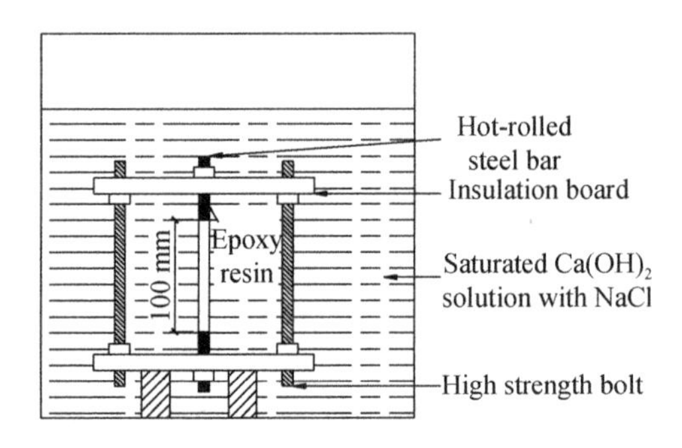

Figure 1 Group 1 specimens

In Group 2, each prestressed hot-rolled ribbed bar was connected with an unstressed hot-rolled plain steel bar to simulated the macro-cell corrosion between dissimilar crossed steel bars in concrete structures, as shown in Figure 2. The details of specimens are shown in Table 1.

Based on the principle of electrochemical corrosion, the polarization equation of a single steel bar for Group 1 can be expressed as Eq. (1).

$$\begin{cases} i = i_{corr}(A - B) \\ A = \exp\left[\dfrac{2.3(E - E_{corr})}{\beta_a}\right] \\ B = \dfrac{\exp\left[-\dfrac{2.3(E - E_{corr})}{\beta_c}\right]}{1 - \dfrac{i_{corr}}{i_L}\left\{1 - \exp\left[-\dfrac{2.3(E - E_{corr})}{\beta_c}\right]\right\}} \end{cases} \tag{1}$$

where E_{corr} is the self-corrosion potential, i_{corr} is the self-corrosion current density, i_L is the limit current density, β_a and β_c are anodic Tafel slope and cathode Tafel slope.

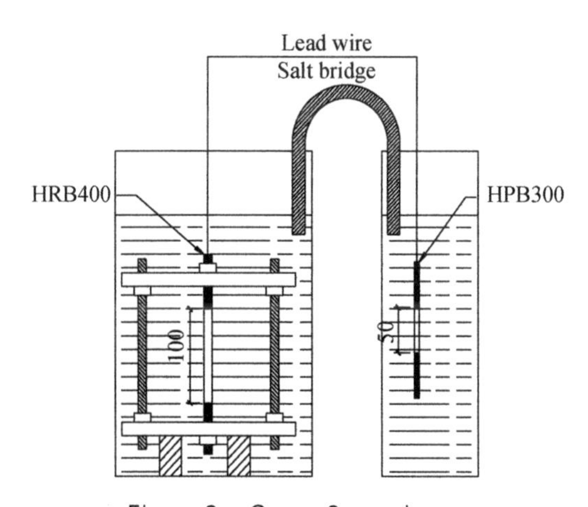

Figure 2 Group 2 specimens

Table 1 The details of the specimens

Group	Specimens code	Tensile stress /MPa	Macro-cell corrosion
Group 1	1-0	0	No
	1-60	60	
	1-120	120	
	1-180	180	
Group 2	2-0	0	Yes
	2-60	60	
	2-120	120	
	2-180	18	

Considering the macro-cell corrosion between dissimilar steel bars, the polarization equation of the specimens in Group 2 can be expressed as Eq. (2).

$$\begin{cases} \beta_{a1}\lg\dfrac{i_{a1}}{i_{corr1}} - \beta_c\lg\dfrac{i_{c1}}{i_{corr1}} + \\ \qquad \dfrac{2.3RT}{zF}\lg\dfrac{i_{L1} - i_{corr1}}{i_{L1} - i_{c1}} = 0 \\ I_g = (i_{a1} - i_{c1})S_1 \\ \dfrac{i_{c1}}{i_{L1}} = \theta \\ \dfrac{i_{a1}}{i_{corr1}} = \gamma \end{cases} \tag{2}$$

where i_{a1} and i_{c1} are the current density of anodic reaction and cathode reaction of the hot-rolled steel bar, respectively. I_g is the macro-cell corrosion current. S_1 is the surface area of the anodic steel bar.

Several electrochemical measurements were performed on the specimens to get the polarization curve and other data. Based on Eq. (1) and Eq. (2), Matlab was used to fit the data.

Results indicate that steel specimens immersed in chloride-contaminated solution under tensile loadings corroded more rapidly compared with the unstressed specimen. When the stress level gets higher, the corrosion current density gets bigger. Besides, the stress level has a great impact on macro-cell corrosion between dissimilar steel bars. The value of the macro-cell corrosion current changes with the stress level.

Acknowledgements

This study was financially supported by

the National Natural Science Foundation of China (Grant No. 51320105013).

References

[1] CORONELLI D, GAMBAROVA P. Structural Assessment of Corroded Reinforced Concrete Beams: Modeling Guidelines [J]. Journal of Structural Engineering, 2004, 130(8): 1214-1224.

[2] LIU X D, FRANKEL G S, ZOOFAN B, et al. Effect of applied tensile stress on intergranular corrosion of AA2024-T3[J]. Corrosion Science, 2004, 46(2): 405-425.

[3] GAO K, LI D, PANG X, et al. Corrosion behavior of low-carbon bainitic steel under a constant elastic load[J]. Corrosion Science, 2010, 52: 3428-3434.

[4] YANG W J, YANG P, LI X M, et al. Influence of tensile stress on corrosion behavior of high-strength galvanized steel bridge wires in simulated acid rain [J]. Materials and Corrosion, 2012, 63(5): 401-407.

[5] ZHANG S, PANG X L, WANG Y B, et al. Corrosion behavior of steel with different microstructures under various elastic loading conditions[J]. Corrosion Science, 2013, 75: 293-299.

[6] FENG X G, TANG Y M, ZUO Y. Influence of stress on passive behavior of steel bars in concrete pore solution [J]. Corrosion Science, 2011, 53(4): 1304-1311.

[7] ZHANG Y, POURSAEE A. Passivation and Corrosion Behavior of Carbon Steel in Simulated Concrete Pore Solution under Tensile and Compressive Stresses[J]. Journal of Materials in Civil Engineering, 2015, 27 (8): 04014234.

Simulation and analysis of pipe jacking excavation process based on Hongxu-Hongmei Project

S. J. Wang & Z. X. Zhang

Department of Geotechnical Engineering, Tongji University, Shanghai 200092, China

Abstract

With the development of urbanization, the construction of municipal underground pipeline is increasingly important. Based on the Hongxu- Hongmei rainwater pumping station and head pipe project, Flac 3D is employed to simulate and analyze the additional stress and surface deformation of surrounding soils caused by excavation in this paper. Then several examples are given to illustrate the influence of pipe jacking, which can help to understand the effect of construction process in more detail. In summary, this study results will benefit in-situ engineers to take suitable measures in real-life engineering.

The afore-mentioned numerical simulation of pipe jacking was widely carried out around the world, e.g. Huang and Hu[1] simulated the mechanical effect of pipe jacking. Fang et al.[2] focused on the difference between theoretical and measured analysis. Wei et al.[3] analyzed the soil disturbance during pipe jacking.

In order to fully reflect the influences of pipe jacking, this study mainly analyzed the lateral settlement, vertical settlement, plastic zone and support surface pressure. Furthermore, parameter sensitivity analysis during pipe jacking was also implemented. The results are as follows (Figure 1–Figure 6).

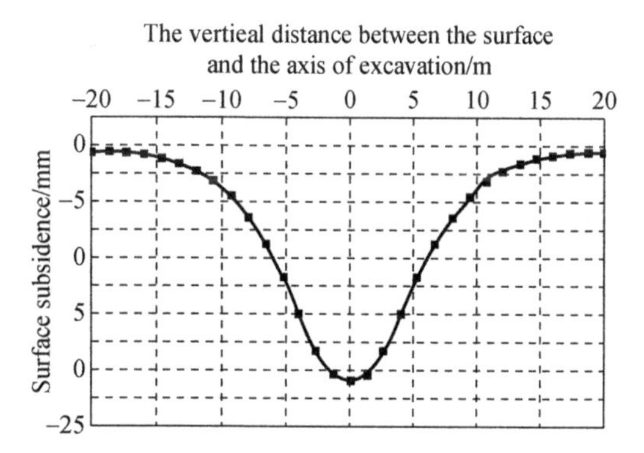

Figure 1 Lateral surface subsidence curve ($Y = 60$ m)

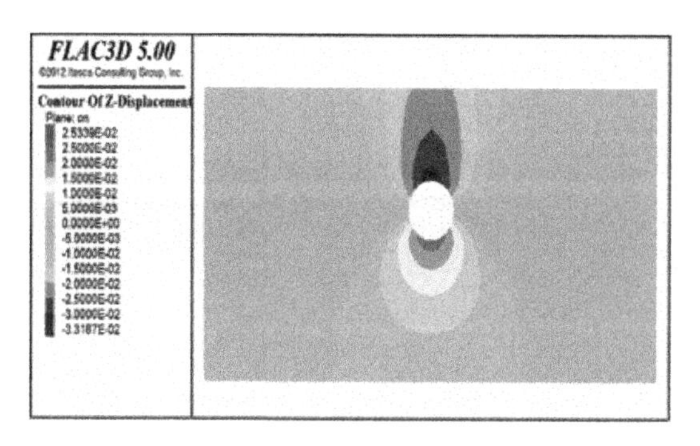

Figure 2 Lateral displacement cloud map

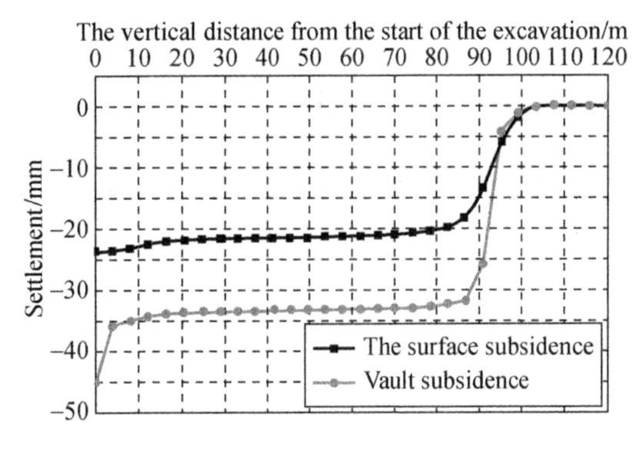

Figure 3 Vertical surface subsidence curve ($X = 0$)

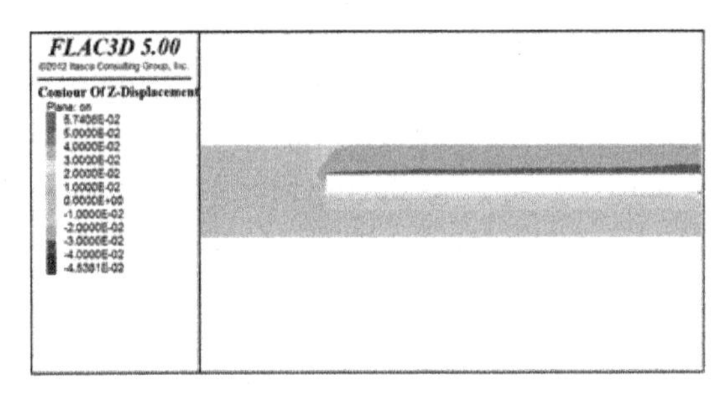

Figure 4 Vertical displacement cloud map ($X = 0$)

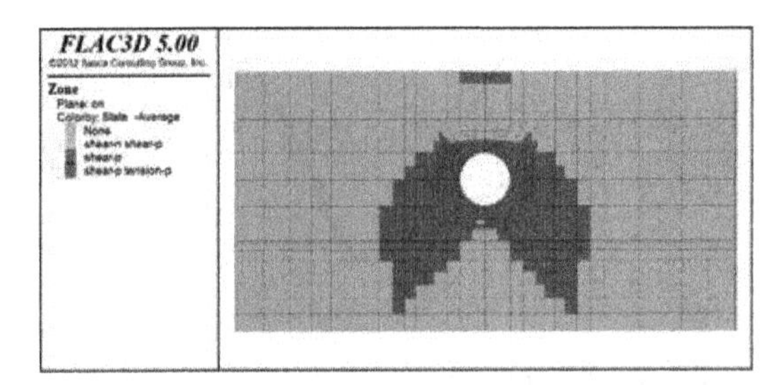

Figure 5 Transverse section of plastic zone of soil ($Y = 60$ m)

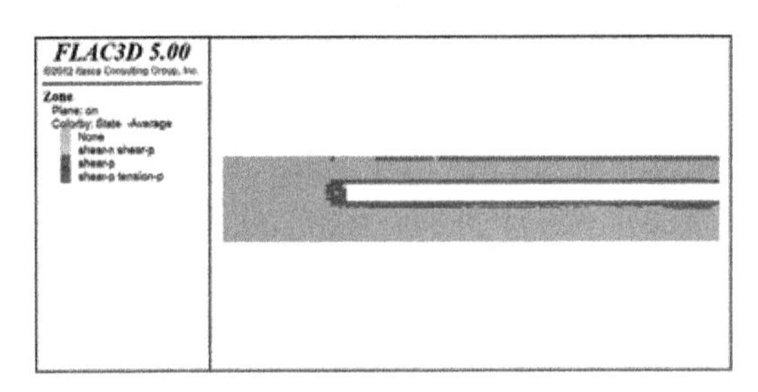

Figure 6 Vertical section of plastic zone of soil ($X = 0$)

Based on the simulation results, the following conclusions can be drawn.

(1) The curve of lateral subsidence trough is normally distributed, which is consistent with the Peck empirical formula.

(2) The results show that the trend of surface longitudinal settlement curve decreases gradually along the axis of the tube body, which obeys the distribution law of surface longitudinal settlement.

(3) The increase of support pressure will reduce the surface settlement in the excavation section, which mainly affects the soil state within 2D distance from the central axis. However, too much support pressure may lead to the surface uplift in the trenchless section.

(4) Lateral friction mainly affects the disturbance of soil mass in the construction process. By reducing the lateral friction between the pipe and soil, the disturbance of soil can be effectively reduced and the fluctuation of surface settlement along the excavation axis can be reduced greatly.

Combined with the actual situation of the project and the numerical simulation results, it shows that increasing the support pressure properly and reducing the lateral friction resistance between the pipe and soil can effectively reduce the disturbance of the soil state and reduce the settlement. Therefore, it is important to control the support pressure reasonably, and reduce the lateral friction resistance in the construction process by using appropriate technology to control the surface settlement.

References

[1] HUANG H W, HU X. Numerical simulation analysis of mechanical effect of pipe jacking [J]. Chinese Journal of Rock Mechanics and Engineering, 2003(3): 400-406.

[2] FANG Y G, MO H H, ZHANG C Y. Theoretical and measured analysis of soil deformation in disturbed area of pipe jacking [J]. Chinese Journal of Rock Mechanics and Engineering, 2003(4): 601-605.

[3] WEI G, XU R Q, TU W. Theoretical analysis and experimental study on soil disturbance caused by pipe jacking [J]. Chinese Journal of Rock Mechanics and Engineering, 2004(3): 476-482.

Numerical simulations on damage and dynamic responses of masonry infilled wall under close-in range explosion

J. H. Hu & H. Wu

Department of Disaster Mitigation for Structures, College of Civil Engineering, Tongji University, Shanghai 200092, China

Abstract

Masonry infilled walls have been widely used in civil structures, due to its poor blast resistance, especially the masonry fragments generated under the close-in range vehicle or suicide bomb explosions will lead to the secondary damage, which seriously threaten the safety of people and equipment inside the buildings. Under the close-in range explosion, the wall is mainly suffered from local damage and disintegrated into more debris, which is greatly different from the global failure of walls in the far-range explosion scenarios, and the related studies are still insufficient.

The explosion load is mainly controlled by the scaled distance Z, where $Z = R/W^{1/3}$, R is the standoff distance, W is the equivalent TNT weight. Orton et al.[1] pointed out that when scaled distance $Z \leqslant 0.4\ \mathrm{m/kg^{1/3}}$, the explosion belonged to the near-field explosion. Henrych and Major[2] used $R_w < R < 10R_w$ as the standard of near-field explosion, where R_w was the radius of charge.

Varma et al.[3] reported a series of blast test data, including reflected pressure, reflected impulse, damage level and maximum deflection of twenty-seven brick panels with different thickness. Based on Varma's tests, Wei and Stewart[4] carried out numerical simulations to estimate the response and damage of unreinforced brick masonry walls subjected to explosive blast loading. Michaloudi and Gebbeken[5] conducted the experimental and numerical study of masonry walls under far-field and contact detonations. In addition, the Air Force Laboratory[6-9] carried out systematic experiments and numerical simulations on the anti-explosion performance of hollow CMU walls and spray polyurea reinforced CMU walls. Wang et al.[10, 11] carried out the explosion test and numerical simulation of the spray polyurea reinforced clay brick walls. However, the researches on the blast resistance of masonry wall mainly focused on the far-field explosion, further study on damage and dynamic responses of masonry wall under close-in range explosion is needed.

Based on the close-in range explosion test on clay brick infilled wall by Shi et al.[12], the Load Blast method, ALE (Arbitrary Lagrangian Euler) method and impulse method are adopted to simulate the blast loadings and predict the damage and dynamic response of masonry infilled wall. Figure 1 and Figure 2 respectively illustrate the finite element model of ALE method and how to apply the impulse produced by explosion to the wall. By comparing with the test data, the applicability of the impulse method is verified. As what Table 1 shows, under close-in range explosion, the peak overpressures predicted by ALE method can agree well with the data from experiment and more accurate than the data predicted by Load Blast method. Figure 3 also shows that ALE method and impulse method accomplish the better simulations of wall's damage and dynamic response under close-in range explosion.

Table 1 Peak overpressure of each measuring point under 1kg TNT explosion (MPa)

Method	Pre1	Pre2	Pre3	Pre4
Experiment	6.4	>10.0	>10.0	3.4
ALE	7.5	11.1	13.3	3.8
Load Blast	9.4	16.2	16.2	4.8

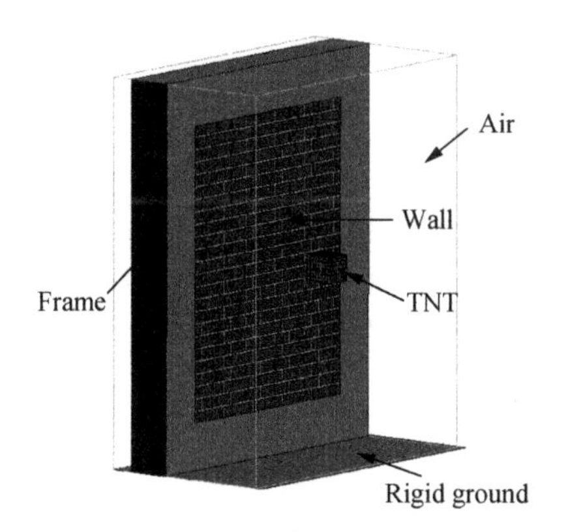

Figure 1 Finite element model of ALE method

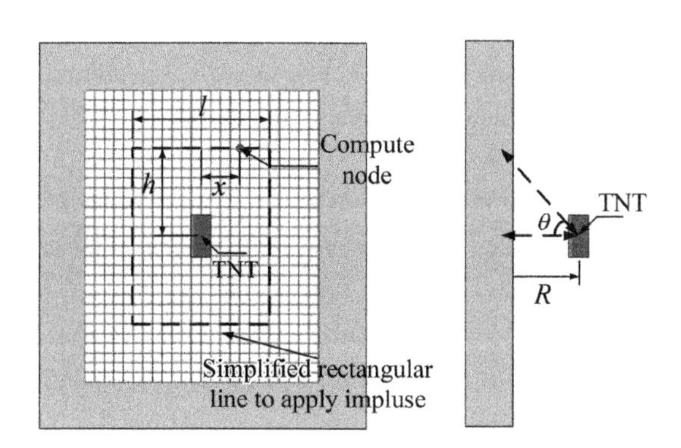

Figure 2 Diagrams of calculation for impulse method

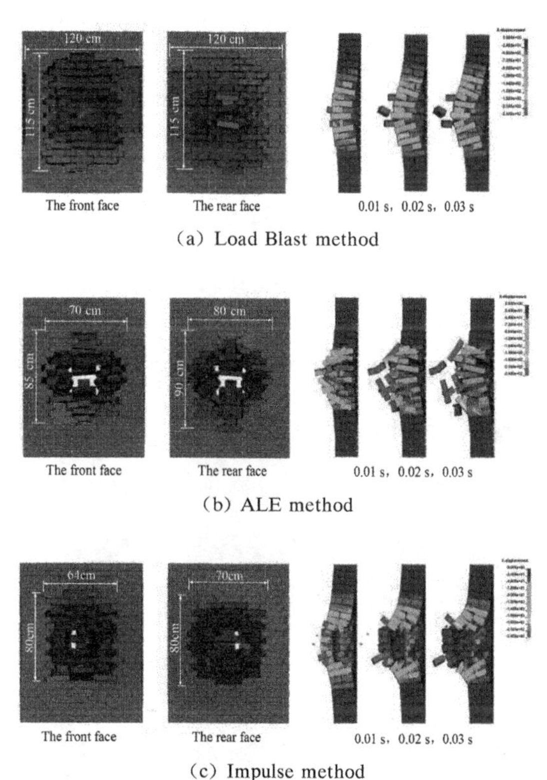

Figure 3 Terminal deformation and displacement of wall under 6 kg TNT explosion predicted by three methods

Then, the influences of the standoff distance, bonding strength between mortar and bricks, as well as the material constitutive models of bricks are discussed.

Figure 4 and Figure 5 respectively show the damage of wall under different standoff distance and identical scaled standoff distance and the terminal damage of wall predicted by different material models. It indicates that, under the close-in range explosion, the damage level of wall gets much serious with the increase of the standoff distance under the identical scaled standoff distance, the corresponding failure mode is transferred from local damage to the global collapse failure; the blast resistance of the wall is generally strengthened with enhancing the bonding strength between brick and mortar, and

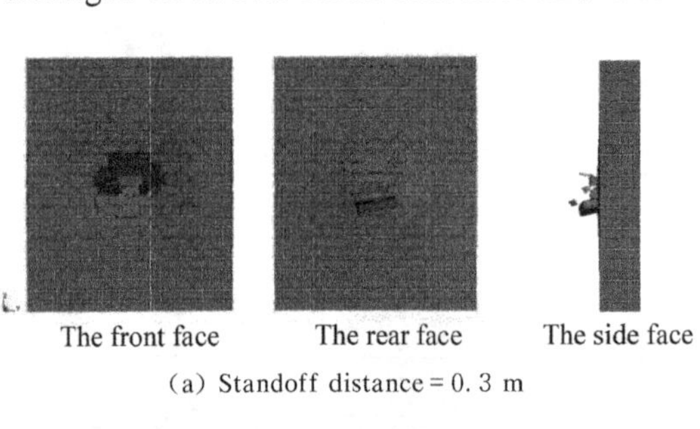

The front face The rear face The side face

(a) Standoff distance = 0.3 m

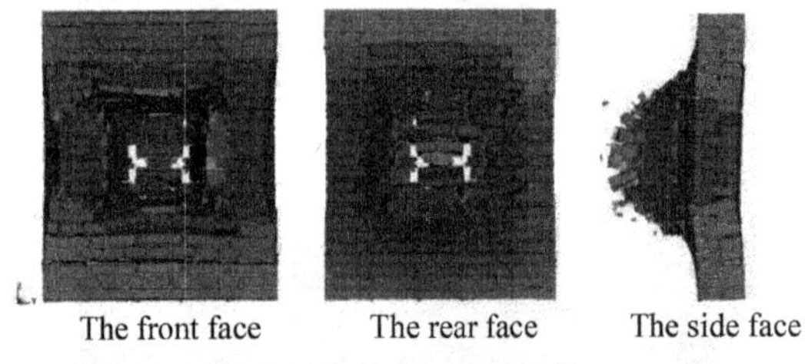

The front face The rear face The side face

(b) Standoff distance = 0.5 m

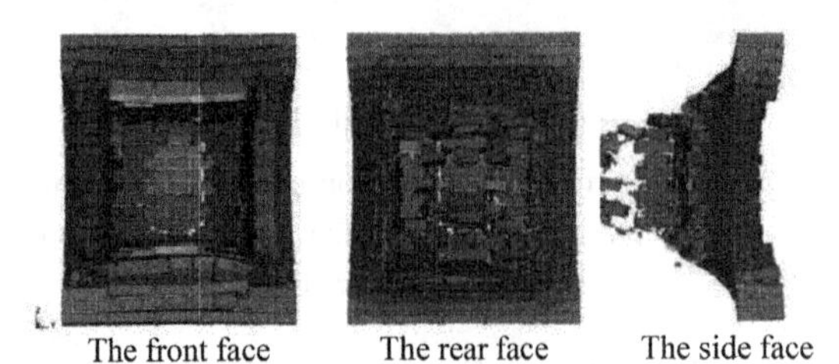

The front face The rear face The side face

(c) Standoff distance = 0.6 m

Figure 4 Damage of wall under different standoff distance and identical scaled standoff distance of 0.22 m/kg$^{1/3}$

mainly influenced by the shear failure stress; both the MAT _ BRITTLE _ DAMAGE and MAT _ WINFRI-TH _ CONCRETE models can give better predictions than MAT _ SOIL _ AND _ FOAM model.

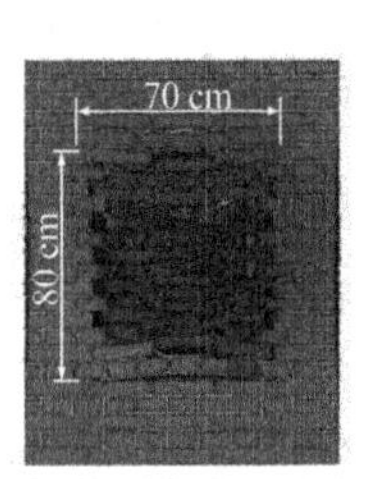

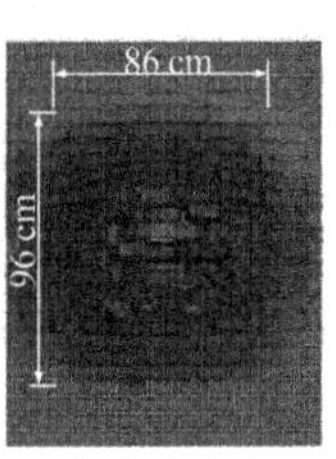

The front face The rear face The side face

(a) MAT_WINFRITH_CONCRETE

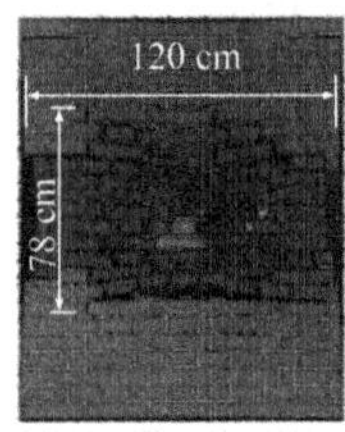

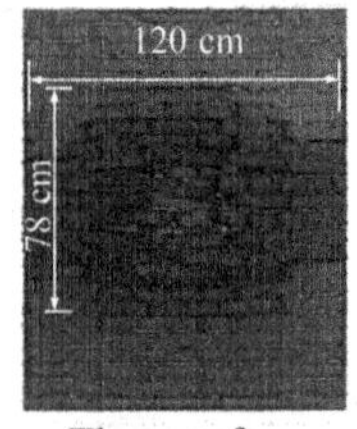

The front face The rear face The side face

(b) MAT_SOIL_AND_FOAM

Figure 5 Terminal damage of wall predicted by two material models under 6 kg TNT explosion

References

[1] ORTON S L, CHIARITO V P, MINOR J K, et al. Experimental testing of CFRP-Strengthened reinforced concrete slab elements loaded by close-in blast[J]. Journal of Structural Engineering, 2014, 140(2): 1612-1625.

[2] HENRYCH J, MAJOR R. The dynamics of explosion and its use [M]. Amsterdam: Elsevier, 1979.

[3] VARMA R K, TOMAR C P S, PARKASH S, et al. Damage to brick masonry panel walls under high explosive detonations[C]// Presented at the ASME Pressure Vessels and Piping Conference. Orlando, Florida, 1997,

351: 207-216.

[4] WEI X Y, STEWART M G. Model validation and parametric study on the blast response of unreinforced brick masonry walls [J]. International Journal of Impact Engineering, 2010, 37(11): 1150-1159.

[5] MICHALOUDIS G, GEBBEKEN N. Modeling masonry walls under far-field and contact detonations[J]. International Journal of Impact Engineering, 2019, 123: 84-97.

[6] DAVIDSON J S, PORTER J R, DINAN R J, et al. Explosive testing of polymer retrofit masonry walls[J]. Journal of Performance of Constructed Facilities, 2004, 18(2): 100-106.

[7] DAVIDSON J S, FISHER J W, HAMMONS M I, et al. Failure mechanisms of polymer-reinforced concrete masonry walls subjected to blast[J]. Structure Engineering, 2005, 131(8): 1194-1205.

[8] DAVIDSON J S, MORADI L, DINAN R J. Selection of a material model for simulating concrete masonry walls subjected to blast [R]. DTIC Document, 2004.

[9] DAVIDSON J S, SUDAME S, DINAN R J. Development of computational models and input sensitivity study of polymer reinforced concrete masonry walls subjected to blast [R]. DTIC Document, 2004.

[10] WANG J G, REN H Q, WU X Y, et al. Blast response of polymer-retrofitted masonry unit walls[J]. Composites Part B: Engineering, 2017, 128: 174-181.

[11] WANG J G. Experimental and numerical investigation of clay brick masonry walls strengthened with spray polyurea elastomer under blast loads[D]. Hefei: University of Science and Technology of China, 2017.

[12] SHI Y C, XIONG W, LI Z X, et al. Experimental studies on the local damage and fragments of unreinforced masonry walls under close-in explosions[J]. International Journal of Impact Engineering, 2016, 90: 122-131.

Study on BIM application for the first project of Xiong'an municipal infrastructure

N. Xu & J. L. Zhao & F. Xie & X. L. Tan
China MCC20 Group Corp. LTD, Shanghai 201900, China

Abstract

The planning and construction of Xiong'an New District is a historic project to ease the function of Beijing's non-capital and promote the coordinated development of Beijing, Tianjin and Hebei. The project is also a millennium plan and a national event.

With the principle of "world vision, international standards, Chinese characteristics, high-point positioning", in order to realize a high-level socialist modern city, BIM (Building Information Modeling) technology was used in the construction of Xiong'an New District.

1. BIM requirements of Xiong'an construction

BIM management platform (Phase Ⅰ) of Xiong'an New District has started to construct. This platform will realize the record, control and management of the whole process of the life of Xiong'an New Area, and supply a solid foundation for the construction of a green and smart city in the new district. Planning and construction of BIM management platform (Phase Ⅰ) includes data layer, application support layer, application layer, covering the ser- vices of display, query, interaction, approval, decision making for current space, overall planning, detailed planning, design plan, engineering construction, project completion.

In addition to the construction of BIM management platform, the new district will further explore the application depth of GIS and BIM, and give full play to the effec-tiveness of the BIM management platform (Phase Ⅰ). Moreover, the new district will form the "Digital Xiong'an Planning and Construction Management Data Standards" of plan, construction, and municipalities by cooperating with relevant units.

2. Overview of the first project of Xiong'an municipal infrastructure

Xiong'an Rongdong Area is one of the first construction districts of Xiong'an New District. It is located in the east of Rongcheng County and the planned land area is about 12.7 km^2. This project includes the construction of the municipal road trunk line of the F community in the Rongdong area (Figure 1), the integrated pipe gallery and drainage network system along with the trunk road. It also includes the construction and application of digital BIM and CIM (City Information Modeling) model.

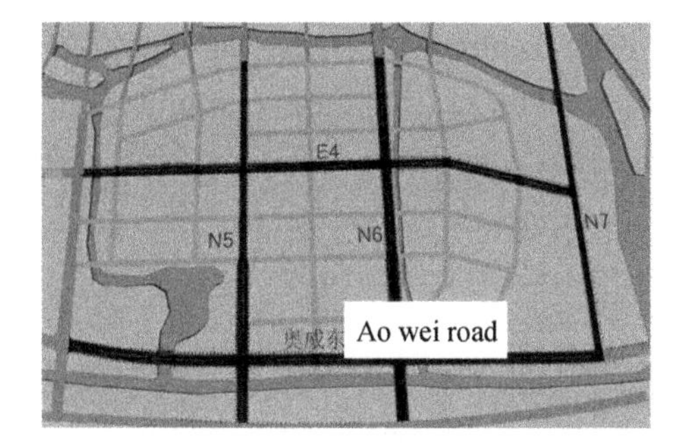

Figure 1 F community in the Rongdong area

3. BIM application of the project

Using GIS data as the underlying architecture, the surrounding environment and building information model will combine by tilt photography. Then build BIM and MCC wisdom site platform to assist project collaborative management. The data of GIS, BIM, IoT was deep integrated, and then imported into the construction management platform of digital Xiong'an. The CIM platform realized digital and intelligent management through advanced information technology, and created a Xiong'an excellent project.

3.1 Technology application of BIM

The UAV (Unmanned Aerial Vehicle) is used for surveying and mapping at various stages of the construction, and the actual scenes in the scene are left in three dimensions to reveal the entire construction process. During the process of modeling, single-professional three-dimensional calibration and multi-professional coordination are carried on to achieve efficient professional integration and coordination, reduce conflicts between various professions and design changes, reduce the changes caused by functional layouts and architectural space collisions that cannot be found in 2D drawings. Detailed design and construction of complex nodes were also carried on at the beginning of the project. With the 3D BIM disclosure, the information of each participant in each stage is shared and symmetric, which avoided the change or rework caused by poor information communication[1]. The 3D site layout ensured the reasonable division of machinery and materials during the construction process. Then simulate the difficult schemes, optimize and visualizes the scheme. With the whole process of virtual construction, the entire construction state can be adjusted, which ensured the rationality of the schedule. With the project quantity acquired from the BIM model, the cost management efficiency improved[2].

The application points include tilt photography (Figure 2), model establishing (Figure 3), 3D review, construction

Figure 2 Results of tilt photography of the project

Figure 3 Intersection model of the project

deepening, site layout, visualization disclosure, project simulation (Figure 4), virtual construction (Figure 5), engineering lifting, etc. Partial results are as follows.

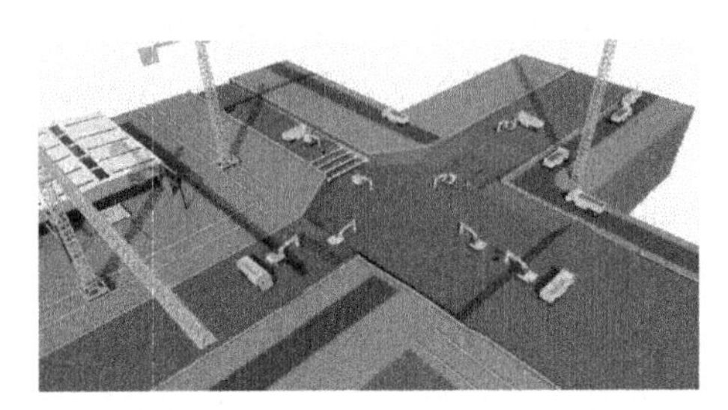

Figure 4 Project simulation of E4 and N6 of the project

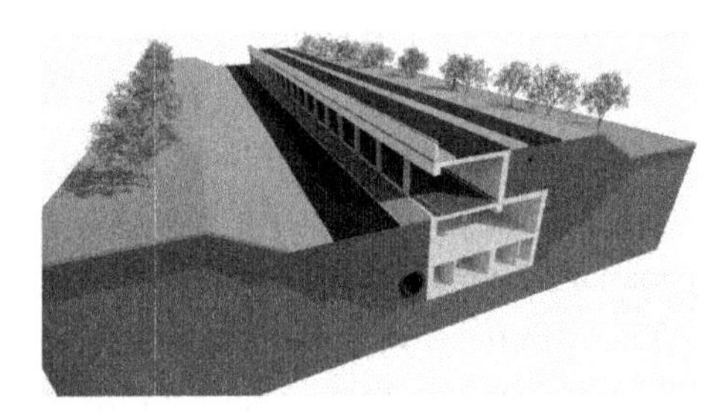

Figure 5 Utility tunnel virtual construction of the project

3.2 Management application of BIM

Through building a municipal BIM platform independently, the GIS tilt photography data and BIM model integrated[3]. With the integrated application of "three-end one cloud", integrated connection of basic data of computer end achieved. Using the mode that is daily on-site management data input of mobile phone, automatic statistics on webpage analysis, cloud automatic data storage and other requirements, multi-party can be collaborative participation. Then on-site management issues leave the mark, and the project quality, safety, schedule, cost and other aspects of refined management improved.

The application points include quality management, safety management, safety inspection, schedule management (Figure 6), engineering quantity management. Partial results are as follows.

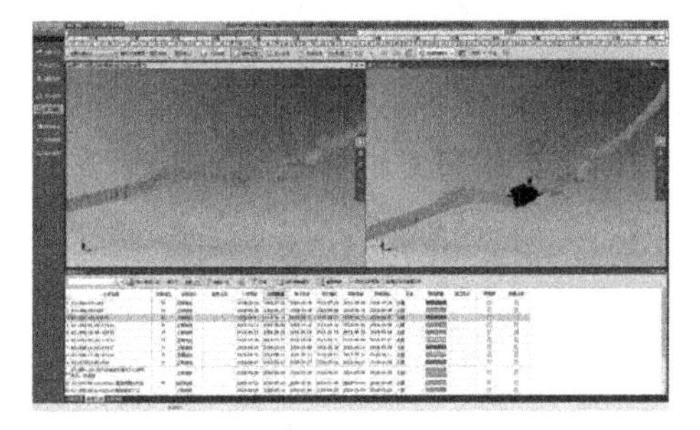

Figure 6 Schedule management of the project

3.3 Information application of BIM

Build the BIM and wisdom site platform, and transfer BIM model and monitoring system to digital construction management platform of Xiong'an, intelligent monitoring system and CIM platform to ensure the delivery of full life data and model. Moreover, establish long-term monitoring during construction period to realize engineering Digital, intelligent management.

The application points include platform management, multi-platform data docking, and completion model.

4. Conclusions

Under the overall framework of the Xiong'an New District, the project department actively promoted the construction of relevant smart city application projects, and formed a promotion mechanism with unified leadership, reasonable division of labor, clear responsibilities, and smooth operation. The group company strongly supported the development of BIM and CIM.

Meanwhile, through improving the BIM organization, policies, funds, talents and publicity measures, the development goals and main tasks of the municipal infrastructure project realized, which reflected the core connotation of the characteristic smart city of Xiong'an New District.

References

[1] WANG Y X, HE Q, TIAN Y L, et al. Analysis on the development trend of municipal BIM[J]. Informatization of China Construction, 2019(2): 40-43.

[2] XU Z Y. Discussion on the application of BIM technology in municipal engineering[J]. Intelligentialize and Informatization, 2018 (4): 261-263.

[3] ZHANG H L. Application research of BIM in municipal infrastructure projects [D]. Beijing: Peking University, 2018.

Unstressed passive wireless sensors for structural health monitoring

L. Y. Xie & Z. R. Yi
Department of Disaster Mitigation for Structures, Tongji University, Shanghai 200092, China
S. T. Xue
Department of Architecture, Tohoku Institute of Technology, Sendai, Japan

Abstract

Structures could be deteriorated by aging effects of strength and stiffness[1]. With the increasing of the service time, structural health monitoring for aging structures is becoming significantly important[2], therefore various sensors have been designed to detect the deformation or strain of structures.

Wired sensors for structural health monitoring use cables to supply power and transmit data. These wired sensors usually show a good performance in stability and accuracy. However, they are limited by the instrumentation time and system cost. Eliminating cables for wired sensors, passive wireless antenna sensors are easier to deploy with less cost using antennas as sensing and transmitting units. This is a promising technology for distributed sensing network over a wide area with dense deployment.

In recent years, there has been an increasing interest in the development of antenna sensors for various physical quantities. Due to the advantage of simple configuration and multimodality, patch antenna has been studied extensively. In general, the physical qualities can be obtained by analyzing resonant frequencies of the patch antenna, which would be shifted by the deformation or environment changes of the monolithic patch antenna. Yi. et al.[3] presented a wireless strain sensor based on monolithic patch antenna. The antenna sensor is attached on the surface of structures and the strain of the structure is obtained by measuring the resonant frequency shifting of the patch antenna. Huang et al.[4] demonstrated a patch antenna sensor that can detect crack propagation with sub-millimeter resolution by analyzing the relationship between the crack width and resonant frequency shifting in longitudinal and transverse direction. Mohammad et al.[5] proposed a patch antenna-based shear sensor. Tchafa et al.[6] proposed a patch antenna-based sensor to determine the changes in strain and temperature from the normalized antenna resonant frequency shifts.

However, the sensing unit of the one-piece antenna is designed to be stressed for deformation measurement. For these stressed antennas, the issues of incomplete strain transfer ratio, insufficient bonding strength, and randomness of crack propagation will compromise the sensor sensitivity and complicate the calibration process. To address these issues completely, we proposed several unstressed antenna sensors, which are shown as follows.

A helical antenna based passive displacement sensor is proposed[7]. This antenna sensor is consisted of a normal mode helical antenna and an inserted silicon rod, which is shown in Figure 1.

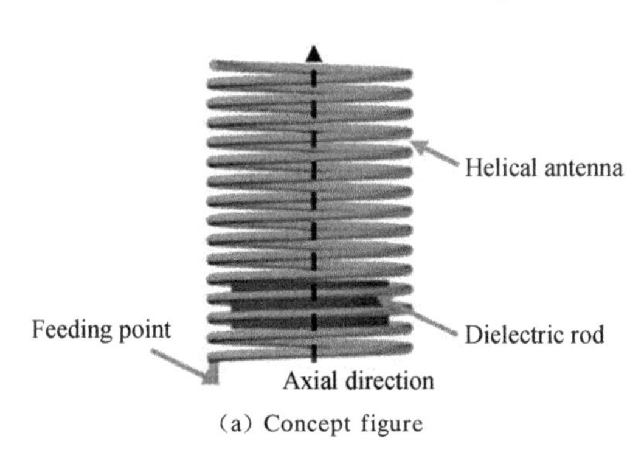

(a) Concept figure

(b) Manufactured antenna sensor

Figure 1 Figures of a helical antenna-based displacement sensor

The electromagnetic field can be altered when the silicon rod dislocated, which would lead to the resonant frequencies shifting of the helical antenna. Hence, the location of the silicon rod can be certified by analyzing the resonant frequency shifting without exerting stress to the antenna. The simulation and experimental results are shown in Figure 2, which suggested a maximum effective measuring range of 7 mm with an average sensitivity of 0.616 MHz/mm.

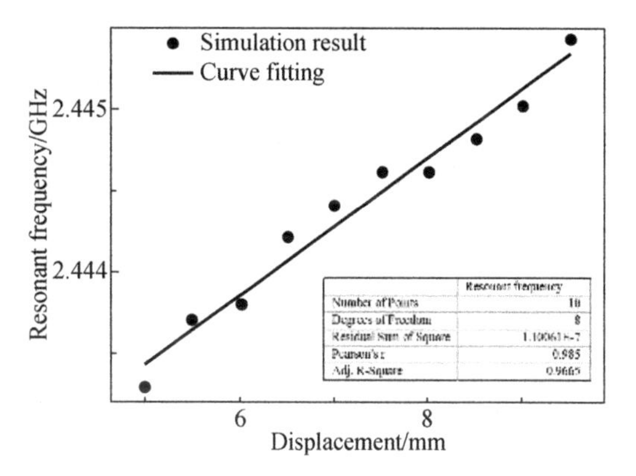

(a) Simulation result

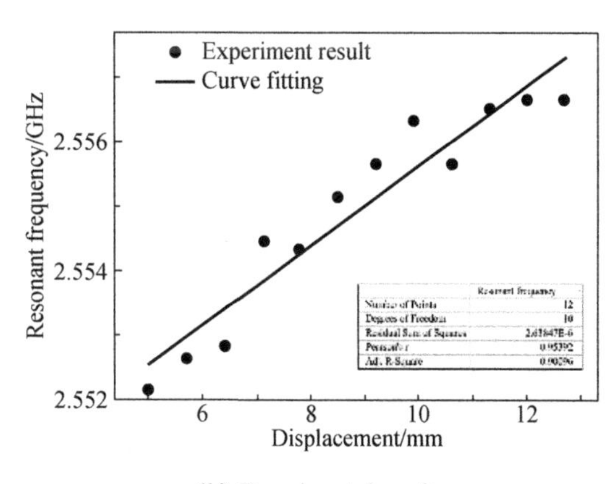

(b) Experimental result

Figure 2 Resonant frequency with respect to displacement of the sensor in each group

The resonant frequency of a patch antenna would be affected by the antenna loading. Based on this principle, the authors have proposed a crack sensor by forming a parallel plate capacitor using two microstrip lines as a sensing unit, which is shown in Figure 3[8]. As the relative movement of two microstrip lines represents the deformation of the monitoring object, the sensing antenna is free of stress.

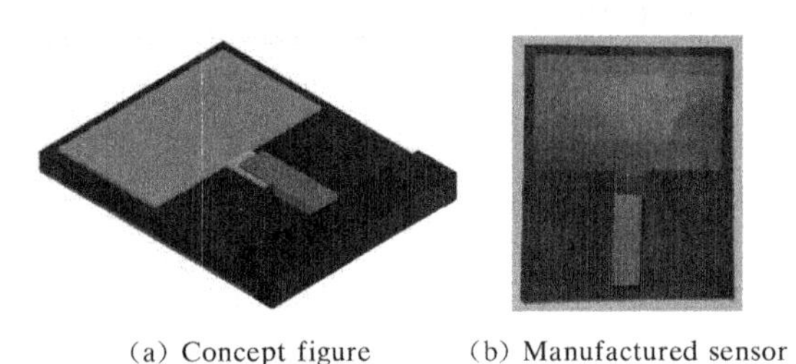

(a) Concept figure (b) Manufactured sensor

Figure 3 Crack sensor based on patch antenna fed by capacitive microstrip line

Another unstressed crack sensor was proposed by combining a monolithic patch antenna with a movable radiation patch, which is shown in Figure 4[9]. The total length of the combined radiation patch would be altered by the relative movement between the patch antenna and the dielectric board, leading to a shift of resonant frequency in the sensing system.

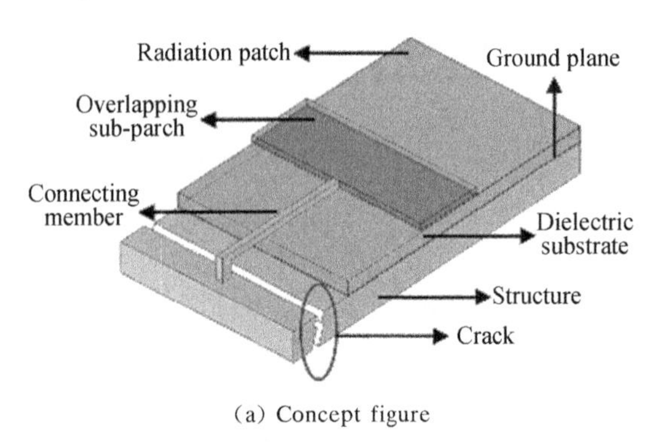

(a) Concept figure

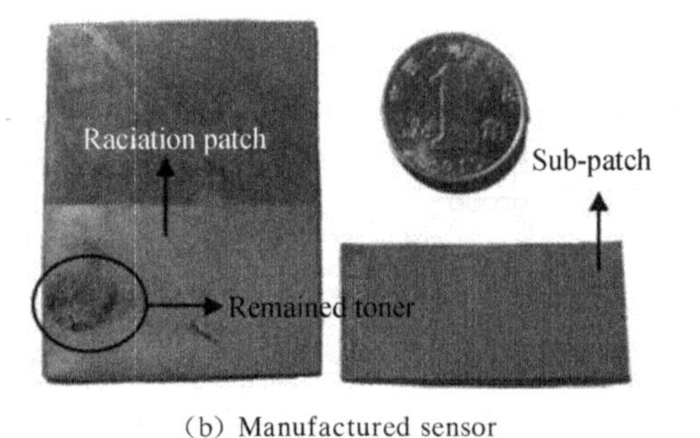

(b) Manufactured sensor

Figure 4 Crack sensor based on patch antenna with overlapping sub-patch

Theory calculation, simulation and experimental results show an effective measuring range of 1.5 mm with a sensitivity of 120.24 MHz/mm on average, which is presented in Figure 5.

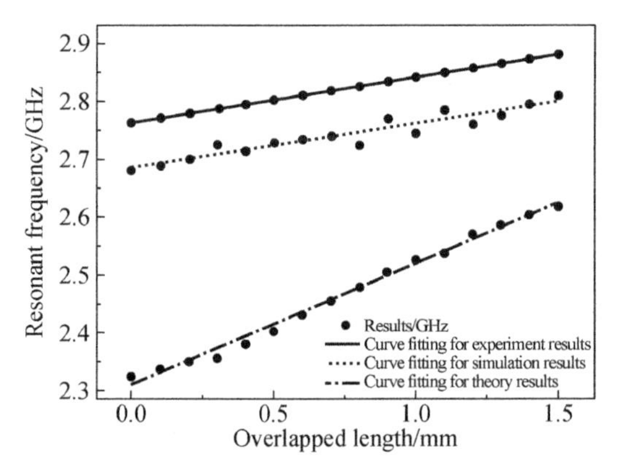

Figure 5 The relationship between resonant frequency and overlapped length in theory calculation, numerical simulation, and in the experiment

Acknowledgements

This research was funded by the Key Program of Intergovernmental International Scientific and Technological Innovation Cooperation in China (Grant No. 2016YFE0127600), the Key Laboratory of Performance Evolution and Control for Engineering Structures (Tongji University), the Ministry of Education of the People's Republic of China (Grant No. 2018KF-4), and the Fundamental Research Funds for the Central Universities.

References

[1] CHENG Y, GAO F, HANIF A, et al. Development of a capacitive sensor for concrete structure health monitoring [J]. Construction and Building Materials, 2017, 149: 659-668.

[2] TOORES M A, RUIZ S E. Structural reliability evaluation considering capacity degradation over time [J]. Engineering Structures, 2007, 29(9): 2183-2192.

[3] YI X H, WU T, WANG Y, et al. Passive wireless smart-skin sensor using RFID-based folded patch antennas [J]. International Journal of Smart and Nano Materials, 2011, 2(1): 22-38.

[4] MOHAMMAD I, GOWDA V R, ZHAI H, et al. Detecting crack orientation using patch antenna sensors[C]//Proceedings of SPIE, 2011.

[5] MOHAMMAD I, HUANG H Y. Shear sensing based on a microstrip patch antenna [J]. Measurement Science and Technology, 2012, 23(10): 105705.

[6] TCHAFA F M, HUANG H. Microstrip patch antenna for simultaneous strain and temperature sensing[J]. Smart Materials and Structures, 2018, 27(6): 065019.

[7] XUE S, YI Z, XIE L, et al. A displacement sensor based on a normal mode helical antenna[J]. Sensors, 2019, 19(17).

[8] XUE S, XU K, XIE L, et al. Crack sensor based on patch antenna fed by capacitive microstrip lines [J]. Smart Materials and Structures, 2019, 28: 085012.

[9] XUE S, YI Z, XIE L, et al. A Passive Wireless Crack Sensor Based on Patch Antenna with Overlapping Sub-Patch [J]. Sensors, 2019, 19(19): 4327.

Influence of image scale on segmenting weak interlayer of rock tunnel face based on deep learning

T. J. Yang

Southwest Transportation Construction Group Co. Ltd., Kunming 650031, China

J. Y. Chen & D. M. Zhang & H. W. Huang

Key Laboratory of Geotechnical and Underground Engineering of Minister of Education, Shanghai 200092, China & Department of Geotechnical Engineering, Tongji University, Shanghai 200092, China

Abstract

By quantitatively measuring the region of the weak interlayer on sites to assess the structure stability of rock tunnel face has formed a significant part under tunneling construction[1, 2]. However, the scale of images collected from Mengzi-Pingbian highway tunnel (MPHT) in Yunnan, China varies owing to the different sampling distances between the tunnel face and photographic system. It may lead unsatisfactory segmentation results in overall task. A pixel-level segmentation[3] deep learning model (Figure 1) based on the convolutional neural network (CNN), known as DeepLab V3 +[4, 5], was proposed in this paper. The basic architecture of the algorithm was then modified, trained and tested in this paper. Meanwhile, a database containing 32 040 images of

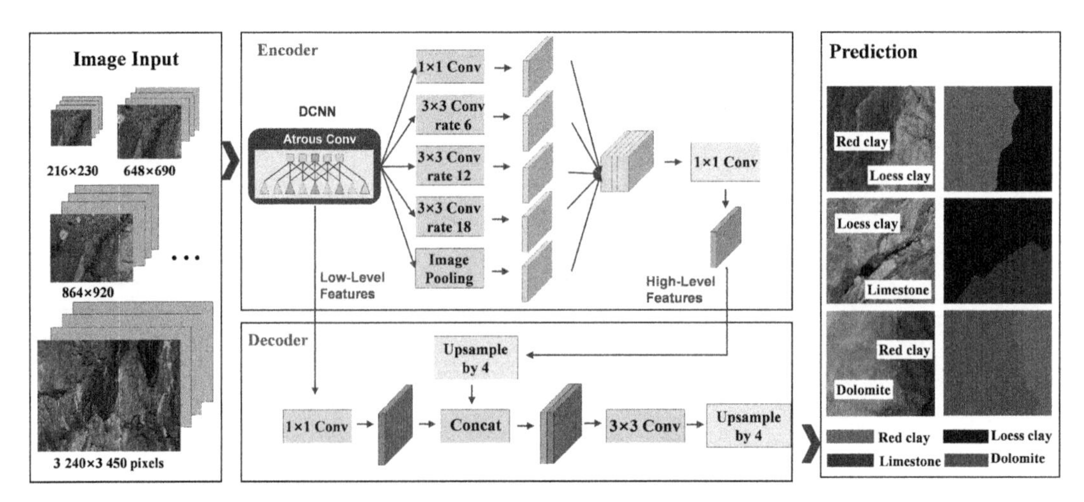

Figure 1 Schematic diagram of the proposed DeepLab V3 + framework with a revised encoder-decoder[6, 7] and depthwise separable convolution[8] structure

limestone, dolomite, red clay and loess clay were labeled manually. Then, images with multiple sizes were applied into the CNN model to verify the robustness and applicability of the model.

Compared the mean pixel accuracy (MPA) and mean intersection of union (MIoU)[9, 10] by testing the multiple image scale (Table 1) from the minimum size (216 × 230 pixels) to the maximum (3 240 × 3 450 pixels), the proposed model exhibited a better performance for the small-size images than the large one. In this task, the image size from 216×230 pixels to 864 × 920 pixels provided higher MPA and MIoU (Figure 2), achieving 88.24% and 78.24%. The image size larger than 1 512×1 380 pixels had a poor performance in term of the boundary segmentation and the noise points.

Table 1 Size statistics of testing images for deep learning

No.	1	2	3	4	5
Image pixels	216×230	432×460	648×690	864×920	1 080×1 150
No.	6	7	8	9	10
Image pixels	1 296×1 380	1 512×1 610	1 728×1 840	1 944×2 070	2 160×2 300
No.	11	12	13	14	15
Image pixels	2 376×2 530	2 592×2 760	2 808×2 990	3 024×3 220	3 240×3 450

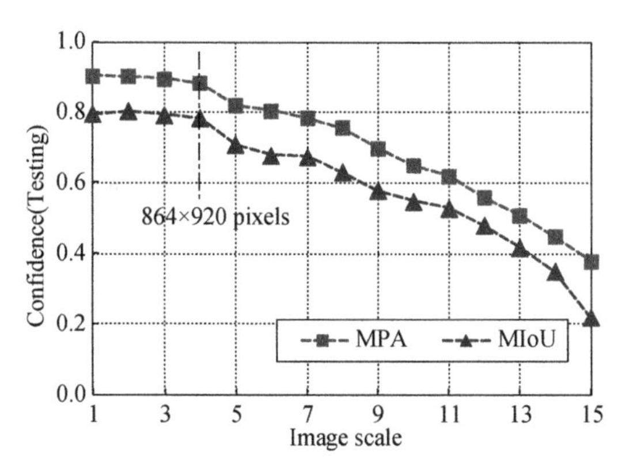

Figure 2 The MPA and MIoU of different image sizes tested in the deep learning model

In summary, the results revealed that the proposed model can accurately segment the weak inter layers for under-construction rock tunnels, and exhibit a strong robustness for the small scale samples. Nevertheless, the inevitable errors and noise points caused by large size images testing are urgent to be reduced in the future study.

Acknowledgements

The research presented in this paper is supported by the Natural Science Foundation Committee Program of China (Grant No. 1538009, Grant No. 51778474), Science and Technology Project of Yunnan Provincial Transportation Department (No. 25 of 2018), the Fundamental Research Funds for the Central Universities of China (Grant No. 0200219129) and key innovation team program of innovation talents promotion plan by MOST of China (No. 2016RA4059).

References

[1] HUANG H W, LI Q T, ZHANG D M. Deep learning based image recognition for crack and leakage defects of metro shield tunnel [J]. Tunnelling and Underground Space Technology, 2018, 77: 166-176.

[2] FANG W L, DING L Y, ZHONG B T, et al. Automated detection of workers and heavy equipment on construction sites: A convolutional neural network approach[J]. Advanced Engineering Informatics, 2018, 37: 139-149.

[3] ZHANG A, WANG K C P, LI B, et al. Automated pixel-level pavement crack detection on 3D asphalt surfaces using a deep-learning network[J]. Computer-Aided Civil and Infrastructure Engineering, 2017, 32(10): 805-819.

[4] CHEN L C, PAPANDREOU G, KOKKINOS I, et al. Deeplab: Semantic image segmentation with deep convolutional nets, atrous convolution, and fully connected crfs [J]. IEEE transactions on pattern analysis and machine intelligence, 2017, 40 (4): 834-848.

[5] LIU C, CHEN L C, SCHROFF F, et al. Auto-deeplab: Hierarchical neural architecture search for semantic image segmentation[C]// Proceedings of the IEEE Conference on Computer Vision and Pattern Recognition. 2019: 82-92.

[6] RONNEBERGER O, FISCHER P, BROX T. U-net: Convolutional networks for biomedical image segmentation [C]// International Conference on Medical image computing and computer-assisted intervention. Springer, Cham, 2015: 234-241.

[7] BADRINARAYANAN V, KENDALL A, CIPOLLA R. Segnet: A deep convolutional encoder-decoder architecture for image segmentation [J]. IEEE transactions on pattern analysis and machine intelligence, 2017, 39(12): 2481-2495.

[8] ZHANG X Y, ZHOU X Y, LIN M X, et al. Shufflenet: An extremely efficient convolutional neural network for mobile devices [C]// Proceedings of the IEEE Conference on Computer Vision and Pattern Recognition, 2018: 6848-6856.

[9] JI J S, WU L J, CHEN Z C, et al. Automated pixel-level surface crack detection using U-Net[C]// International Conference on Multi-disciplinary Trends in Artificial Intelligence. Springer, Cham, 2018: 69-78.

[10] LI S Y, ZHAO X F, ZHOU G Y. Automatic pixel-level multiple damage detection of concrete structure using fully convolutional network[J]. Computer-Aided Civil and Infrastructure Engineering, 2019, 34(7): 616-634.

Experimental verification of angle shear connector welded to hat-shaped CFS section

H. Shin & H. S. Oh & T. Kang

Department of Architecture and Architectural Engineering, Seoul National University, Korea

Abstract

In order to provide the steel-concrete synthesis effect of composite beam, a method of welding steel angle shear connectors to hat-shaped cold-formed steel (CFS) section has recently been used. For the design of such composite beam system, Kang et al.[1] proposed the shear strength design equation of angle shear connector welded to hat-shaped CFS section (Eq. (1)) based on AISC 360[2] design equation for channel shear connector welded to wide-flange section. The equation is adopted by AC495[3], which specifies the design, test method, and acceptance criteria for such composite beam system.

$$Q_n = \frac{0.6 \times (100\ \mathrm{mm})^{3/2}(t_f + 0.5t_w)\sqrt{f_c' E_c}}{\sqrt{l_a}} \tag{1}$$

where Q_n is nominal shear strength of each steel angle shear connector (N), t_f is thickness of horizontal leg of steel angle (mm), t_w is thickness of vertical leg of steel angle (mm), f_c' is specified concrete compressive strength (MPa), E_c is modulus of elasticity of concrete (MPa), and l_a is CFS structural beam web-to-web clear distance (mm).

In this study, to validate the AC495 design equation, a series of push-out tests (Figure 1) and beam flexural tests (Figure 2) were carried out with the shear connector spacing and direction as the main variables. From the tested 10 push-out specimens, the measured angle shear connector strengths exceeded 2.5 to 4 times the nominal strength determined by the AC495 design equation (Table 1). In addition, shear connector related failure modes did not appear in the beam flexural specimen with shear connector spacing three times larger than the spacing determined by AC495 design equation, which verifies the push-out test results (Figure 3). The current shear connector design equation in AC495 was found to underestimate shear connector strength significantly, because the shear strength provided by concrete itself is not considered in the design equation. That conservative estimation on the shear connector strength can lead to excessively short shear connector spacing, which makes construction difficult and costly. Thus, it is necessary to improve the equation for more reasonable design of angle shear connectors.

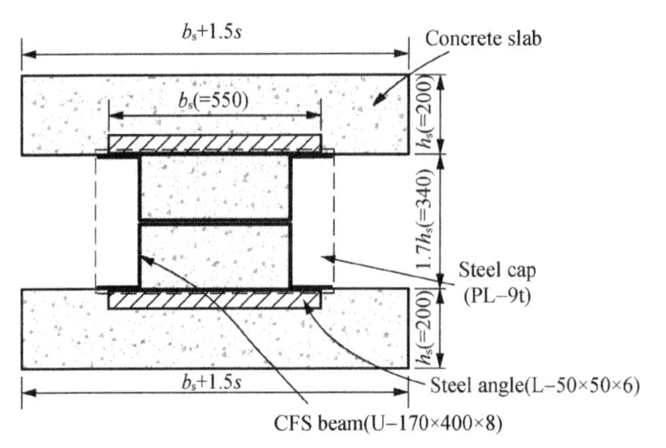

(a) Push-out specimen: Plan

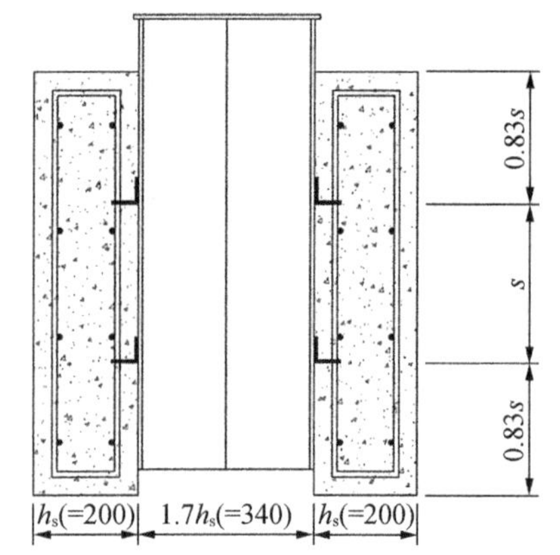

(b) Push-out specimen: Elevation

(c) Push-out specimen setup

Figure 1 Push-out test

Table 1 Push-out test results

Specimen	Angle direction	Shear connector spacing, s/mm	Measured peak strength per shear connector, Q_p/kN	Nominal strength per shear connector, Q_n/kN
PT-R175		175	685	
PT-R225		225	751	186
PT-R300	The right direction (┐ ┌)	300	744	
PT-R400		400	846	
PT-R550		550	822	179
PT-I175		175	624	
PT-I225		225	669	186
PT-I300	Inversed direction (┘ └)	300	760	
PT-I400		400	701	
PT-I550		550	649	179

Figure 2 Composite beam flexural test

(a) FT-F175 specimen (shear connector spacing in accordance with AC495 design equation)

(b) FT-F550 specimen (three times the spacing required by AC495 design equation)

Figure 3 CFS beam-concrete interface slip after the flexural test

Acknowledgements

This work was supported by the National Research Foundation (NRF) of Korea (Grant No. 2015R1A5A1037548) and POSCO.

References

[1] HKK Thomas, K Kim, S Kim, et al. Development of Design Equation for Steel Angle Anchors Welded to Hat-Shaped Cold-Formed Steel Section[C]// The 2017 World Congress on Advances in Structural Engineering and Mechanics, Korea, 2017.

[2] AISC 360-16, Specification for Structural Steel Buildings [S]. Chicago: American Institute of Steel Construction, IL, 2016.

[3] ICC-ES AC495, Cold-Formed Steel Structural Beams with Steel Angle Anchors Acting Compositely with Cast-In-Place Concrete Slabs [S]. CA: ICC Evaluation Service, LLC, Brea, 2018.

Mechanical properties of butt weldments made with Q345B–E5015 at elevated temperature

Y. Liu & Z. Guo & Y. B. Liu & X. R. Wang

School of Mechanics & Civil Engineering, China University of Mining and Technology, Xuzhou 221116, China

Abstract

This study fulfilled steady-state tensile tests and transient-state tests to investigate the mechanical properties of butt weldments, composed of Q345B base metal and E5015 electrodes, subjected to elevated temperatures. The results include the elastic modulus, yield, and ultimate strength of the butt weldments at temperatures from 20 ℃ to 800 ℃. Comparison of the results shows that the welding heat affected zone (HAZ) is vulnerable to fracture when subjected to reheating above 500 ℃. It's more reasonable to use the results from the transient-state tests to deduce the reduction factors of elastic modulus of weldments subjected to 200 ℃ or higher. When heating over 400 ℃, these reduction factors are basically in line with those of the proposal of Eurocode 3-1-2[1] and AISC: 2005[2]. The reduction factors of yield strengths of weldments, from 100~300 ℃ and 600~800 ℃, should be considered the effects of welding and use the results in transient-state tests. When the temperature was between 300 ~ 600 ℃, the reduction factors could apply the proposal of ASCE[3]. The order of heating and loading influences the material mechanical properties subjected to high temperature. In other words, the couple of physical stress fields and thermal induced stress fields is affected not only by the quantity of superposition, but also by the time series.

Based on the stress-strain curves, this paper, additionally provided the logical reduction factors of these design parameters. Furthermore, these parameters were compared with the recommendation in the current design standards, such as BS5950[4], Eurocode 3-1-2[1], AISC: 2005[2], ASCE[3], AS 4100[5], and CECS 200: 2006[6]. The following conclusions could be drawn.

(1) It's more reasonable to use the results from the transient-state tests to deduce the reduction factors of elastic modulus of weldments subjected to 200 ℃ or higher. At 100 ℃, the reduction factors, when strains exceeded 0.2%, should be obtained from the steady-state results. When heating over 400 ℃, these reduction factors are basically in line with those of the proposal of Eurocode 3-1-2 and AISC: 2005.

(2) The effects of initial stresses are more important on the fire resistance of metal. The order of heating and loading influences the material mechanical properties subjected to high temperature.

(3) The reduction factors of yield strengths of weldments, from 100~300 ℃ and 600~800 ℃, should be considered the effects of welding and use the results in transient-state tests. When temperature between 300 ~ 600 ℃, the reduction factors could apply the proposal of ASCE. And more, the current design standards are not applied to predict the reduction factors of ultimate strengths of weldments that influenced by the HAZ.

Acknowledgements

The research in this paper was supported by the Key Project of Research and Development Program of Xuzhou, China. The authors would like to appreciate its support of this area of research as one of the plan items of application and innovation, which is numbered as KC18220.

References

[1] EN1993-1-2, Eurocode 3: Design of steel structures, Part 1-2: general rules-structural fire design [S]. Brussels: European Committee for Standardization (CEN), 2005.

[2] AISC 360-05, Specification for Structural Steel Buildings American Institute of Steel [S]. Construction I (AISC), 2005.

[3] Structural fire protection [S]. New York, USA: American society of civil engineering, 1992.

[4] BS 5950, Structural use of steelwork in building-Part 8: code of practice of fire resistant design [S]. British: British Standard, 2003.

[5] AS 4100, Steel structures [S]. Sydney, Australia: Australian standards, 1998.

[6] CECS 200-2006, Technical code for fire safety of steel structure in buildings [S]. Beijing, China: Chinese Plan Press, 2006.

Numerical analysis of the effect of isolation pile construction on soil in deep soft ground

Q. Zhang & Z. X. Zhang
Department of Geotechnical and Underground Engineering, Tongji University, Shanghai 200092, China
C. C. Fu
Shaoxing Keqiao Rail Transit Group Co., Ltd., Shaoxing 312030, China

Abstract

In soft soil ground, the soil will be disturbed during the excavation or construction of a foundation pit, resulting in large settlement on the surface of surrounding area. When the surface settlement exceeds a certain range of values, it will cause settlement and inclination of the surrounding building foundation. Material will even crack so that safety of the building cannot be guaranteed. In order to prevent adverse effects on adjacent buildings during the construction process, isolation piles are usually arranged between the protected buildings and the construction site to meet the protection requirements in the construction of a place closer to the building. As one of the main methods for stratum reinforcement during construction and reducing the disturbance of construction to the surrounding environment, the isolation pile has been widely used in case of foundation pit excavation, shield construction and surcharge construction projects. Research on isolation piles has been getting widespread attention around the world.

Most of the current researches focus on the effect and mechanism of the piles after the piles are construed completely, but this research concentrating on the surrounding soil disturbance during the construction of the pile is relatively few. During the construction of the pile, the surrounding soil or structure will be deformed resulting from construction disturbance or improper construction control.

Based on the Wanxiu Road Vehicle Base Station project near Hangzhou-Ningbo Passenger High-speed Railway Line, numerical models of single pile and row pile are established. Through the numerical simulation of Flac3D, analysis on the disturbance law of the surrounding piles caused by different single pile construction parameters and construction of row pile are conducted. Furthermore, appropriate construction control methods are suggested.

The main research contents and approaches are given below.

1. **Single pile model and effect of different construction parameters on soil**

Single pile numerical model is generated as shown in Figure 1.

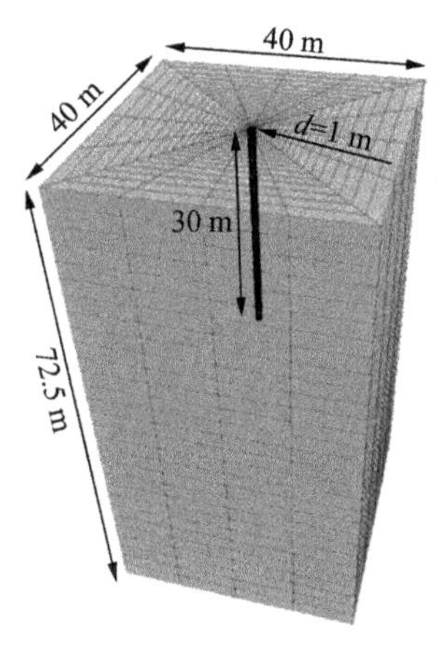

Figure 1 Numerical model of single pile

By using FLac3D, the pile drilling process is simulated by null zone. The effect of the soil of the wall mud should be equivalent to the still water based on the existing references, whose pressure is applied to the circumferential side of the hole in the form of normal uniform load. A certain calculation step is set to simulate the hole-empty and stress release time. The interface is established on the surface of the soil around the hole to represent interaction between pile and soil. New pile solid zone is generated by several segments, and then move into the hole to simulate process of pile-forming.

The effects of wall mud density, hole-empty time, drilling speed and pile-forming speed on surface settlement were compared through different construction conditions.

2. Row pile model and distribution patterns of soil parameters after construction

On the basis of single pile construction simulation, the construction process of 60 row isolation piles was simulated by sequential construction method. The row pile construction model is shown in Figure 2.

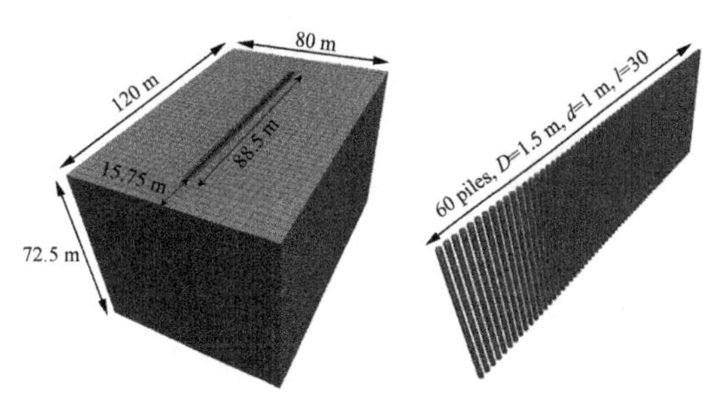

Figure 2 Numerical model of row pile

Using the built-in FISH language of Flac3D, a composite construction cycle function is created to realize the simulation of row pile construction. The function calculation process is shown in Figure 3.

The results (Figure 4) show that the maximum vertical displacement occurs at the 2/3 total length of the isolation pile construction direction, and large soil settlement change occurs near the pile which is under construction.

3. Changing patterns of soil parameters considering construction process

Along the direction of the row pile, there are 9 monitor points surrounding the piles, which is shown in Figure 5. Regarding complete construction of each single pile as a construction condition, which is, a total of 60 conditions. An analysis on changing patterns of the influence of 60 construction conditions on the soil around the pile is conducted.

As Figure 6 shows, since the construction process such as drilling and pile formation will produce disturbance to the soil around the pile, seven monitor points C to H will have sudden changes in soil settlement at the beginning of the construction of the nearest pile. Settlement value maintains high growth rate in 10

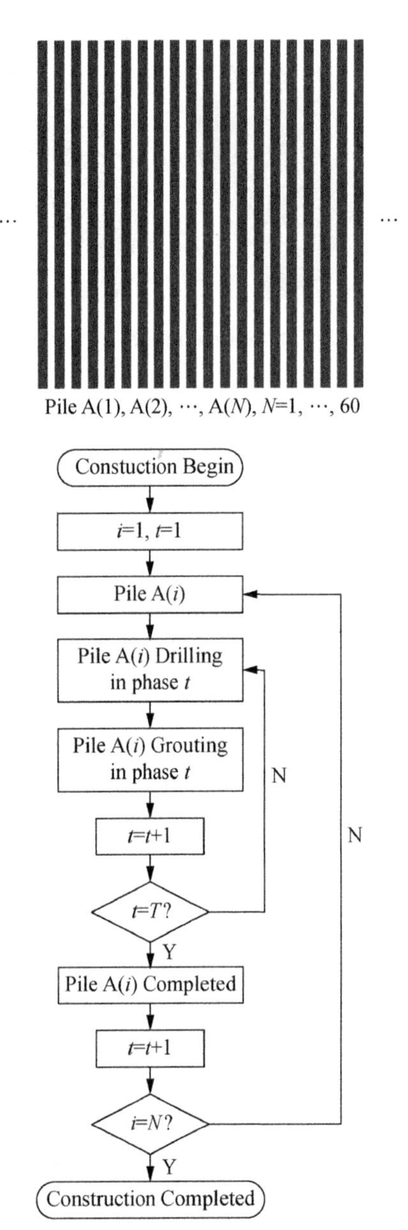

Figure 3　Calculation process of construction cycle function

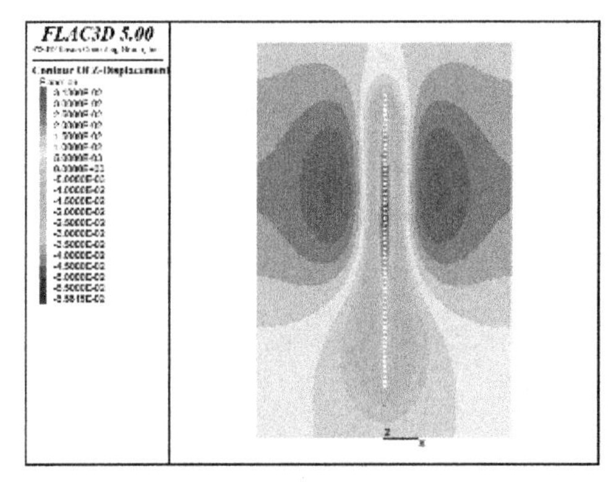

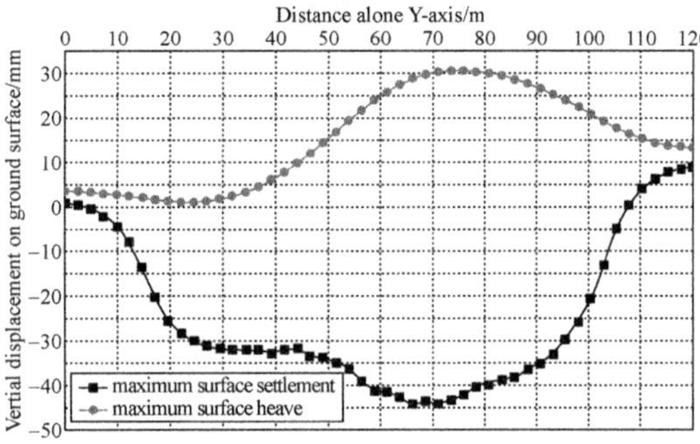

Figure 4　Vertical displacement on ground surface

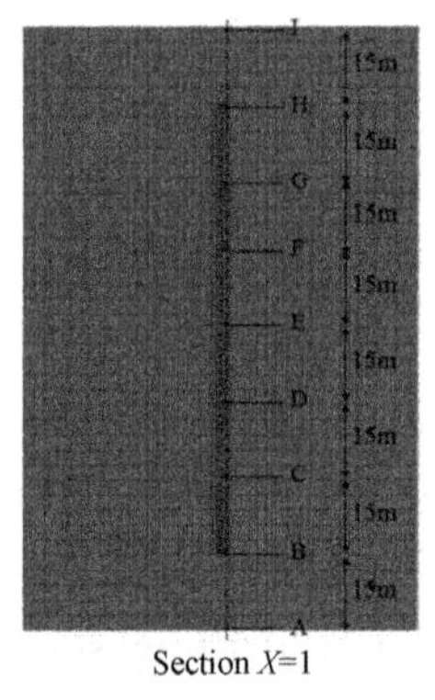

Figure 5　Distribution of monitor points

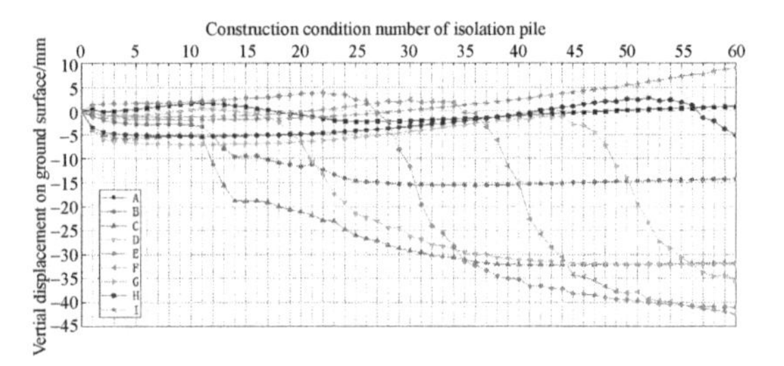

Figure 6　Vertical displacement of monitor points

following construction conditions with a maximum of 66%. Among all the monitor points, the settlement points of the monitor points E and F at the section of the cross-section $Y = 60$ m and $Y = 70$ m are the largest, which is correspond with the settlement distribution law at the end of construction mentioned before. That is, considering construction process, a large settlement also occur at the 2/3 length of the isolation pile construction direction. So more attention should be paid to the monitoring and protection of structure surrounding piles during the construction process.

References

[1] SHI S. Analysis of influence of isolation pile construction on deformation of pile foundation of adjacent high speed railway viaduct[J]. World Bridges, 2012, 40(5): 54-58.

[2] LIU B L. Influence of isolation pile construction on the displacement of adjacent high-speed railway piers[J]. Urban Mass Transit, 2015, 18(12): 80-83.

[3] WANG Z Q. Numerical Simulation of Precast Pile Installation Based on the LS-DYNA[D]. Liaoning: Liaoning University of Technology, 2016.

Experimental investigation on axial compression behavior of columns strengthened with textile reinforced concrete

S. Y. Li & Q. Zhang
College of Civil and Transportation Engineering, Hohai University, Nanjing 210098, China
Y. S. Zhao
Master, Faculty of Civil Engineering and Mechanics, Jiangsu University, Zhenjiang 202013, China

Abstract

Columns were considered to play an especially important role in bearing vertical loads and resistance to lateral action in the structures, which would directly determine the safety of the structures. However, in recent years, the existing columns according to old design standards could not satisfy usage requirements due to more strict seismic design requirements in the new seismic codes. Therefore, in such case building should be strengthened by the proper method. The application of externally bonding fiber reinforced polymer jackets have been evidenced as a convenient and feasible solution. Numerous experimental studies have shown that adopt the FRP as a confinement system could increase both compressive strength and deformation capacity of concrete member under vertical and lateral loads[1, 2]. However, a few drawbacks and limitations have been discovered result from the presence of organic resin[3]. A new composite material named textile reinforced concrete (TRC) system consisted of fiber textile meshes combined with high performance mortar was developed to solute these issues[4].

In this paper, the experimental study was carried out on the concrete square columns strengthened with TRC subjected to axial compression, taking into account the influence of the number of textile layers, concrete strength and glass short fiber on confinement effect of TRC system, with the aim to evaluate and analyze the effectiveness of different confinement scheme in increase compression strength and deformation capacity of TRC confined columns.

A two-dimensional knitted fabric that was made of carbon fiber bundles and alkali-free glass fiber bundles was used in this experiment. A total of thirteen square specimens consist of three unconfined columns and ten columns confined with TRC were designed and casted in the present investigation. For the specimen series, PC denoted plain concrete, L and the following number denoted the number of textile layers. N and F denoted with and without short fibers respectively, and C and the following number denoted the concrete strength. The experimental results were also discussed and analyzed in the following sections, mainly including the failure modes and load-axial deformation curves of specimens.

The unconfined columns occurred failure immediately once reaching the peak compressive strength, showing an obvious brittle feature. The finally failure mode was characterized by vertically splitting failure along a main crack. However, the typical failure pattern of columns strengthened with TRC was that multiply hair cracks instead of a major crack initialed at the mid-height of the column. Once the peak strength was achieved, a significant lateral deformation together with a progressive cracking of the cement-based mortar could be observed. The failure was characterized by a combination of the interfacial zone debonding in the overlapping zone and the textile rupture; moreover, the textile rupture was rather dominant. It was noteworthy that the thinner and denser cracks at the surface of TRC strengthening layer with short glass fibers added into for confined columns were observed. Figure 1 depicted the failure patterns of confined and unconfined specimens.

Figure 2 presents the load-axial deformation relationship of confined columns and unconfined columns. The load-axial deformation response for each confinement configuration is characterized

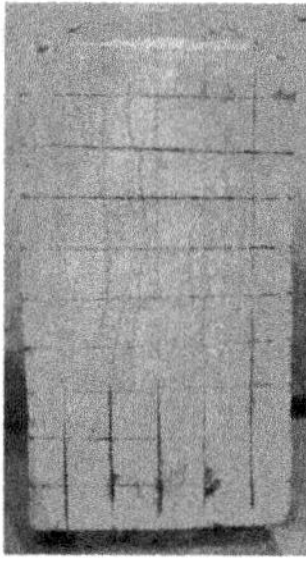
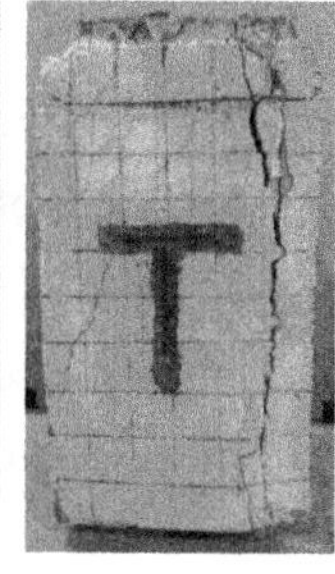

(a) PC-L0-C30 (b) PC-L2N-C30 (c) PC-L2F-C30

Figure 1 Failure mode of specimens

by three different branches. Firstly, a parabolic response with a similar slope to that of unconfined specimens is observed in the initial loading. Then the curve of second stage is represented by a non-linear ascending trend up to the peak strength of the confined specimens. The curve in the last stage is represented by a soft descending branch that dropped when the confinement fracture occurred. In addition, it can be seen that a significant gain in compressive strength and deformation capacity and a higher curves for square columns confined with TRC is obtained in comparison with unconfined columns. The TRC system is proved to be very effective in increasing peak strength and axial stiffness of the unstrengthened specimens. The increment is related to the number of textile layers and short fibers. The ultimate loads of TRC confined columns exhibit an increased gradually trend with the increase of the number of textile layers. A most high compressive strength will be achieved when the specimens strengthened with four layers of TRC. According to Figure 2(b), a certain increase in both ultimate load and axial deformation is obtained when added short glass fibers into TRC strengthened layer compared to that of without short glass fibers, indicating the short fibers also is the parameter affecting the confinement effect of TRC system.

In conclusion, TRC is a very promise composite material as confinement system to improve compressive strength and deformation capacity of concrete members.

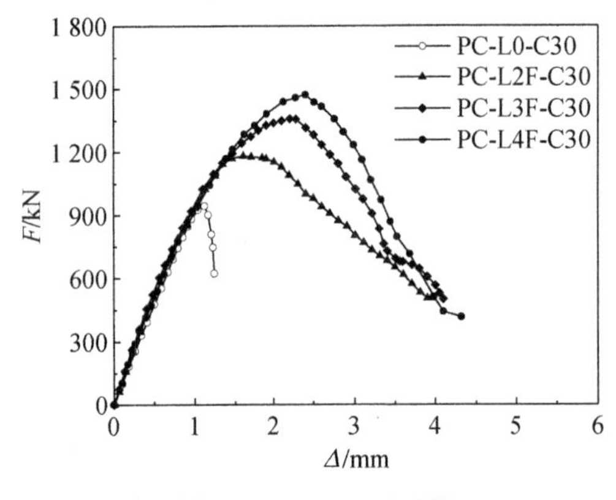

(a) Concrete strength C30

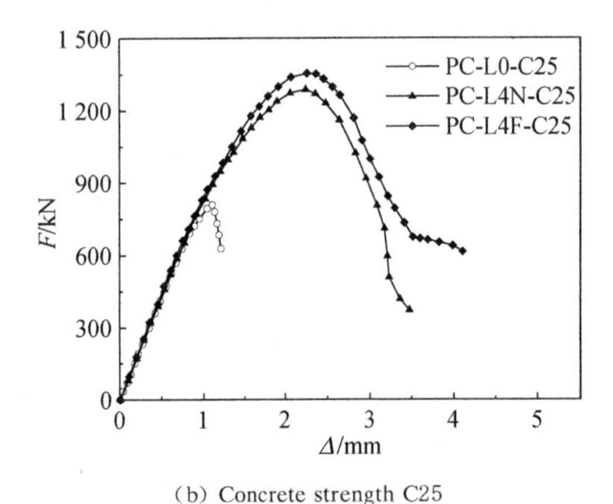

(b) Concrete strength C25

Figure 2 Load-axial deformation relationship of columns

Acknowledgements

The authors gratefully acknowledge the financial support from the National Natural Science Foundation of China (Grant No. 51508154) and Natural Science Foundation of Jiangsu Province (Grant No. BK20150803). The experimental work described in this paper was conducted at the structural laboratory of Yancheng Institute of Technology. Helps during the testing from staffs and students at laboratory are greatly acknowledged.

References

[1] NANNI A, BRADFORD N M. FRP jacketed concrete under uniaxial compression [J]. Construction and Building Materials, 1995, 9 (2): 115-124.

[2] TOUTANJI H. Stress-strain characteristics of concrete columns externally confined with advanced fiber composite sheets [J]. ACI Materials Journal, 1999, 96(3): 397-404.

[3] COLAJANNI P, De Domenico F, RECUPERO A, et al. Concrete columns confined with fibre reinforced cementitious mortars: experimentation and modelling[J]. Construction and Building Materials, 2014, 52: 375-384.

[4] BOURNAS D A, LONTOU P V, PAPANICOLAOU C, et al. Textile-reinforced mortar versus fiber-reinforced polymer confinement in reinforced concrete columns[J]. ACI Structural Journal, 2007, 104(6): 740-748.

Seismic behavior and plastic hinge position moving up of steel fiber ductile concrete piers (SFDCP)

Y. J. Wang & R. D. Chen & J. Y. Wang & T. Y. Zhang
College of Civil Engineering and Architecture, Zhejiang University, Hangzhou 310058, China

Abstract

In the seismic damage of bridges, it is common for the cracking damage of the plastic hinge area of the pier to cause the overall collapse of the bridge structure. How to avoid premature failure of the plastic hinge area becomes one of the keys to the design of the pier. All over the world, the lateral restraint reinforcement of the plastic hinge are specified from the aspects of reinforcement ratio and structural measures, and the core concrete is restrained to improve the compressive strength and meet the ductility required for the design[1]. On this basis, many scholars have carried out many researches to solve the inherent defects of the concrete material in the plastic hinge area.

As listed in Table 1, six full-scale bridge pier specimens are designed, and the test piers (SFDCP) are partially reinforced with steel fiber fine stone concrete, which avoids the damage of the lower plastic hinge region with the most unfavorable force, and the bearing capacity is improved by 18%. We verified the feasibility of using steel fiber fine stone concrete replacing 25% reinforcement of the plastic hinge area on the foundation of the common piers. Finally, a simplified calculation formula is proposed to estimate the reasonable plastic hinge length of the SFDCP which match well with the test results. The SF-1 specimen is shown in Figure 1.

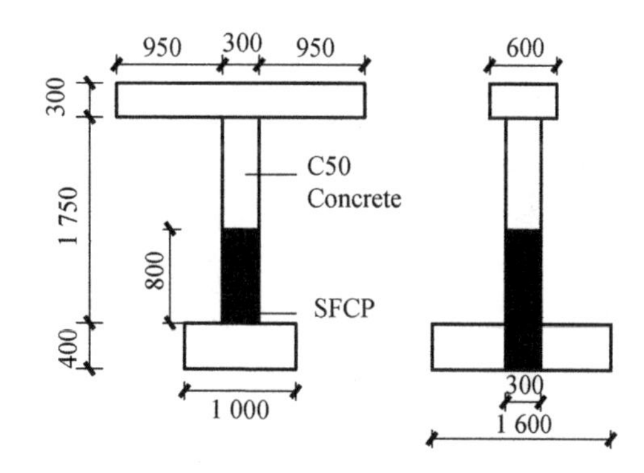

Figure 1 Dimension and reinforcement bars of specimen (unit: mm)

Table 1 Test parameters

No.	h/mm	RRRS%
C-1	0	0.686
SF-1	800	0.686
SF-2	800	0.514
SF-3	800	0.343
SF-4	500	0.686
SF-5	300	0.686

RRRS: clamping ratio of replacing section.

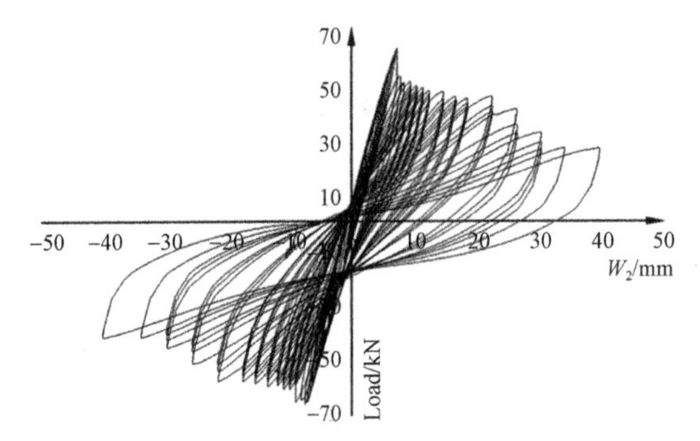

Figure 2 Hysteresis curve of SF-1

The test parameters and hysteresis curve of SF-1 are both shown in Table 1 and Figure 2.

The SF-1 exerts a good restraining torsion effect in the plastic hinge region, the torsional peak load reached 55.84 kN · m, comparing to 34.58kN · m of the ordinary pier C-1.

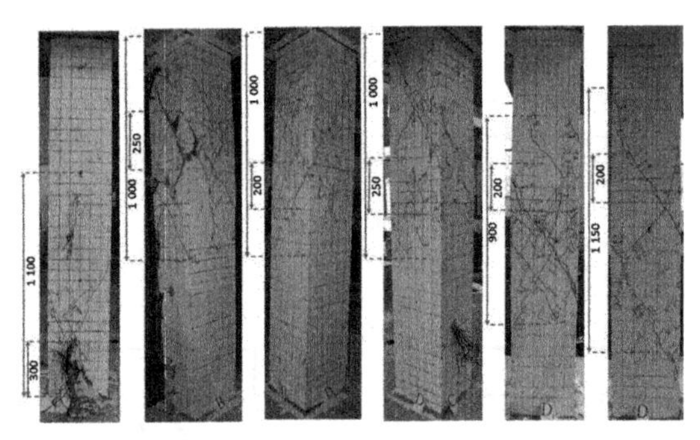

Figure 3 Schematic diagram of equivalent plastic hinge length

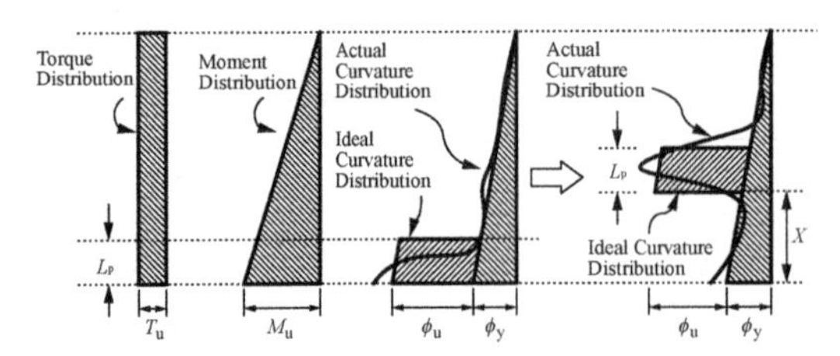

Figure 4 Schematic diagram of the equivalent plastic hinge moving up[3]

In this paper, the torsion-bending ratio of the test is 0.47, and the superposition of torsion under small torsion-bending ratio has little impact on the position and length of the plastic hinge of the ordinary concrete specimen[4], that is, the evenly distributed torque has less influence on the bending curvature of the section, and the plastic hinge moves up in Figure 3 and Figure 4. There is the following formula:

$$\Delta_u^- = \frac{\Phi_y L^2}{3} + \left(\Phi_u - \Phi_y \cdot \frac{L-X}{L}\right) \cdot L_p \cdot (L - X - 0.5L_p) \quad (1)$$

Then we find a formula that can estimate the length of the SFDCP different from some norms[5] which only considers the ordinary piers.

$$L_P = 0.08(L - t) + 0.022 f_y d_b \quad (2)$$

And a new formula to estimate the plastic hinge position of the SFDCP:

$$X = t + 0.5 \cdot (L - t) \cdot \rho_v \quad (3)$$

Both Eq. (2) and Eq. (3) agree well with the test results. Each error is less than 24%.

The plastic hinge moves up in the SFDCP, so that the overall bearing capacity of the pier body is fully exerted, and the torsional bearing capacity is improved by more than 18%.

The 50% reduction of the amount of the reinforcement in the steel fiber fine stone concrete reinforcement area has no obvious influence on the resistance of the SFDCP.

In this paper, for the plastic hinge moving up, the formula for determining the plastic hinge moving amplitude is proposed and the plastic hinge length formula recommended by the specification is further modified. For the test results, a good fitting effect is obtained, which is suitable for steel fiber fine stone concrete replaced pier design.

Acknowledgements

This study was supported by Zhejiang Provincial Key Research and Development Program of China (Grant No. 2018C03045) and the Fundamental Research Funds for the Central Universities (Grant No. 2019FZA4018). The authors appreciated the great assistance of Cheng-bin LIU, Yu PENG and Zhi-hua HU in the execution of experiments.

References

[1] Ministry of Housing and Urban-Rural Development of the People's Republic of China. Code for seismic test of buildings: JGJ/T 101—2015 [S]. Beijing: China Building Industry Press, 2015.

[2] FOSTER S J, ATTARD M M. Strength and ductility of Fiber-Reinforced High-Strength concrete columns[J]. Journal of Structural Engineering, 2001, 127:28-34.

[3] PRIESTLEY M J N, PARK R. Strength and ductility of concrete bridge columns under seismic loading[J]. ACI Structural Journal, 1987, 84(1): 61-76.

[4] HAN Q, DU X L, LIU J B, et al. Seismic damage of highway bridges during the 2008 Wenchuan earthquake [J]. Earthquake Engineering and Engineering Vibration: English, 2009, 8(2): 263-273.

[5] Caltrans. Caltrans Seismic design criteria[S]. Version 1.6. Sacramento: California Department of Transportation, 2010.

Experimental and numerical studies on hysteretic behavior of wood in the parallel-to-grain direction

J. Y. Tang & X. B. Song

Key Laboratory of Performance Evolution and Control for Engineering Structures of Ministry of Education, Tongji University, Shanghai 200092, China & Department of Structural Engineering, College of Civil Engineering, Tongji University, Shanghai 200092, China

Abstract

Timber structures have many advantages, such as eco-friendliness, high stiffness-to-weight ratio and easy assembling, and are widely used around the world. Natural disasters, such as earthquake, will cause damage in structures. Understanding the mechanical properties of timber members and connections and their hysteretic behavior is crucial to behavior analysis and prediction of timber structures under earthquakes.

Recently, continuum damage mechanics has been successfully adopted to predict the mechanical behavior of wood[1-6]. However, little research has been taken on the hysteretic behavior of wood. Dinwoodie[7] provided that the tensile strength would loss 6%～15.3% if wood had been compressed to the max compress stress. However, no detailed results of the stress and strains were reported.

This paper focuses on the experimental and numerical analysis of the mechanical behavior of timber under cyclic loading in the parallel-to-grain direction.

1. Experimental Study

The sizes of the specimen are shown in Figure 1. Particularly, the width of the specimen is 10 mm. There were totally 5 specimens and No. 1 graded Canadian spruce-pine-fir lumber was used to make the specimens.

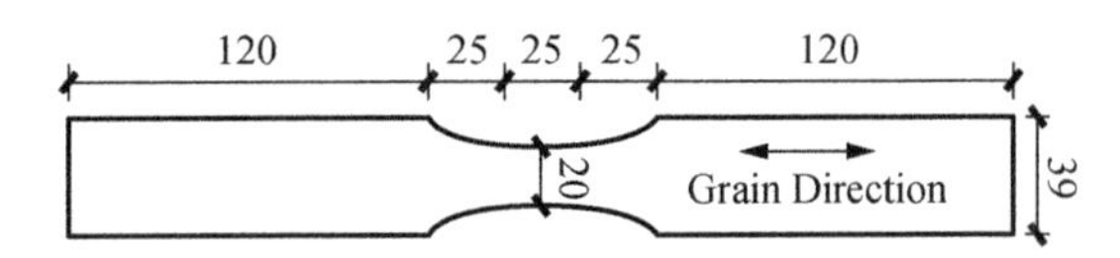

Figure 1 Dimension of the specimens (unit: mm)

Reversed cyclic loading was applied with a stress-displacement mixed controlled cyclic protocol. When the specimen was under tension, the loading was controlled by the stress rate to make sure that the maximum tensile stress was 25 MPa. When the specimen was under compression, the loading was controlled by displacement to make sure that the maximum compressive strain was larger than the one in the previous loading cycle.

The load speed was 0.8 mm/min. Typical test result is shown in Figure 2.

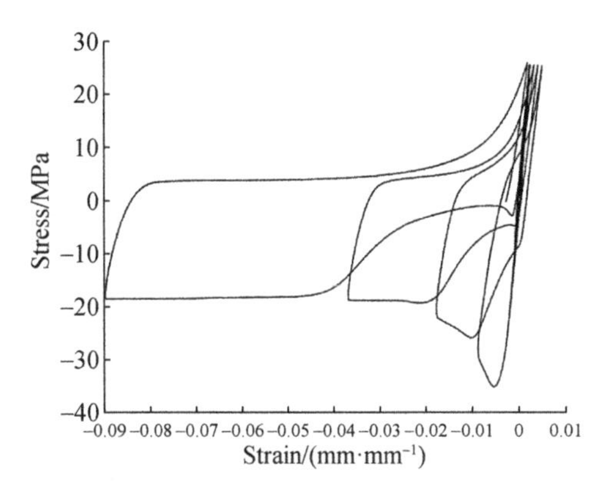

Figure 2 Stress-strain curves from cyclic loading tests

The test and data measurement device is shown in Figure 3 (a). When the specimen was in large compressive damage, there was an obvious kinking band[8] on the middle part of the specimen, as shown in Figure 3(b). This indicated that the wood fiber was kinked. When the stress turned in tension (positive) from compression (negative), there was a low stiffness segment on the stress-strain curves and the kinking bands disappeared gradually. When the strain turned in positive, the wood fiber was straightened totally and the stiffness recovered.

(a) Test and data measurement device

(b) The kinking bands on the specimen

Figure 3 Experiment device and specimen

2. Numerical Study

Based on the experiment phenomenon and the three-dimensional combined elastic-plastic and damage model established by Wang et al.[5], in this study, the effective stress was divided into a tensile and a compressive components by spectral factorization[9]:

$$\bar{\boldsymbol{\sigma}} = \bar{\boldsymbol{\sigma}}^{+} + \bar{\boldsymbol{\sigma}}^{-} \tag{1}$$

where $\bar{\boldsymbol{\sigma}}$ is the effective stress tensor; $\bar{\boldsymbol{\sigma}}^{\pm}$ is the effective tensile (or compressive) stress tensor.

The elastic-plastic model was developed using the Hill's yielding criterion in Voigt's notation as:

$$f^{\pm} = \sqrt{\bar{\boldsymbol{\sigma}}^{\pm T} \boldsymbol{H}^{\pm} \bar{\boldsymbol{\sigma}}^{\pm}} - \bar{\sigma}_{\text{iso}}^{\pm} = 0 \tag{2}$$

where $\boldsymbol{H}^{\pm}$ is the Hill's strength parameters matrix; and $\bar{\sigma}_{\text{iso}}^{\pm}$ is the isotropic hardening equivalent effective tensile (compress) stress.

The variation of the tensile stiffness caused by compressive damage was assumed based on a new hardening rule:

$$\bar{\sigma}_{\text{iso}}^{+} = -s_1 \left[\ln\left(\frac{1}{s_2 \bar{\varepsilon}^{p+} + s_3} \right) - s_4 \right] \tag{3}$$

where $\bar{\varepsilon}^{p+}$ is equivalent plastic tensile strain; and s_1, s_2, s_3, s_4 are shape parameters.

Based on the algorithm in Eq. (1) to Eq. (3), a three-dimensional model of the timber specimens was developed, and the loading protocol was considered to run the numerical analysis. The model and the simulation results (in comparison with the test results) are shown in Figure 4 and Figure 5, respectively.

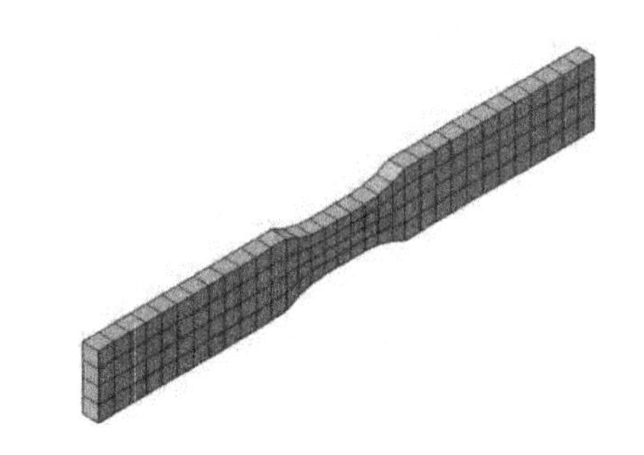

Figure 4 The finite element model of the specimen

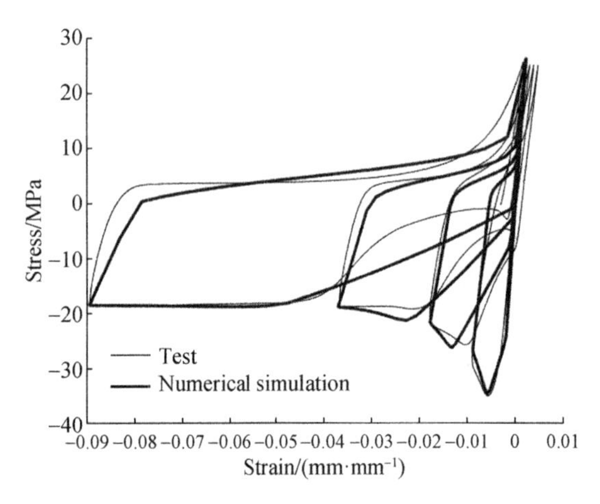

Figure 5 Stress-strain curves from experiment and model prediction

It was found that not only the irreversible deformation and the softening behavior of wood under compression in the parallel-to-grain direction could be traced by the model, but also the variation of tensile stiffness caused by compressive damage could be predicted by the model with acceptable accuracy.

3. Conclusions

In this study, cyclic loading tests were conducted on the timber specimens along the parallel to grain direction to establish a complete stress-strain model. The effective stress tensor was divided into positive and negative components, and Hill's yielding criteria together with a new hardening rule was adopted in a numerical study. It achieved great agreement.

Acknowledgements

This project was supported by National Natural Science Foundation of China (PI) (Grant No. 51878477).

References

[1] XU B H, BOUCHAIR A, TAAZOUNT M, et al. Numerical and experimental analyses of multiple-dowel steel-to-timber joints in tension perpendicular to grain [J]. Engineering Structure, 2009, 31(10): 2357-2367.

[2] SANDHAAS C. Mechanical behavior of timber joints with slotted-in steel plates[D]. Delft, Netherlands: Delft University. of Technology, 2012.

[3] KHELIFA M, KHENNANE A. Numerical analysis of the cutting forces in timber[J]. Journal of Engineering Mechanics, 2013, 140(3): 523-530.

[4] SIRUMBAL-ZAPATA L F, MALAGA-CHUQUITAYPE C, ELGHAZOULI A Y. A three-dimensional plasticity-damage constitutive model for timber under cyclic loads[J]. Computers and Structures, 2018, 195(2018): 47-63.

[5] WANG M Q, SONG X B, GU X L. Three-Dimensional Combined Elastic-Plastic and Damage Model for Nonlinear Analysis of Wood[J]. Journal of Structural Engineering, 2018, 144(8): 04018103.

[6] OUDJENE M, KHELIFA M. Elasto-plastic constitutive law for wood behavior under compressive loadings[J]. Construction and Building Materials, 2009, 23(11): 3359-3366.

[7] DINWOODIE J M. Failure in Timber Part 3:

The Effect of Longitudinal Compression on Some Mechanical Properties [J]. Wood Science and Technology. 1978, 12(1978): 271-285.

[8] POULSEN J S, MORAN P M, SHIH C F, et al. Kink band initiation and band broadening in clear wood under compressive loading[J]. Mechanics of Materials, 1997, 25(1997): 67-77.

[9] WU J Y, LI J, FARIA R. An energy release rate-based plastic-damage model for concrete [J]. International Journal of Solids and Structures, 2006, 43(2006): 583-612.

Study on seismic performance and energy demand of RC frame constructed with lead viscoelastic damper under strong earthquake

W. Y. Huang, C. Zhang, F. Shi, Y. Zhou
School of Civil Engineering, Guangzhou University, Guangzhou 510006, China

Abstract

To build a more resilient city against strong earthquakes[1, 2], it is quite significant to construct seismic resilient buildings[3, 4]. Lead viscoelastic damper (LVD)[5], demonstrated in Figure 1, was deemed as a novel resilient damper due to its self-recovery material properties (i.e. dynamic recrystallization of lead cords and superelasticity characteristic of viscoelastic materials), thus it can be incorporated in the buildings to improve their seismic resilience[6, 7].

This paper aims to assess the seismic performance, collapse-resistant performance and structural energy demands of reinforced concrete (RC) frame equipped with LVD subjected to strong earthquake. The investigations were applied to a 6-story 22.2 meter-height traditional RC frame and LVD-damped frame designed based on current Chinese codes[8-10] and simulated by Opensees[11]. The simulated models of RC frame and LVD-damped frame were depicted in Figure 2 and Figure 3.

Time history analysis of the traditional RC frame (RCF), LVD-damped frame (LVDF) and its bare frame (i.e. without LVDs) were carried out to investigate the damping effect of LVDs

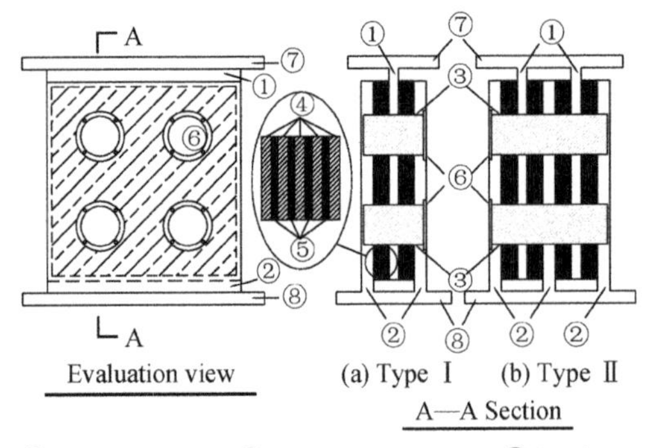

Figure 1 Configurations of LVD

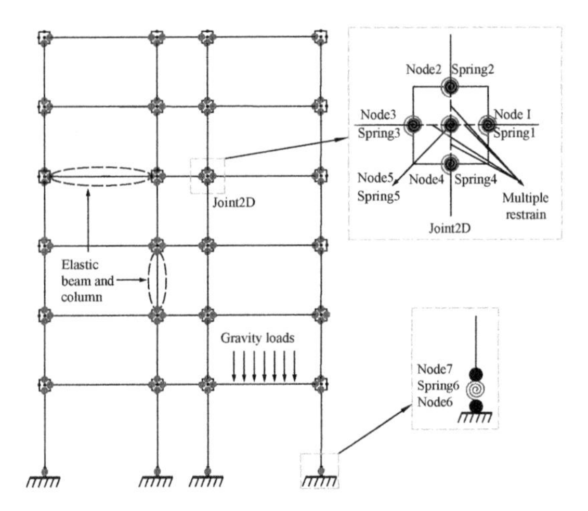

Figure 2 Analytical RC frame model

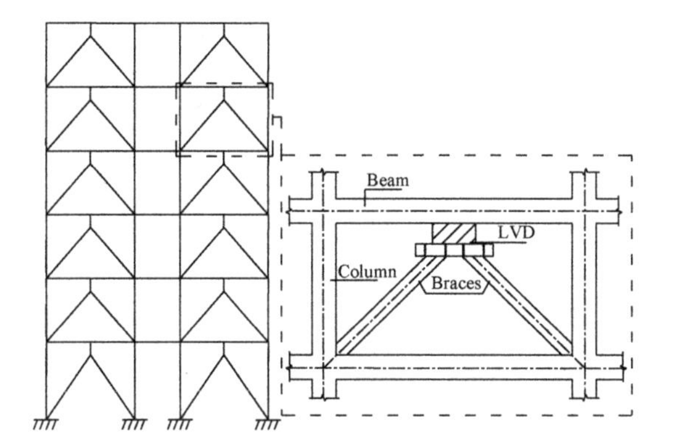

Figure 3 Numerical model of LVD frame

installed in the structure. To assess structure seismic performance systematically, incremental dynamic analysis (IDA)[12] using ATC-63 recommended 22 far-field ground motion records[13], seismic fragility analysis and collapse assessment through the IDA results were conducted for the traditional RC frame and LVD-damped frame[14]. In addition, to survey the contribution of LVDs on reforming structural energy dissipation mechanism, energy demands of RC frame and LVD-damped frame were statistically analyzed[15]. Results of time history response investigation show that both the traditional RC frame and LVD-damped frame were available to meet the structural design requirements under designed earthquake. However, by adding LVDs, structure seismic performance was greatly improved compared with conventional RC frame and LVD-damped bare frame. Besides, as illustrated in Figure 4, the IDA results demonstrate that structural dynamic responses would be availably suppressed by installing LVDs, which could also reduce the undesirable dispersion of structural response making analysis results more reliable.

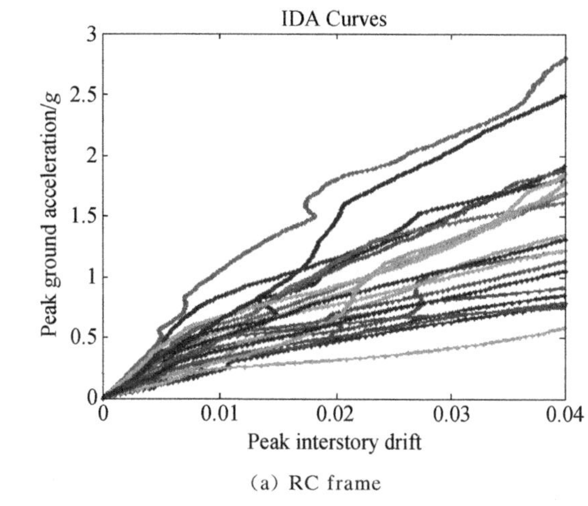

(a) RC frame

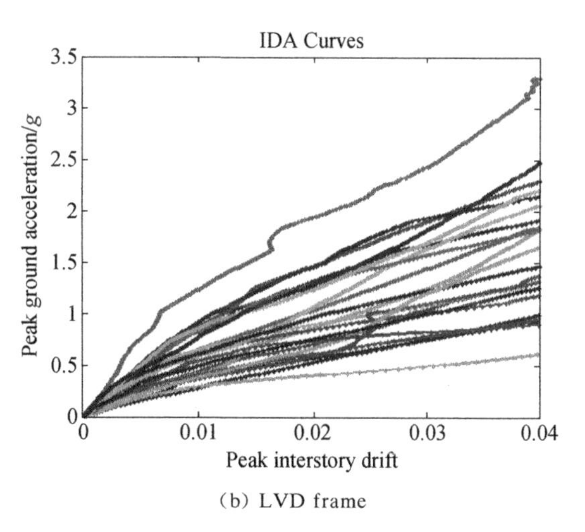

(b) LVD frame

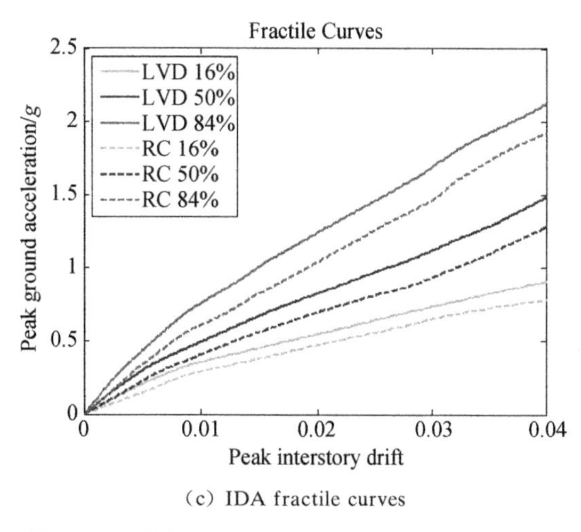

(c) IDA fractile curves

Figure 4 IDA results of RC frame and LVD frame

Structural damage evaluation consequences are shown in Figure 5 and Table 1. The analysis results reveal that exceedance probabilities of LVD-damped frame within different damage levels is lower than that of RC frame, showing that LVD designed frame can satisfyingly mitigate structural damage and increase collapse margin ratio (CMR) by 15% approximately.

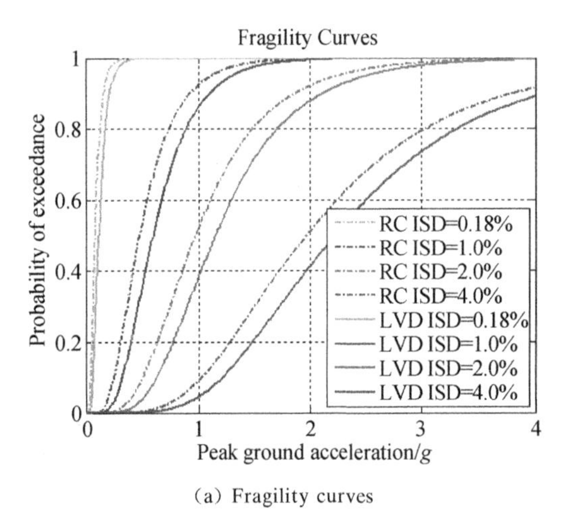

(a) Fragility curves

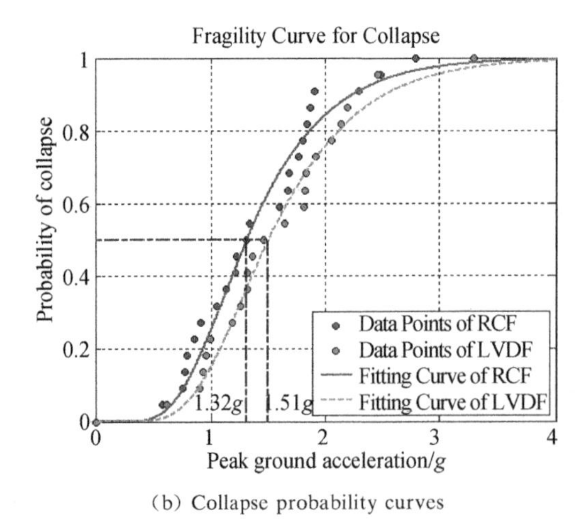

(b) Collapse probability curves

Figure 5 Fragility and collapse probability curves

Specifically, the CMR was calculated based on ATC-63 method as follow:

$$CMR = IM_{50\%\text{collapse}} / IM_{\text{MCE}} \tag{1}$$

Table 1 CMR calculate results

	RCF	LVDF
$IM_{50\%\text{Collapse}}/g$	1.32	1.51
IM_{MCE}/g	0.40	0.40
CMR	3.29	3.78

In addition, it is also summarized from Table 2 that by dissipating a great percentage of structural input seismic energy under frequency earthquake (FE), design earthquake (DE), rare earthquake (RE) and very rare earthquake (VRE), constructing LVDs would be beneficial to generously reduce the plastic hinge rate, cumulative energy consumption and energy consumption ratio of the plastic hinge areas. In other words, furnishing LVDs could effectually mitigate structural damage of the RC frame and have a more stable structural energy dissipation mechanism under strong earthquake.

Table 2 Energy dissipation proportions of structural components

		Beam	Column	LVD
RCF	FE	0	0	—
	DE	16.61%	0	—
	RE	42.70%	0	—
	VRE	71.80%	9.67%	—
LVDF	FE	0	0	64.25%
	DE	0	0	63.63%
	RE	9.56%	0	44.08%
	VRE	14.62%	0	35.93%

Acknowledgements

The authors gratefully acknowledged the generous support of this work by the

National Natural Science Foundation of China (No. 51878195) and Guangzhou University Postgraduate Foundation Innovation Project (No. 2018GDJC - M43).

References

[1] CIMELLARO G P, REINHORN A M, BRUNEAU M. Framework for analytical quantification of disaster resilience [J]. Engineering Structures, 2010, 32(11): 3639-3649.

[2] BRUNEAU M, CHANG S E, EGUCHI R T, et al. A Framework to Quantitatively Assess and Enhance the Seismic Resilience of Communities[J]. Earthquake Spectra, 2003, 19(4): 733-752.

[3] ZHOU Y, LU X L. State-of-the-art on rocking and self-centering structures [J]. Journal of Building Structures, 2011, 32(9): 1-10.

[4] WANG W, FANG C, ZHAO Y S, et al. Self-centering friction spring dampers for seismic resilience [J]. Earthquake Engineering and Structural Dynamics, 2019, 48(9): 1045-1065.

[5] SHI Fei. Performance and Application Research of (Lead) Viscoelastic Damper [D]. Guangzhou: Guangzhou University, 2012.

[6] ZHANG C, HUANG W Y, XU X, et al. Comprehensive design and seismic performance analysis of retrofitted frame equipped with sector lead viscoelastic damper [J]. Journal of building structures, 2018, 39 (S1): 94-99.

[7] HUANG W Y, ZHANG C, WANG G P. Influence of Different Design Parameters of Sector Lead Viscoelastic Dampers on the Seismic Performance of Retrofitted Frame Structures[J]. China Earthquake Engineering Journal, 2019, 41(3): 638-644.

[8] Ministry of Housing and Urban-Rural Development of the People's Republic of China. Code for design of buildings: GB 50011—2010 [S]. Beijing: China Architecture and Building, 2010.

[9] Ministry of Housing and Urban-Rural Development of the People's Republic of China. Code for design of concrete structuresl: GB 50010—2010 [S]. Beijing: China Architecture and Building, 2010.

[10] Ministry of Housing and Urban-Rural Development of the People's Republic of China. Technical specification for seismic energy dissipation of buildings: JGJ 297—2013 [S]. Beijing: China Architecture and Building, 2013.

[11] MCKENNA F. OpenSees: A Framework for Earthquake Engineering Simulation [J]. Computing in Science and Engineering, 2011, 13(4): 58-66.

[12] VAMVATSIKOS D, CORNELL C A. Incremental dynamic analysis [J]. Earthquake Engineering and Structural Dynamics, 2002, 31: 491-514.

[13] FEMA-P695. Quantification of building seismic performance and factors [S]. Washington, DC: Federal Emergency Management Agency. June 2009.

[14] RUIZ-GARCÍA J, NEGRETE M. Drift-based fragility assessment of confined masonry walls in seismic zones [J]. Engineering Structures, 2009, 31(1): 170-181.

[15] UANG C M, BERTRERO V V. Evaluation of seismic energy in structures [J]. Earthquake Engineering and Structural Dynamics, 1990, 19(1): 77-90.

Seismic behavior of traditional timber-masonry frames with different types of diagonal bracings

C. Salamone

Department of Structural Engineering, Tongji University, Shanghai, 200092, China & Department of Architecture and Engineering, University of Bologna, Ravenna 48121, Italy

X. B. Song

Department of Structural Engineering, Tongji University, Shanghai 200092, China

Abstract

Traditional constructions that combine wood with a filler material are widely used throughout the world. The geography and the context, in which they are built, have influenced their design[1]. Several countries developed different techniques that evolved over the years. These structures have always been a topic of interest in engineering for the interaction between the timber and the masonry. The understanding of these buildings is essential to figure out the solutions to preserve historic values. Modern societies must find the right strategies to conserve these timeless heritages[2].

The purpose of this study is to compare different construction techniques used in China and Europe, and try to find the optimal solution among the timbered-frame structures. The comparison of samples under cyclic static loading tests is performed. Five specimens are investigated: bare timber frame without infill (F1, Chuan-dou main frame) then retrofitted with masonry infill (F1R, retrofitted Chuan-dou main frame); timber frame with masonry infill (F2, Chuan-dou main frame with masonry infill); timber frame with masonry infill and the single timber cross-bracing (F3, Italian Baraccato system); timber frame with masonry infill and the zigzag timber bracings (F4, Italian Maso system); timber frame with masonry infill and the double timber cross-bracing (F5, Portuguese Pombalino system), as shown in Figure 1.

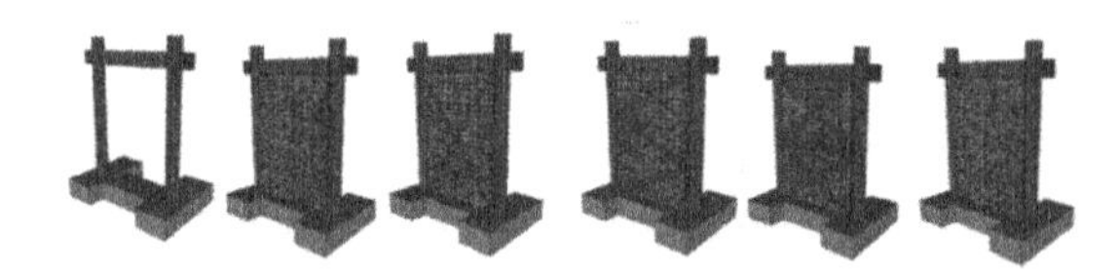

Figure 1 Prospective view of the specimens

For each specimen, the main frame is assembled with the mortise and tenon joints without the application of metal fasteners. Moreover, the wooden diagonals elements follow the traditional European construction techniques with the use of half-lapping joints and screws. The specimens were made in the State Key Laboratory of Disaster Reduction in Civil Engineering at Tongji University,

Shanghai. The main forces applied during the experiments were: one horizontal load by a hydraulic actuator and concentrated vertical loads by two hydraulic jacks. The CUREE loading protocol was used, as shown in Figure 2.

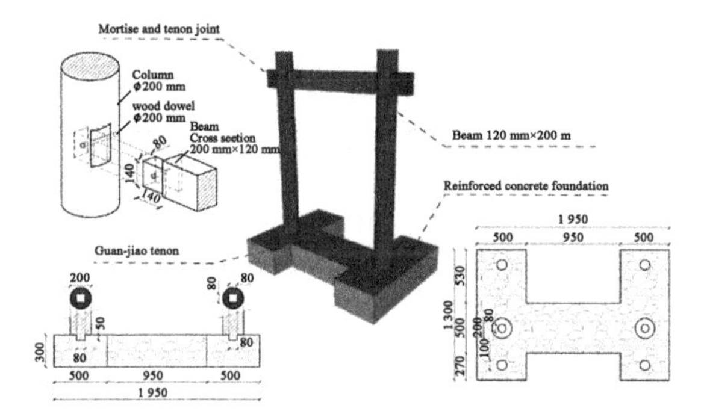

Figure 2 Wooden members of each bare frame

The specimens show a nonlinear and inelastic behavior through the hysteresis curves (Figure 3). The mortise and tenon joints were subjected to slight rotation, which showed a semi-rigid behavior, without any structural damages. The results demonstrated that the combination of timber frame with masonry infill and diagonal bracings could enhance the Chinese traditional timber buildings. From these experiments, the resistance of traditional constructions to historical seismic activity could be understood, compared to newly reinforced concrete structures. These traditional technologies could be an eco-friendly alternative in the construction industry. Focusing on their mechanisms, weaknesses and strengths, it is possible to keep using this traditional construction to improve their structural behavior for the future generations.

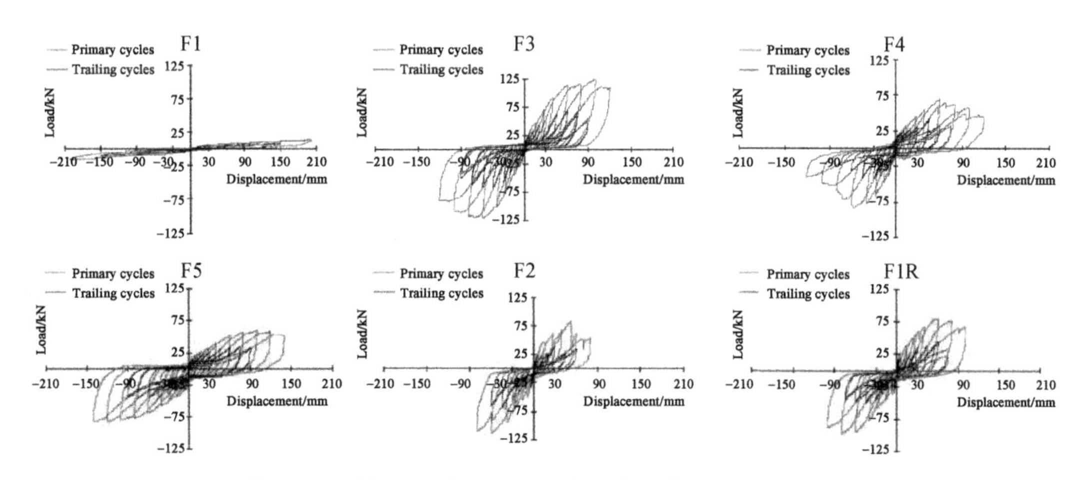

Figure 3 Hysteresis curves of various frame specimens

References

[1] QU Z, DUTU A, ZHONG J R, SUN J J. Seismic Damage to Masonry-Infilled Timber Houses in the 2013 M7.0 Lushan, China, Earthquake[J]. Earthquake Spectra, 2015, 31(3): 1859-1874.

[2] LANGENBACH R. Learning from the past to protect the future: Armature Crosswalls [J]. Engineering Structures, 2008, 30(8): 2096-2100.

Recycling of cathode ray tube glass in ultra-high performance concrete

H. N. Wei & T. J. Liu & A. Zhou

School of Civil and Environmental Engineering, Harbin Institute of Technology, Shenzhen, Shenzhen 518055, China

Abstract

Cathode ray tube (CRT) has been utilized to produce monitor displays in the past decades, and it is classified as a hazardous material because it contains lead[1-3]. With the development of new display technology, the amount of obsolete CRT is growing dramatically. Most waste CRTs are disposed in landfill sites, and only approximately 26% are reused[4-5]. The disposal of waste CRTs in landfills has significant drawbacks, as lead leakage can pollute water and soil and result in oral intake of lead by humans[6-8]. Hence, waste CRTs should thus be carefully treated, and the proper recycling method of waste CRTs has attracted particular attention. However, the current recycling methods involve with either environmental pollution or limited recovery scale, bringing main challenges for waste CRT management.

In this paper, an innovative cementitious material[9-10], ultra-high performance concrete (UHPC) was introduced to combine and recycle hazardous waste CRT. Crushed CRT funnel glass was utilized to substitute river sand in various ratios in UHPC. The mechanical properties and toxic heavy metal leachability of UHPC were investigated, and the microstructure of UHPC was also observed. Test results indicate that the addition of CRT funnel glass can increase the flowability of UHPC, but decrease the compressive and flexural strength of UHPC. When the replacement ratio of CRT funnel glass is 25%, the flowability improves significantly while the mechanical properties reduce slightly (about 3%) compared to those of the reference group. Analyses by X-ray diffraction and electron microscopy scan demonstrate that CRT funnel glass inhibits cement hydration and weakens the interfacial transition zone, resulting in the reduced strengths of UHPC. Meanwhile, the toxicity characteristic leaching procedure of UHPC containing CRT funnel glass at 160 days was conducted, and the leached lead (Pb) concentration was displayed in Table 1. The leached lead concentration was only 3.35 mg/L even the replacement ratio of CRT funnel glass was 100%, which is below the regulatory limit (5 mg/L) of United States Environmental Protection

Agency. It can be concluded that the lead leaching of hazardous CRT glass was effectively inhibited in UHPC. The mechanism behind was that the dense microstructure and low permeability coefficient of UHPC significantly restrained the lead leaching process. Moreover, the leached concentrations of other heavy metals were also listed in Table 1, including Cr, Zn, Ba, Cd, Ni, Cu and As. It was shown that the leached concentrations of those heavy metals at 160 days were far below the regulation limit.

Table 1 Concentrations of leached heavy metals of UHPC incorporated with CRT glass at 160 days (unit: mg/L)

Replacement ratio	Pb	Cr	Zn	Ba	Cd	Ni	Cu	As
25%	0.20	0.06	0.06	1.91	—	—	—	—
50%	0.44	0.08	0.15	1.99	—	—	—	—
75%	0.89	0.09	0.19	2.28	—	—	—	—
100%	3.35	0.12	0.28	2.40	—	—	—	—

Note: — indicates below the detection limit (< 0.001 mg/L).

The material design (UHPC and hazardous waste CRT) in this paper provided an effective alternative to recycling hazardous waste CRT on a large scale without limitations on the replacement ratio and safety concern, as well as paves the way for a sustainable UHPC.

Acknowledgements

This work was supported by the National Key Research and Development Plan of China (Grant No. 2018YFC0705400); National Natural Science Foundation of China (Grant No. 51878225, 51678200); and Program of Shenzhen Science and Technology Plan (Grant No. JCYJ2017081116 0514862).

References

[1] LIU T J, QIN S S, ZOU D J, et al. Experimental investigation on the durability performances of concrete using cathode ray tube glass as fine aggregate under chloride ion penetration or sulfate attack [J]. Construction and Building Materials, 2018, 163: 634-642.

[2] OKADA T. Water-soluble lead in cathode ray tube funnel glass melted in a reductive atmosphere [J]. Journal of Hazardous Materials, 2016, 316: 43-51.

[3] LING T C, POON C S. Utilization of recycled glass derived from cathode ray tube glass as fine aggregate in cement mortar[J]. Journal of Hazardous Materials, 2011, 192 (2): 451-456.

[4] LONG W J, GU Y C, ZHENG D, et al. Utilization of graphene oxide for improving the environmental compatibility of cement-based materials containing waste cathode-ray tube glass[J]. Journal of Cleaner Production, 2018, 192: 151-158.

[5] LU J X, DUAN Z H, POON C S. Combined use of waste glass powder and cullet in architectural mortar [J]. Cement and Concrete Composites, 2017, 82: 34-44.

[6] SINGH N, LI J H. An efficient extraction of

lead metal from waste cathode ray tubes (CRTs) through mechano-thermal process by using carbon as a reducing agent [J]. Journal of Cleaner Production, 2017, 148: 103-110.

[7] ZHAO H, POON C S. A comparative study on the properties of the mortar with the cathode ray tube funnel glass sand at different treatment methods [J]. Construction and Building Materials, 2017, 148: 900-909.

[8] SINGH N, LI J H, ZENG X L. Solutions and challenges in recycling waste cathode-ray tubes [J]. Journal of Cleaner Production, 2016, 133: 188-200.

[9] ARORA A, AGUAYO M, HANSEN H, et al. Microstructural packing- and rheology-based binder selection and characterization for Ultra-high Performance Concrete (UHPC) [J]. Cement and Concrete Research, 2018, 103: 179-190.

[10] MENG W N, KHAYAT K. Effects of saturated lightweight sand content on key characteristics of ultra-high-performance concrete [J]. Cement and Concrete Research, 2017, 101: 46-54.

Behavior of large-size square PEN FRP-concrete-steel hybrid multi-tube concrete columns under axial compression

Q. Q. Li & Y. L. Bai

Key Laboratory of Urban Security and Disaster Engineering of Ministry of Education, Beijing University of Technology, Beijing 100124, China

Abstract

Hybrid fiber-reinforced polymer (FRP)-concrete-steel multi-tube concrete columns (MTCCs) are a new tape of hybrid columns recently formed at the University of Wollongong. An MTCC consists of a number of inner tubes made of steel and an outer tube made of FRP, with the space inside all the tubes filled with concrete. In MTCCs, the combination of three materials (i.e. FRP, concrete, and steel) possesses several important advantages not available with existing columns including its excellent corrosion resistance as well as excellent ductility and ease for construction. The use of small circular steel tubes in MTCCs eliminates the difficulties connected with the manufacture, transportation and installation of large steel tubes. Furthermore, these steel tubes filled with concrete form a rigid wall to confine the concrete surrounded by them. Although large-size square/rectangular hybrid MTCCs may be needed for practical applications, due to aesthetic and other reasons, the existing experimental studies have only focused on the behavior of small-size circular and square concrete specimens confined with glass fiber-reinforced polymer (GFRP) composites[1-3], which are referred to as conventional FRP composites. The data available for large-size square MTCCs columns are very limited and the previous studies have never focused on the usage of an FRP jacket made of Large Rupture Strain (LRS) FRP composites. Aiming to fill such a knowledge gap, this paper presents the results of an experimental investigation on the performance of large-size square MTCCs wrapped with Large Rupture Strain (LRS) FRP composites, namely, Polyethylene Naphtholate (PEN) FRP composites[4, 5], which possesses a large rupture strain (usually larger than 5%) and is much cheaper and more environmentally friendly than conventional FRP composites.

A total of 4 large-size square MTCCs were manufactured and tested under monotonic axial compression (Figure 1). Details of the test columns are shown in Table 1. All specimens had a 1 500 mm height and a 500 mm square cross-section (measured from the inner side of the FRP tube) with a corner radius of 50 mm (Figure 1). The key parameters which were examined included the number of layers of FRP wrap, and the type and configuration of internal steel tubes. The 4 large-size square MTCC specimens covered two types of steel tubes (Type A and Type B, Table 2) and two steel tube configurations (i. e. eight-tube square configuration and four-tube con square figuration, Figure 1). For ease of reference, each specimen is given a name, which starts with a number 8 or 4 to represent the number of steel tubes in the specimens, followed by a letter A or B to represent the type of steel tubes. Another number 2 or 4 is to represent the number of plies of fiber sheets. The last Roman numeral at the end of "8A - 2" is to distinguish two nominally identical specimens.

During the test, the load kept increasing for "8A-2-Ⅰ" specimen until hoop rupture of the PEN FRP jacket occurred and the specimen lost its structural integrity with noticeable noises (Figure 2). For the sake of laboratory safety, the other three specimens terminated loading before the rupture of FRP jacket.

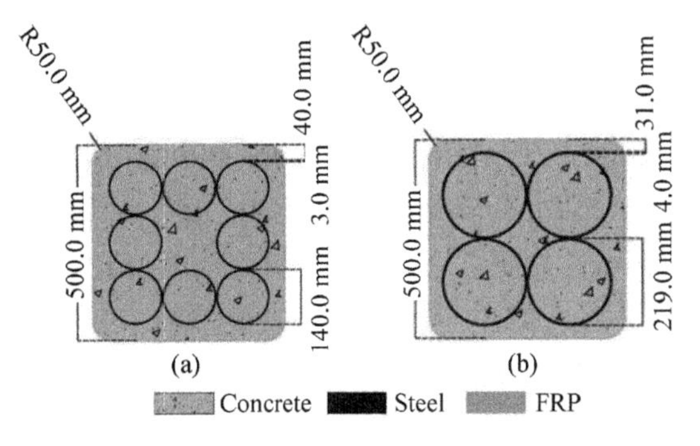

Figure 1 Cross-sections of test specimens

Table 1 Details of specimens

Specimen	FRP type	Number of layers of fiber sheets
8A-2-Ⅰ, Ⅱ	PEN	2
8A-4	PEN	4
4B-2	PEN	2

Table 2 Dimensions of steel tube

Steel tube type	Diameter /mm	Thickness /mm	Diameter-to-thickness ratio
A	140	3	46.67
B	219	4	54.75

Figure 2 Failure modes of typical specimen

The key results of all the test specimens are summarized in Table 3. In this table, P_u is the ultimate load of the specimen from the test; ε_{cu} is the ultimate axial strain of the specimen from the test; ε_{cu} is normalized by the axial strain at the peak stress of unconfined concrete ε_{co} (0.002 7). It is evident that the ε_{cu} value is up to around 9 times the ε_{co} value, which indicates the highly ductile behavior of the 8A-2-I specimen.

Table 3 Key results of compression test

Specimen	P_u/kN	ε_{cu}/%	$\varepsilon_{cu}/\varepsilon_{co}$
8A-2-I	11 409	2.45	9.1
8A-2-II	11 752	1.54	5.7
8A-4	12 972	2.43	9.0
4B-2	11 664	1.38	5.1

The axial load-strain curves of all specimens are shown in Figure 3. Due to the use of a thinner PEN FRP jacket, all specimens displayed a post-peak descending branch and the maximum axial load capacity was reached before FRP rupture. The eight-tube MTCC specimens with four plies of fiber sheets had a higher load capacity than the corresponding specimens with two plies of fiber sheets.

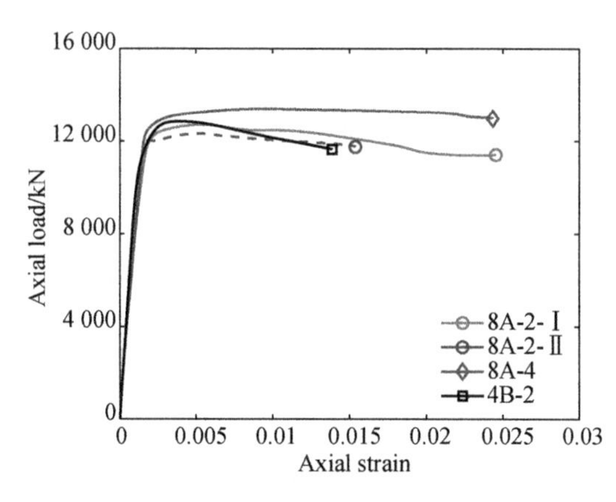

Figure 3 Axial load-axial strain curves

Acknowledgements

The authors are grateful for the financial support received from the National Key Research and Development Program of China (2017YFE0103000).

References

[1] YU T, CHAN C W, TEH L H, et al. Hybrid FRP-concrete-steel multi-tube concrete columns: stub column tests[C]// Proceedings of the 24th Australian Conference on the Mechanics of Structures and Materials. Leiden, The Netherlands: CRC Press/Balkema, 2016: 1731-1736.

[2] YU T, CHAN C W, TEH L H, et al. Hybrid FRP-concrete-steel multitube concrete columns: concept and behavior[J]. Journal of Composites for Construction, 2017, 21(6): 04017044.

[3] CHAN C W, YU T, ZHANG S. Compressive behavior of square fibre-reinforced polymer-concrete-steel hybrid multi-tube concrete columns[J]. Advances in Structural Engineering, 2018, 21(8): 1162-1172.

[4] DAI J G, BAI Y L, TENG J G. Behavior and modeling of concrete confined with FRP composites of large deformability[J]. Journal of composites for construction, 2011, 15 (6): 963-973.

[5] BAI Y L, DAI J G, TENG J G. Cyclic compressive behavior of concrete confined with large rupture strain FRP composites[J]. Journal of Composites for Construction, 2013, 18(1): 04013025.

Dynamic tensile mechanical properties of polyethylene terephthalate (PET) fiber bundle

Z. W. Yan & Y. L. Bai

Key Laboratory of Urban Security and Disaster Engineering of Ministry of Education, Beijing University of Technology, Beijing 100124, China

J. G. Dai

Department of Civil and Environmental Engineering, the Hong Kong Polytechnic University, Hong Kong, China

Abstract

The polyethylene terephthalate (PET) FRP is a new type of FRP composite. It can be made from the waste plastic bottles, which is environmentally friendly. Owing to the fact that this new material has a tensile rupture strain of more than 5%, it is also called the large-rupture-strain (LRS) FRP composite. Besides, it has a bilinear stress-strain relationship, which is different from the linear stress-strain relationship exhibited by the conventional FRP (carbon FRP, glass FRP, aramid FRP and basalt FRP). When this material was in the initial linear stress-strain portion, it could provide the strength enhancement for the reinforced concrete (RC) structure with the relative large elastic modulus. When it was in the second linear stress-strain portion, it has a great energy absorption capacity with a relative small elastic modulus and a large tensile strain. Therefore, some studies have reported to date on the seismic strengthening of the RC structure with the LRS FRP as the external jackets[1-8]. Nevertheless, the RC structures suffer not only the seismic loading but also the impact loading such as the vehicle impact in the bridge pier during their service life[9]. The investigations of the impact resistance of the RC structure strengthened with FRP are necessary for the application of the PET FRP in the impact-resistant strengthening of the RC structure.

Considering the fact that the FRP composite mainly bears the tension in the strengthening of the RC structure and the fiber bundle is the main load-carrying element in FRP composites. Therefore, it is of vital importance to investigate the dynamic tensile mechanical properties of the PET fiber bundle.

In this study, a total of 20 PET fiber bundle specimens with a gauge length of 25 mm (Figure 1) were tested at the displacement rate of 1, 2, 3 and 4 m/s using an Instron drop-weight impact system, as shown in Figure 2(a). The corresponding strain rates were calculated with the value of 40, 80, 120 and 160 s^{-1} according to the definition of the strain rate that is the ratio of the displacement rate to the corresponding gauge length of the specimen. For the comparison purpose, 5 specimens were tested at the displacement rate of 2.5 mm/min with the corresponding strain rate of 1/600 s^{-1} using an MTS testing machine, as shown in Figure 2(b).

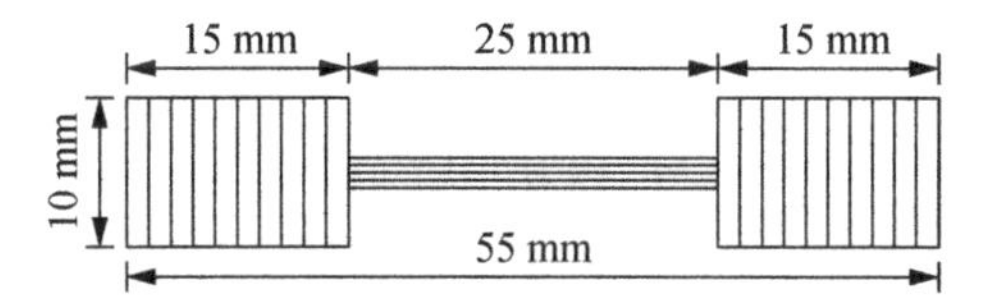

Figure 1 Schematic diagram of the PET fiber bundle

(a) Instron drop-weight impact system (b) MTS testing machine

Figure 2 Testing machine

Figure 3 shows the failure modes of the PET fiber bundle specimens at the strain rate of 1/600 (quasi-static loading) and 160 s^{-1} (dynamic loading). As can be seen in the figure, the fiber bundle failed in a chaotic manner under quasi-static loading whereas the failure locations of the filaments were relatively concentrated with a trim fractography. This phenomenon might be explained as follows. Under the quasi-static loading, the defects of the filament could be fully developed, followed by the successive rupture of the filament. Under the dynamic loading, there was no enough time for the defects to develop, resulting in the failure of the fiber bundle as a whole at the weakest part.

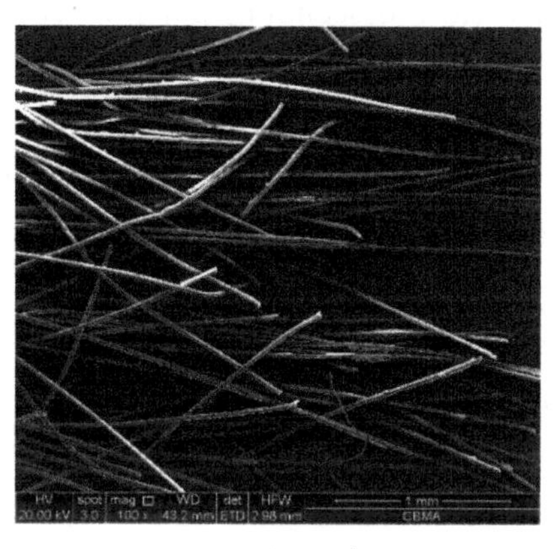

(a) 1/600 s^{-1}

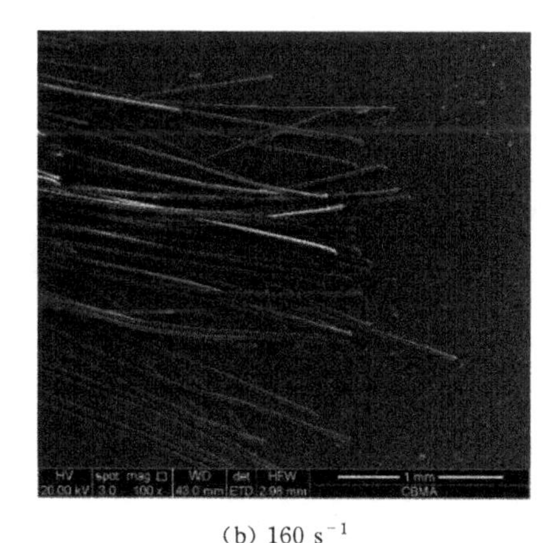

(b) 160 s^{-1}

Figure 3 Failure modes at different strain rate

Figure 4 shows the stress-strain curves at the strain rate of 1/600 and 160 s^{-1}. The curves exhibit a bilinear stress-strain relationship, initialing with a linear elastic stage. Then, the curve went to the second linear stage until the peak of the curve with the slope of the second linear portion smaller than that of the former.

After the peak of the curve, the curve had a drop from peak to zero, which shows a brittle failure of the PET fiber bundle. As can be seen in the figure, the descending portions of the curve under the quasi-static and dynamic loading were different from each other: the former had a relative slow descending portion whereas the latter had a sharp one. This is due to the fact that the successive failure of the filament could be well recorded under quasi-static loading whereas under the dynamic loading, it is difficult to be recorded due to the sudden rupture of the specimen.

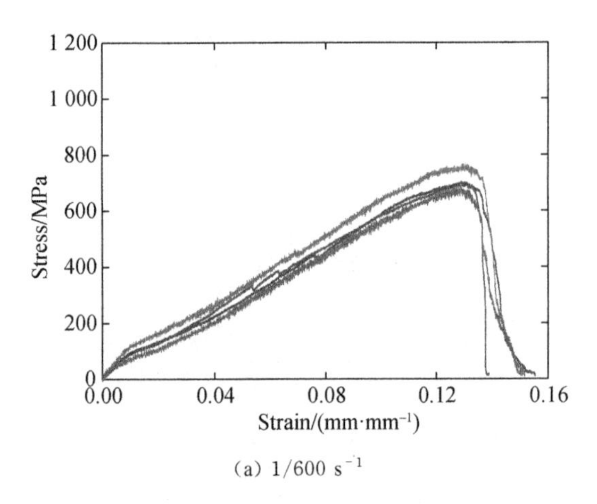

(a) $1/600\ s^{-1}$

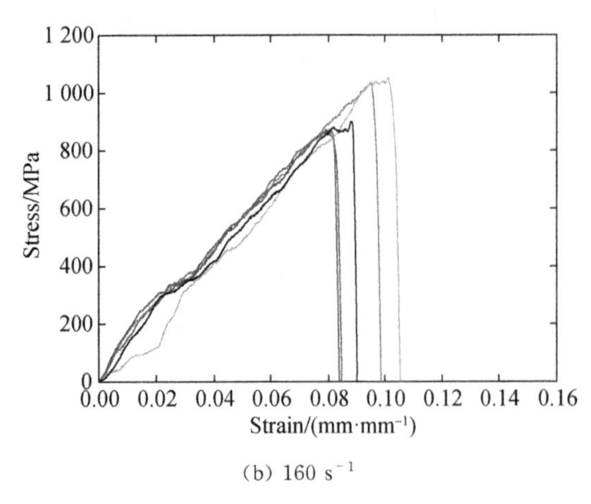

(b) $160\ s^{-1}$

Figure 4 Stress-strain curves at different strain rate

From the stress-strain curve in Figure 4, the tensile strength, failure strain, elastic modulus and toughness can be obtained, whose average values are listed in Table 1. The tensile strength is the peak stress of the curve and the failure strain is the strain corresponding to the peak stress. When the strain rate increased from 1/600 to 160 s^{-1}, the tensile strength had an increase from 708 MPa to 941 MPa whereas the failure strain decreased from 13.0% to 9.11%. The elastic modulus has two values, including the initial elastic (E_1) and second elastic modulus (E_2). The former is defined as the slope of the initial linear elastic portion, and the latter is defined as the slope of the second linear portion. The initial and second elastic modulus increased from 11.0 GPa and 5.83 GPa to 17.3 GPa and 10.8 GPa, respectively, with an increase in the strain rate from 1/600 to 160 s^{-1}. The toughness is defined as the area enclosed by the stress-strain curve and strain axial. An increase in the strain rate from 1/600 to 160 s^{-1} led to a decrease in the toughness from 57.1 MPa to 46.7 MPa.

Table 1 Dynamic mechanical properties at different strain rates

Strain rate /s^{-1}	Tensile strength /MPa	Failure strain /%	E_1 /GPa	E_2 /GPa	Toughness /MPa
1/600	708	13.0	11.0	5.83	57.1
160	941	9.11	17.3	10.8	46.7

Acknowledgements

The authors are grateful for the

financial support received from the National Key Research and Development Program of China (2017YFE0103000), the National Natural Science Fund of China (Grant No. 51678014, 51778019).

References

[1] BAI Y L, DAI J G, TENG J G. Cyclic compressive behavior of concrete confined with large rupture strain FRP composites[J]. Journal of Composites for Construction, 2014, 18(1): 04013025.

[2] DAI J G, BAI Y L, TENG J G. Behavior and modeling of concrete confined with FRP composites of large deformability[J]. Journal of Composites for Construction, 2011, 15(6): 963-973.

[3] DAI J G, LAM L, UEDA T. Seismic retrofit of square RC columns with polyethylene terephthalate (PET) fibre reinforced polymer composites[J]. Construction and Building Materials, 2012, 27(1): 206-217.

[4] SALEEM S, HUSSAIN Q, PIMANMAS A. Compressive behavior of PET FRP-confined circular, square, and rectangular concrete columns[J]. Journal of Composites for Construction, 2017, 21(3): 04016097.

[5] BAI Y L, DAI J G, OZBAKKALOGLU T. Cyclic stress-strain model incorporating buckling effect for steel reinforcing bars embedded in FRP-confined concrete[J]. Composite Structures, 2017, 182: 54-66.

[6] ISPIR M. Monotonic and cyclic compression tests on concrete confined with PET-FRP[J]. Journal of Composites for Construction, 2014, 19(1): 04014034.

[7] ISPIR M, DALGIC K D, ILKI A. Hybrid confinement of concrete through use of low and high rupture strain FRP[J]. Composites Part B: Engineering, 2018, 153: 243-255.

[8] BAI Y L, DAI J G, MOHAMMADI M, et al. Stiffness-based design-oriented compressive stress-strain model for large-rupture-strain (LRS) FRP-confined concrete[J]. Composite Structures, 2019, 223: 110953.

[9] OU Y F, ZHU D J, ZHANG H A, et al. Mechanical properties and failure characteristics of CFRP under intermediate strain rates and varying temperatures[J]. Composites Part B-Engineering, 2016, 95: 123-136.

Experimental study of recycled aggregate concrete confined with recycled polyethylene terephthalate composites

W. Y. Yuan & Q. Han & Y. L. Bai

Key Laboratory of Urban Security and Disaster Engineering of Ministry of Education, Beijing University of Technology, Beijing 100124, China

Abstract

Numerous studies have indicated that FRP-confined recycled aggregate concrete (RAC) is a particularly effective method to overcome the shortcomings of RAC (e.g. low strength, poor durability). The recycled polyethylene terephthalate (PET) composites are relatively new type of environmentally-friendly fibers with bilinear tensile stress-strain relationship and high energy absorption capacity[1-3]. The combination of recycled PET FRP jacket and RAC provides a good solution for the reuse of waste plastics and construction garbage. In this study, the axial compression tests of 24 RAC cylinders (150 mm × 300 mm) externally confined by PET FRP were carried out. Different parameters such as the recycled aggregate (RA) replacement ratio (i.e. 0%, 33%, 67%, 100%), the number of FRP layers (i.e. 1-ply, 2-ply, 3-ply) were analyzed. More details about the mix proportions of RAC and the flat coupon tensile test of PET FRP can be obtained from reference [4], [5], respectively.

As shown in Figure 1, three strain gauges, with a gauge length of 20 mm, were uniformly distributed on the non-overlapping zone to determine the hoop strain of the specimen around the mid height of the column. Four linear variable displacement transducers (LVDTs) were fixed every 90 degree to determine the axial strain of the 150 mm mid-height of the column.

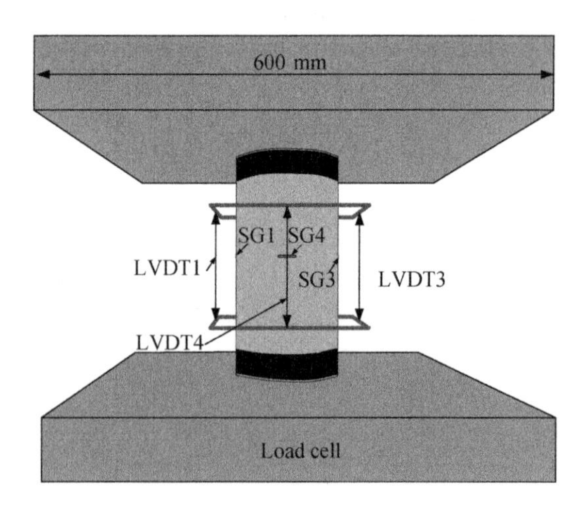

Figure 1 Test setup

Figure 2 displays the typical failure mode of specimens with different RA replacement ratios. When the peak load is reached, the PET FRP jacket seemed to rupture simultaneously in mid-height portions and the coarse aggregates are cleaved outside the overlapping zone. In general, the failure mode has no correlation to the replacement ratio of RA. Similar observations were obtained in the experiments of conventional FRPs (e.g. CFRP, GFRP)-confined RAC (e.g. reference[4], [5], [6]).

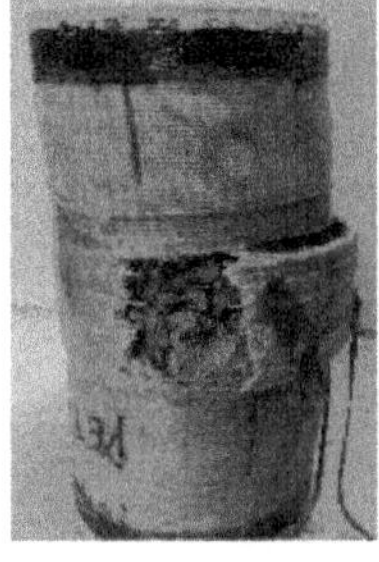

(a) R0-PET-2-a

(b) R33-PET-2-a

(c) R67-PET-2-a

(d) R100-PET-2-a

Figure 2 Failure mode

Table 1 presents the key results of the specimens. Where t_{frp} is the thickness of PET FRP jacket; $\varepsilon_{h,rup}$, ε_{cu} and f'_{cc} are the average actual hoop rupture strain, compressive strength and ultimate axial strain of the wrapped specimens with identical configuration respectively. The compressive strength of the unconfined specimens with RA replacement ratio of 0%, 33%, 67%, 100% is 31.5 MPa, 28.6 MPa, 25.6 MPa, 24.1 MPa, respectively. This proves that the addition of recycled aggregate reduces the strength of concrete.

Table 1 Key test results

Specimen	t_{frp} /mm	$\varepsilon_{h,rup}$ /%	ε_{cu} /%	f'_{cc} /MPa
R0-PET-1-a, b	0.841	6.360	3.790	41.61
R0-PET-2-a, b	1.682	7.100	6.520	69.59
R0-PET-3-a, b	2.523	6.840	8.310	97.98
R33-PET-1-a, b	0.841	6.885	4.665	42.91
R33-PET-2-a, b	1.682	6.820	6.655	70.03
R33-PET-3-a, b	2.523	6.540	8.075	92.44
R67-PET-1-a, b	0.841	8.055	5.745	45.02
R67-PET-2-a, b	1.682	7.700	7.660	72.04
R67-PET-3-a, b	2.523	8.105	9.695	98.06
R100-PET-1-a, b	0.841	7.445	5.255	44.92
R100-PET-2-a, b	1.682	6.720	6.725	61.70
R100-PET-3-a, b	2.523	6.925	8.655	95.69

Note: RX-PET-XX-XXX, where X = the replacement ratio of RA; XX = the number of PET FRP layers (1 represents the specimen externally jacketed with 1ply PET FRP); XXX = the number to differentiate two specimens with identical configuration.

Figure 3 displays that the compressive strength and ultimate axial strain increase significantly with the increase of FRP layers because of the excellent characteristics of large rupture strain for PET FRP jackets. However, Figure 4 shows that the specimens with the same FRP layers and different RA replacement ratios have similar strength and ultimate axial strain, indicating that RA replacement ratio has a marginal effect on the compressive behavior of RAC strengthened with PET FRP. In other words, the confinement effect of LRS FRP can solve the problem of strength reduction caused by the addition of RA.

This observation is consistent with that of conventional FRPs-confined recycled concrete[6-9].

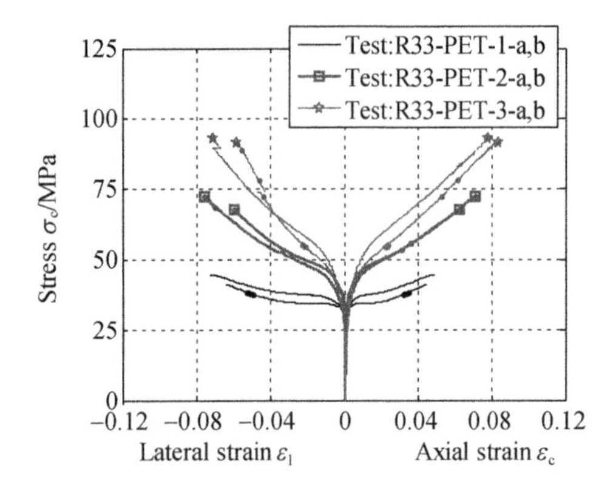

Figure 3 Effect of LRS FRP layer on stress-strain behavior of confined RAC

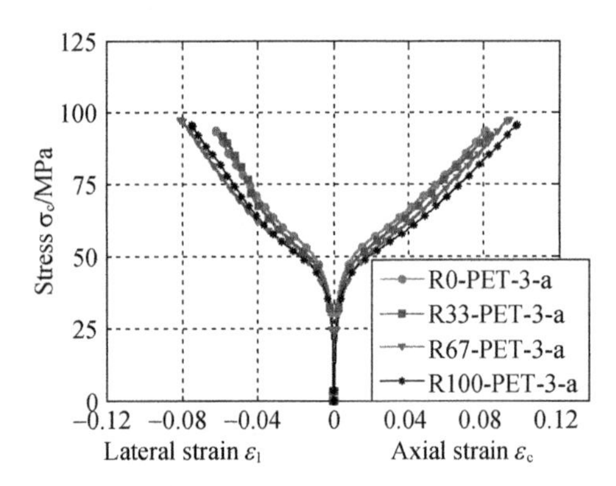

Figure 4 Effect of RA replacement ratio on the stress-strain behavior

The stress-strain curves of PET FRP-confined RAC specimens depict a triple-section pattern with a long smooth transition zone, analogous to the curves of PET FRP-confined natural aggregate concrete (NAC)[5]. The first portion is quite similar to unconfined RAC because the confining effect is negligible due to the insignificant lateral dilation of RAC. A smooth transition zone is observed after the first portion when FRP is activated, followed by a linear segment until the rupture of FRP. The experimental curve is different from the bilinear curves of conventional FRPs-confined RAC. The main reason may be that the bilinear characteristics of PET FRP and the occurrence of RA reduce the expansion property of the specimens.

Acknowledgements

The authors are grateful for the financial support received from the National Key Research and Development Program of China (2017YFE0103000), the National Natural Science Fund of China (51778019).

References

[1] DAI J G, BAI Y L, TENG J G. Behavior and modeling of concrete confined with FRP composites of large deformability[J]. Journal of composites for construction, 2011, 15 (6): 963-973.

[2] DAI J G, LAM L, UEDA T. Seismic retrofit of square RC columns with polyethylene terephthalate (PET) fibre reinforced polymer composites [J]. Construction and Building Materials, 2012, 27(1): 206-217.

[3] ISPIR M. Monotonic and cyclic compression tests on concrete confined with PET-FRP[J]. Journal of Composites for Construction, 2014, 19(1): 04014034.

[4] XIE T, OZBAKKALOGLU T. Behavior of recycled aggregate concrete-filled basalt and carbon FRP tubes [J]. Construction and Building Materials, 2016, 105: 132-143.

[5] BAI Y L, DAI J G, LIN G, et al. Stiffness-based design-oriented compressive stress-strain model for large-rupture-strain (LRS) FRP-confined concrete [J]. Composites Structures, 2019, 223: 110953.

[6] CHEN G M, HE Y H, JIANG T, et al. Behavior of CFRP-confined recycled aggregate concrete under axial compression [J]. Construction and Building Materials, 2016, 111: 85-97.

[7] ZHAO J L, YU T, TENG J G. Stress-strain behavior of FRP-confined recycled aggregate concrete [J]. Journal of Composites for Construction, 2014, 19(3): 04014054.

[8] LI P, SUI L, XING F, et al. Static and cyclic response of low-strength recycled aggregate concrete strengthened using fiber-reinforced polymer[J]. Composites Part B: Engineering, 2019, 160: 37-49.

[9] ZHOU Y, HU J, LI M, et al. FRP-confined recycled coarse aggregate concrete: Experimental investigation and model comparison[J]. Polymers, 2016, 8(10): 375.

Probability distribution model for bearing capacity of reinforced concrete columns with non-uniform corroded stirrups under axial compression

X. Q. Jiang & B. B. Zhou & X. H. Wu

College of Ocean Science and Engineering, Shanghai Maritime University, Shanghai 201306, China

Abstract

Considering the influence of spatial variation of corrosion on cross-sectional areas and mechanical properties of stirrups, the probability distribution model for bearing capacity of corroded reinforced concrete columns subjected to chloride penetration under axial compression is studied in this paper. Introducing the probability distribution model of spatial heterogeneity factor for corroded reionforcement, the spatial distribution of corrosion of stirrup was quantified by discretizing the corroded column into several small elements along the axial direction. Using ABAQUS finite element analysis software, the bearing capacity of reinforced concrete columns with non-uniform corroded stirrups under axial compression was analyzed, and the probability distribution model was established. The calculation results showed that the bearing capacity of reinforced concrete columns with non-uniform corroded stirrups under axial compression could be fitted well by the Gumbel distribution, which was mainly dependent on the spatial variation degree of corrosion for stirrups, as shown in Figure 1. With

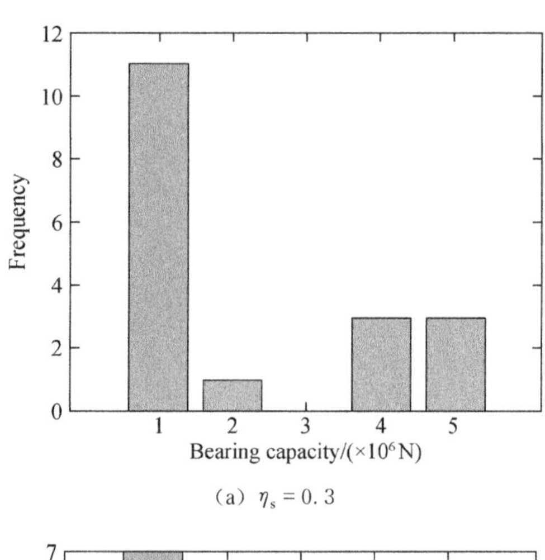

(a) $\eta_s = 0.3$

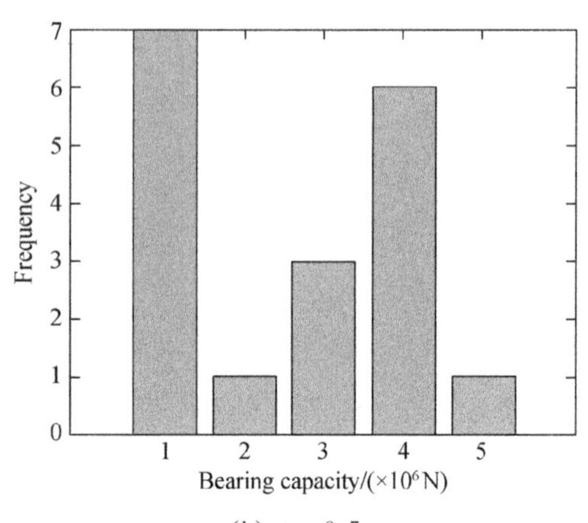

(b) $\eta_s = 0.5$

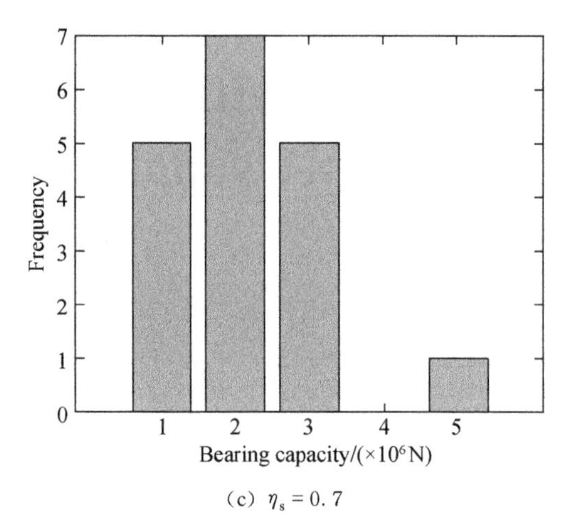

(c) $\eta_s = 0.7$

Figure 1 Probability distribution of bearing capacity of corroded RC Columns under axial compression corresponding to different average corrosion loss ratio

the increases of corrosion degree, the location parameter μ was constant, while the scale parameter σ decreased significantly, as shown in Table 1.

Table 1 Values of μ and σ corresponding to different average corrosion loss ratio

η_s	μ	σ
0.3	3 195 833.26	29 197.06
0.5	3 173 612.40	27 824.79
0.7	3 161 100.18	18 967.81

References

[1] ZHANG W P, ZHOU B B, GU X L, et al. Probability distribution model for cross-sectional area of corroded reinforcing steel bars [J]. Journal of Materials in Civil Engineering, 2014, 26(5): 822-832.

[2] GU X L, GUO H Y, ZHOU B B, et al. Corrosion non-uniformity of steel bars and reliability of corroded RC beams [J]. Engineering Structures, 2018, 167: 188-202.

[3] ESAKLUL K A. Innovative approaches to downhole corrosion control[C]//8th Middle East Oil Show & Conf., Proc., Vol. 1, Society of Petroleum Engineers (SPE), 1993:569-575.

[4] CHOI J. Some mathematical constants[J]. Applied Mathematics and Computation, 2007,187(1): 122-140.

[5] ZHU Z K, YAN R Q, LUO L H, et al. Detection of signal transients based on wavelet and statistics for machine fault diagnosis[J]. Mechanical Systems and Signal Processing, 2009,23(4):1076-1097.

[6] CHOI J S, HONG S, CHI S B, et al. Probability distribution for the shear strength of seafloor sediment in the KR5 area for the development of manganesenodule mine[J]. Ocean Engineering, 2011, 38(17-18): 2033-2041.

Axial performance of push-on joints with rubber gasket of large-diameter water supply pipelines inside utility tunnels

X. Li & Z. Zhong & B. Hou & J. Li & X. Du

Key Laboratory of Urban Security and Disaster Engineering of Ministry of Education, Beijing University of Technology, Beijing 100124, China

Abstract

The segmented ductile-irons pipeline with rubber-gasket push-on joints is one of the most commonly used type of pipeline in the water supply system. For typical nonstructural components, such as the large-diameter pipelines inside the utility tunnels, the seismic dynamic response is affected by both acceleration and displacement[1]. The influence of the stiffness, strength reduction and energy dissipation characteristics of the push-on joints of pipelines on the seismic response of the pipeline during earthquakes cannot be ignored.

A series of pseudo-static tensile tests were performed on the water-filled push-on joints of the DN400 ductile iron water pipelines in the utility tunnel to investigate the axial mechanical behavior and failure mechanism of the joints. Figure 1 illustrates a typical configuration of the push-on joints of ductile iron pipeline, which consists of the bell, the spigot and the rubber-gasket. The water-filled push-on joints of the ductile iron pipelines were subjected to both monotonic and cyclic tensile loadings in their axial direction in the experiments. A displacement control scheme was adopted in the tests with a loading rate of 0.1 mm/s. Figure 2 shows the loading protocol for the cyclic tensile tests. The loading amplitude gradually increases from 2.5 mm up to the failure of the joint, and two loading cycles were performed at each amplitude. During the axil tensile tests, water sealing condition was recorded at the pipe joint. When initial leakage with small water drops form at the joint, it is regarded that the joint suffers a minor damage, which does not seriously affect the water transportation capability of the pipeline. With the further increase of the joint opening, serious water leakage occurs with water inside the pipeline continuously flowing out from the joint. It is regarded that the pipe joint suffers significant damage, which impairs the water delivery capacity of the pipeline. The axial tensile tests were terminated when significant damage to the push-on joint were observed during the tests. A new rubber gasket was replaced in the push-on joint after each set of tests.

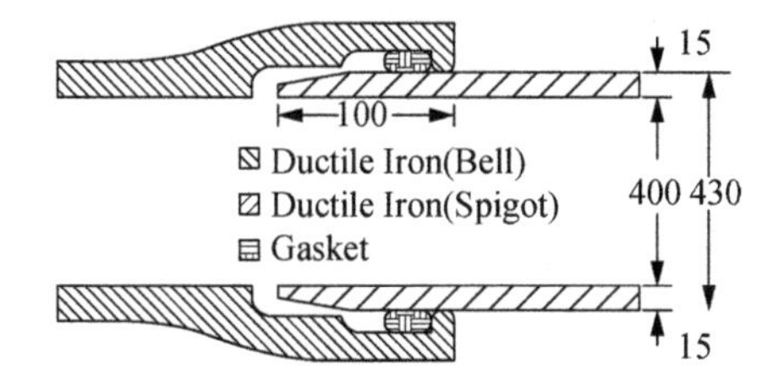

Figure 1 Configurations of push-on joints of ductile iron pipeline (unit: mm)

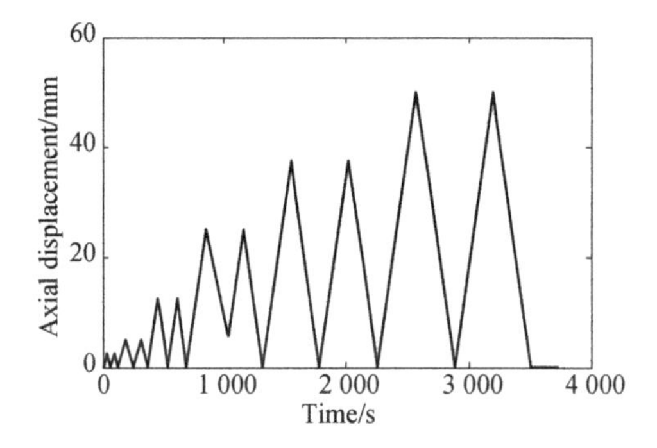

Figure 2 Cyclic loading protocol of axial tensile tests of push-on joint

Figure 3 compares the experimental results from monotonic and cyclic axial tensile tests. It was found that under monotonic and cyclic loading the push-on joint exhibits similar axial stiffness, which primarily attributes to the shear stiffness of the compressed rubber gasket between the spigot and the bell. However, the axial resistance of the push-on joint from the monotonic test is much larger than that from the cyclic test. It is because the lubricant of rubber sealing ring was air drying and crusting before the monotonic tests. In the subsequent cyclic tests, the lubricant remained effective, which had significant influence on the initial axial tensile strength of the push-on joint.

For the axial cyclic tests, it can be seen that the equivalent axial joint stiffness reduces exponentially with the increase of the axial joint opening, as shown in Figure 4.

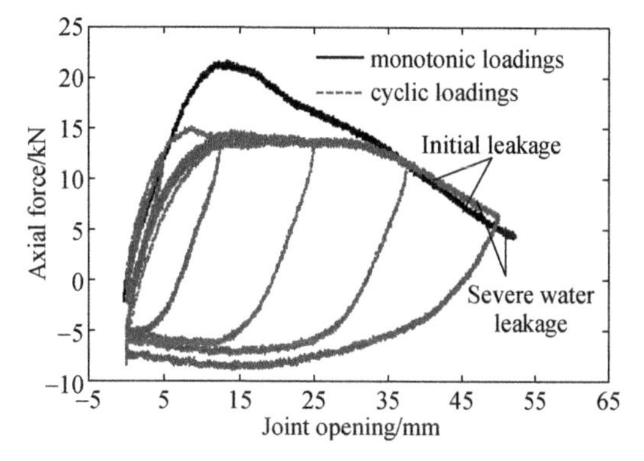

Figure 3 Axial force-joint opening response of push-on joints under different loading protocols

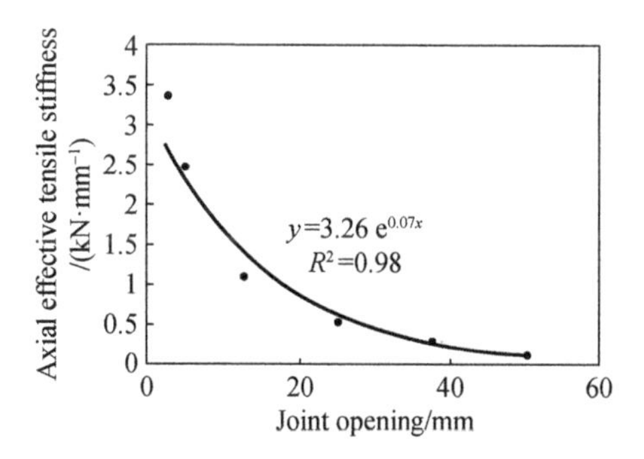

Figure 4 Axial effective tensile stiffness of pipe joint

Besides, the initial and severe water leakage occurred at axial joint openings of 40 mm and 47 mm, respectively, for both monotonic and cyclic tests. The results show that the loading method has insignificant effect on the critical joint openings corresponding to the initial and severe water leakage at the joint as shown in Table 1 and Figure 3. In this paper, the experimental results from this research are also compared with the tests results from domestic and foreign scholars[2-7], as shown in Figure 5. It should be pointed out that Zhong et al.[7] and Meis et al.[2] presented peak joint opening at the moment of "serious water leakage", similar to this study. However, the peak joint openings from other studies are corresponding to the joint axial deformation when the pipe spigot is completely pulled out.

Table 1 Axial test results of push-on joints

Loading method	initial water leakage		severe water leakage	
	displacement /mm	Axial force /kN	displacement /mm	Axial force /kN
Monotonic	45	7.6	52	4.6
cyclic	40	10.4	47	7.5

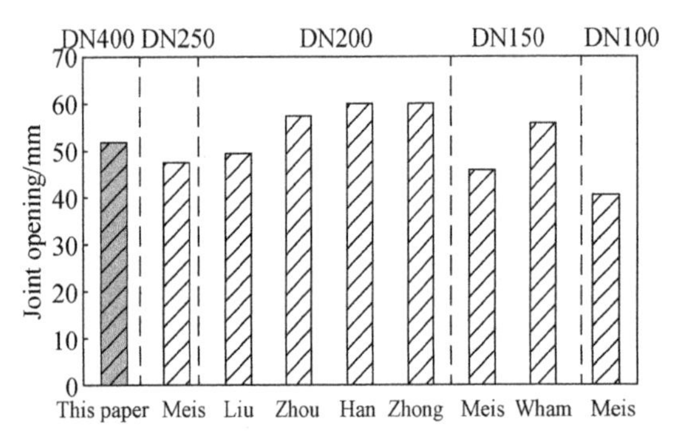

Figure 5 Comparison of peak joint openings from this test with published results

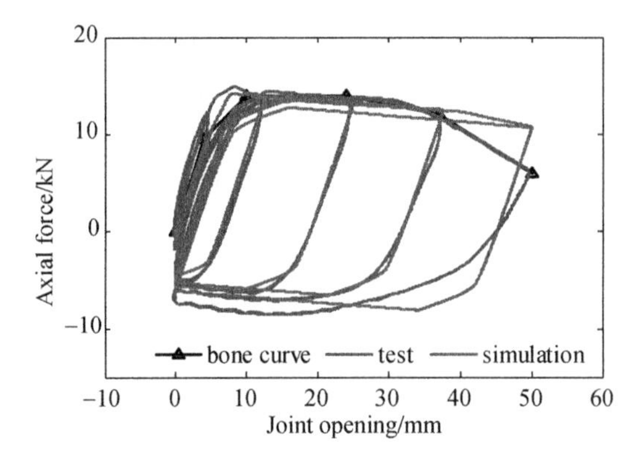

Figure 6 Comparison of axial force-joint opening response between experimental and numerical result

A new finite element model for the push-on joint was developed in OpenSees in this study to simulate its nonlinear behavior. Figure 6 compares the hysteretic axial force-joint opening response of numerical simulation and tests under cyclic loadings. The joint models can generally capture the stiffness degradation and energy dissipation characteristics of the push-on joint under cyclic loadings. The experimental and numerical results can provide a sound basis for the future study on the seismic performance of water pipelines in utility tunnels.

Acknowledgements

This work presented in this paper was supported by the National Natural Science Foundation of China (U1839201) and he National Key Research and Development Program of China (2018YFC1504305). Any opinions, findings and conclusions or recommendations expressed in this paper are those of the authors and do not necessarily reflect the views of the financial supporters. Sincere thanks are extended to the Key Laboratory of Urban Security and Disaster Engineering of Ministry of Education, Beijing University of Technology for providing the computational facilities.

References

[1] FEMA E - 74. Reducing the risks of nonstructural earthquake damage: A practical guide [S]. Federal Emergency Management Agency, 2011.

[2] MEIS R D, MARAGAKIS E M, SIDDHARTHAN R. Behavior of underground piping joints due to static and dynamic loading [R]. Technical Report MCEER-03-0006, State University of New York at Buffalo, November 17, 2003.

[3] WHAM B P, O'ROURKE T D. Jointed pipeline response to large ground deformation[J]. Journal of Pipeline System Engineering and Practice, 2015, 7 (1): 04015009.

[4] LIU W M, SUN S P. Seismic test study of pipeline interface [C]// The 5th National Earthquake Engineering Conference, Beijing, 1998: 970-975.

[5] ZHOU J H. Destructive test study and aseismic analysis of water supply pipeline [D]. Dalian: Dalian University of Technology, 2010.

[6] HAN Y, ZHANG K H, DUAN J F. The pull-out test of ductile cast iron pipe with flexible joint[J]. Sichuan Building Science, 2017, 43(4): 80-83.

[7] ZHONG Z L, WANG S R, DU X L. Experimental and numerical study on axial mechanical properties of pipeline under pseudo-static loading [J]. Engineering Mechanics, 2019, 36(3): 224-230.

Seismic performance assessment of shallowly buried underground structure using endurance time method

L. Zhen & Z. Zhong & Y. Shen & M. Zhao & X. Du

Key Laboratory of Urban Security and Disaster Engineering of Ministry of Education, Beijing University of Technology, Beijing 100124, China

Abstract

Serious damage of the Dakai stations with accompanying life and economic loss in the 1995 Hanshin earthquake in Japan[1] clearly shows that the seismic performance assessment of underground structures is of particular importance. In recent years, there has been an increasing tendency in structural design codes and seismic design guidelines to move toward adoption of performance-based design approaches[2]. Seismic response prediction and assessment of structures is the one of core contents in the performance-based earthquake engineering. At present, the conventional methods for seismic performance assessment of the underground structures include nonlinear static Pushover analysis[3-5] and incremental dynamic analysis (IDA)[6]. However, the pushover method adopts a static approach to analyze a dynamic problem, which fails to capture the dynamic characteristics of the soil-structure interaction system and the uncertainties of the ground motions. The IDA method uses full nonlinear dynamic analyses with its limitations of high computational cost and low efficiency.

The endurance time method (ETM) is an efficient seismic performance evaluation method characterized by developing series of seismic response spectra compatible acceleration time histories whose amplitudes increase with the duration[7-9], as shown in Figure 1. Those endurance acceleration time histories are subsequently used as the input for engineering structures to perform nonlinear dynamic analyses. ETM can effectively capture the entire dynamic response of the structure from elastic to plastic till finally collapse, and can be used as an alternative approach to evaluate the seismic performance of structures.

This paper used Dakai subway station as a case study. Two-dimensional finite element model was established using the open source computer program, OpenSees, considering nonlinear dynamic soil-structure interaction, as shown in Figure 2. The nonlinear beam-column fiber element was employed for the underground structure and four-node plane strain quad element

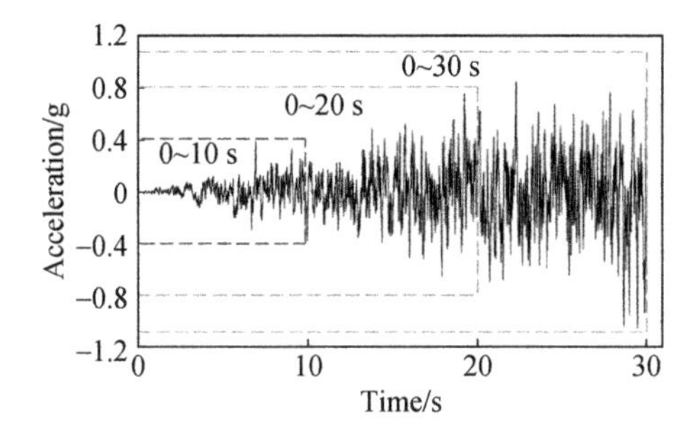

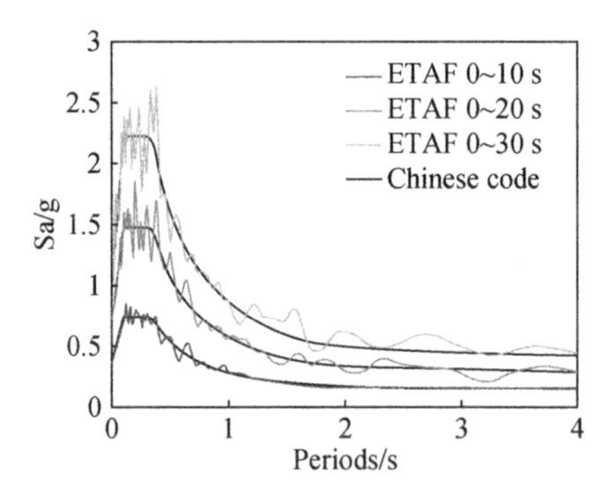

Figure 1 An ET acceleration function (ETAF)

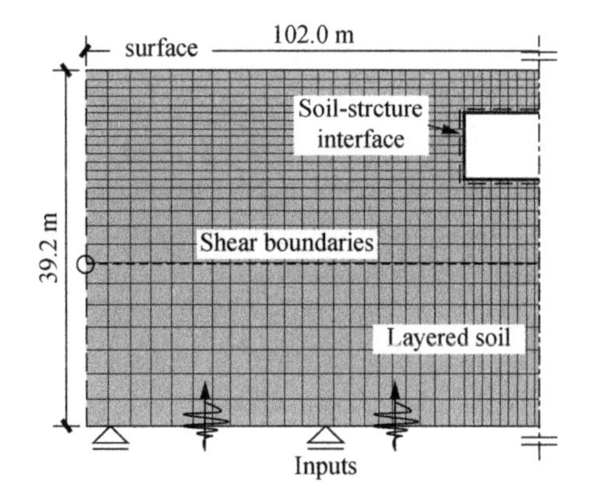

Figure 2 2D integrated finite element model of soil-structure interaction system

was employed for the soil. The mechanical behavior of concrete and steel was modeled using the Kent-Scott-Park and the Giuffré-Pinto constitutive model, respectively. The mechanical behavior of sand and clay was simulated by using the Pressure Depend Multi Yield material and Pressure Independ Multi Yield material in OpenSees, respectively. The soil domain was truncated at a width of 102 m (6 times of the width), where the lateral boundaries are sufficiently far from the underground structure so as to eliminate the influence of the boundary effects on the seismic response of the underground structure. Moreover, horizontal kinematic constraints were introduced to the nodes on two side boundaries to ensure the same horizontal movement of the two nodes at the same burial depth and effectively simulate the shear deformations of the soil layers under upward propagation of in-plane waves.

Three endurance time acceleration functions were generated based on the design response spectra of Chinese seismic design code and a set of 15 real ground motion records were selected from the PEER-NGA database. The seismic response of the Dakai subway station subjected to three ETAFs and 15 real ground motions were compared and investigated in this study.

The numerical results show that the responses of endurance time analyses generally fall between the envelopes of the incremental dynamic analyses of 15 real ground motions. The average response of the subway station using ETM is also in good agreement with the average results using IDA, as shown in Figure 3. Besides, the response spectrum corresponding to the fundamental period of the soil-structure interaction system is more preferable than the peak ground acceleration as the seismic intensity measure for the performance evaluation of the underground structures, as shown in Figure 3.

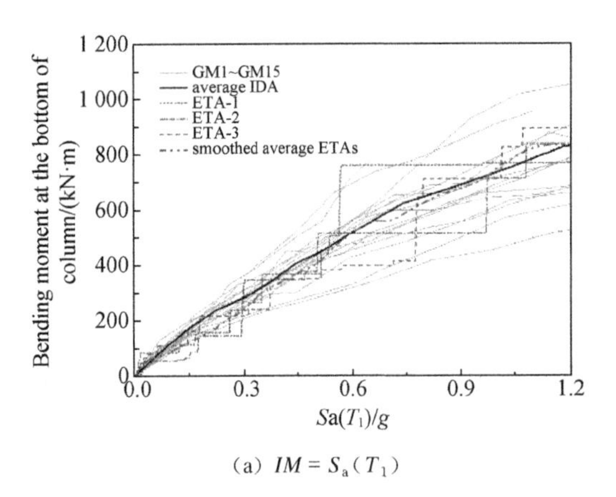

(a) $IM = S_a(T_1)$

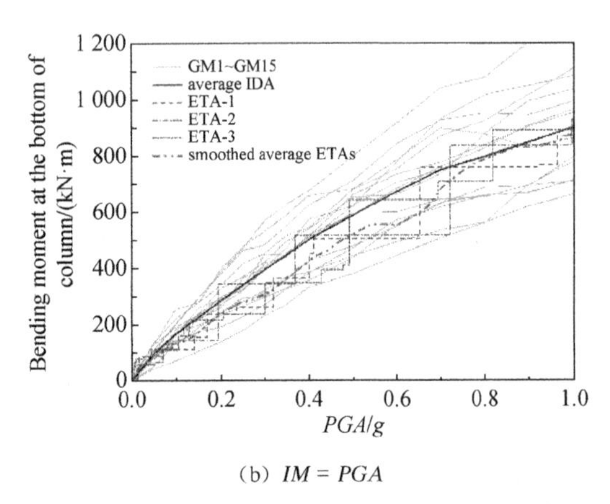

(b) $IM = PGA$

Figure 3 Comparison of ETM and IDA results

Overall, the endurance time analysis provides a new computationally efficient alternative for seismic performance evaluation of the underground structures other than the traditional nonlinear IDA.

Acknowledgements

The work presented in this paper was supported by the National Natural Science Foundation of China (51978020, U1839201) and the National Key Research and Development Program of China (2018YFC1504305).

References

[1] PARRA-COLMENARES E J. Numerical modeling of liquefaction and lateral ground deformation including cyclic mobility and dilation response in soil systems [D]. Corvallis: Oregon State University, 1996.

[2] GHOBARAH A. Performance-based design in earthquake engineering: State of development [J]. Engineering Structures, 2001, 23(8): 878-884.

[3] LIU J B, LIU X Q, LI B. A pushover analysis method for seismic analysis and design of underground structures[J]. China civil engineering journal, 2008, 41(4): 73-80.

[4] LIU J B, LIU X Q, XUE Y L. Study on applicability of a pushover analysis method for seismic analysis and design of underground structures [J]. Engineering mechanics, 2009, 26(1): 49-57.

[5] LIU J B, WANG W H, ZHAO D D, et al. Pushover analysis methed of underground structures under reversal load and its application in seismic damage analysis[J]. China earthquake engineering journal, 2013, 35(1): 21-28.

[6] VAMVATSIKOS D, CORNELL C A. Incremental dynamic analysis [J]. Earthquake Engineering and Structural Dynamics, 2002, 31(3): 491-514.

[7] ESTEKANCHI H E, VAFAI A, SADEGHAZAR M. Endurance time method for seismic analysis and design of structures [J]. Scientia Iranica, 2004, 11(4): 361-370.

[8] HARIRI-ARDEBILI M A, SATTAR S, ESTEKANCHI H E. Performance-based seismic assessment of steel frames using endurance time analysis [J]. Engineering Structures, 2014, 69: 216-234.

[9] ESTEKANCHI H E, VALAMANESH V, VAFAI A. Application of endurance time method in linear seismic analysis [J]. Engineering Structure, 2007, 29(10): 2551-2562.

Ground motion intensity measures for seismic performance assessment of mountain tunnels

C. M. Zhang & Z. L. Zhong & L. B. Zhen & Y. Y. Shen & M. Zhao

Key Laboratory of Urban Security and Disaster Engineering of Ministry of Education, Beijing University of Technology, Beijing 100124, China

Abstract

Since 1989 Loma Prieta earthquake and 1994 Northridge earthquake, performance-based earthquake engineering (PBEE) has been widely recognized in both engineering practice and academic research of seismic analysis of above ground structures[1]. Ground motion intensity measure (*IM*) is a key link between the seismic hazard analysis and structural seismic response analysis, which plays an important role in the PBEE framework. Therefore, how to define a reasonable ground motion *IM* which can represent earthquake intensity and reduce the dispersion of structure response has become a critical topic in the field of earthquake engineering.

In order to find suitable *IMs* for seismic response analysis of mountain tunnels, a two dimensional finite element model is established in this paper to simulate the nonlinear dynamic interaction between mountain tunnel and surrounding rock. The general-purposed finite element code ABAQUS is employed in this study for the nonlinear dynamic time-history analyses of mountain tunnel subjected to earthquake excitation and the numerical model is depicted in Figure 1. The model is 160 m wide and 100 m deep. The modelled tunnel has a circular cross section of radius $r = 5$ m and its center is located 50 m below the ground surface. The lining inner radius is 4. 5 m and the wall thickness is 0. 5 m. The concrete lining of tunnel is simulated by the concrete damaged plasticity constitutive model and the surrounding rock is simulated by the Mohr-Coulomb constitutive model. The boundary conditions of the model are as follows: the boundary at the top of the numerical model is free and the boundary at the model bottom is artificial viscous-spring boundary[2]. Tied degrees of freedom boundary (TDOF) which imposes horizontal kinematic constraints to the nodes with the same burial depth at the two lateral boundaries of model are applied at the lateral boundaries of the model[3]. The whole analysis process is divided into two steps to produce a more realistic simulation of seismic responses of tunnel structures. The first step is to establish geostatic stress equilibrium.

Then, ground motion is applied to the bottom boundary to perform dynamic analyses. The near field ground motion without velocity pulse and far field ground motion recommended by FEMA-P695 are used as the input ground motions for the numerical study[4]. Comparison of capability of the commonly used twenty *IMs* including the efficiency, practicality and sufficiency in assessing the engineering demand of tunnel structures are presented in this paper[5-7].

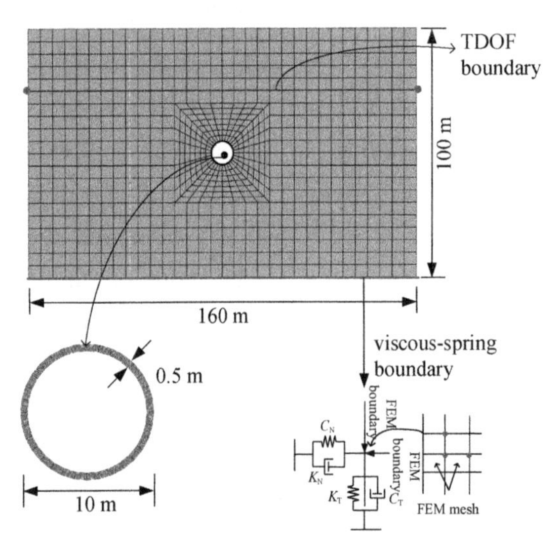

Figure 1 Finite element model

Besides, the overall lining damage indices in compression (*OLDC*) and in tension (*OLDT*) are used as the engineering damage measures (*DMs*) to estimate the lining damage state.

$$OLDC = \sum_i \frac{E_i^e}{\sum_i E_i^e} \cdot DC_i^e \tag{1}$$

$$OLDT = \sum_i \frac{E_i^e}{\sum_i E_i^e} \cdot DT_i^e \tag{2}$$

where E_i^e is the dissipated energy of the ith element; DC_i^e and DT_i^e are the damage index of the ith element in compression and tension, respectively. *OLDC* and *OLDT* reflect the cross-sectional lining damage of the mountain tunnels in single values, which can be easily correlated to scalar seismic *IM*.

Previous researchers found that *DM* - *IM* relationships follow a standard power law as shown in Eq. (3)[8].

$$DM = \mathrm{a}(IM)^b \tag{3}$$

which can be transformed into

$$\ln(DM) = \ln(\mathrm{a}) + b\ln(IM) \tag{4}$$

This transformation allows the constants ln(a) and b to be estimated by simple linear regression of ln(*DM*) and ln (*IM*). The efficiency is characterized in terms of the dispersion of the residuals of linear regression results. The dispersion is quantified by the standard deviation of the logarithm of the residuals, denoted as σ herein and can be calculated by

$$\sigma_{\ln EDP \mid IM} = \sqrt{\frac{\sum[\ln(EDP) - \ln(aIM^b)]^2}{n-2}} \tag{5}$$

The numerical results of efficiency of the selected *IM*s are shown in Figure 2.

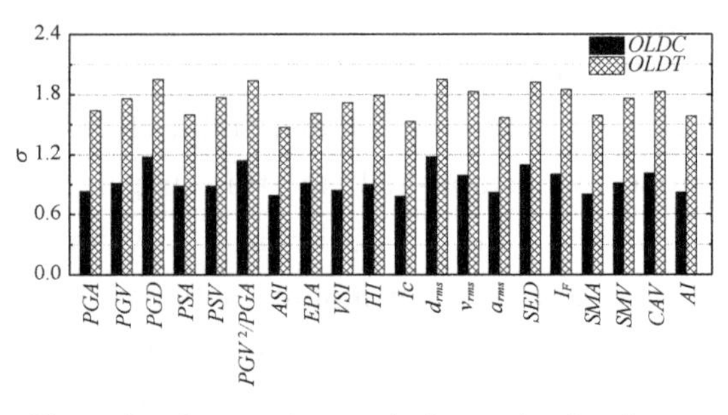

Figure 2 Regression analysis results for the efficiency of all *IM*s

Practicality refers to whether or not there is any direct correlation between an *IM* and the *DM*. If a ground motion *IM* is not practical, there is little or no dependence of the level of structural demand upon the level of the *IM*. Practicality is measured by the regression parameter b in Eq. (4). Larger b value indicates that the *IM* is more practical. The results of practicality of the selected *IM*s are shown in Figure 3.

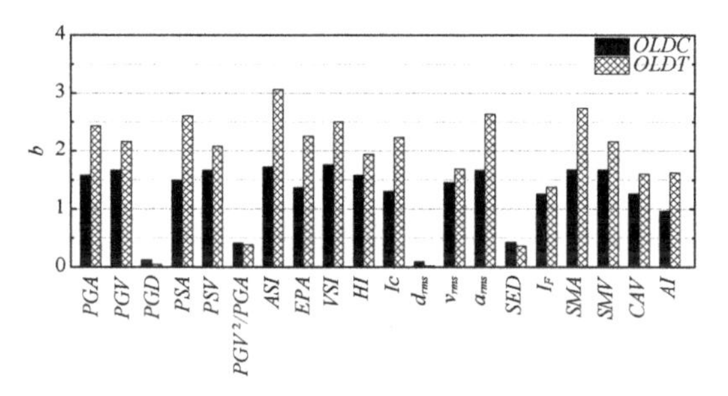

Figure 3 Practicality comparison of candidate *IM*s

In addition to the discussion of efficiency and practicality of *IM*s, the sufficiency of *IM*s is also considered in this study. Sufficiency is determined by the statistical significance of the trend of the residuals from the regression between the *DM* and magnitude (M) or distance (R). The results of sufficiency are shown in Figure 4 and Figure 5.

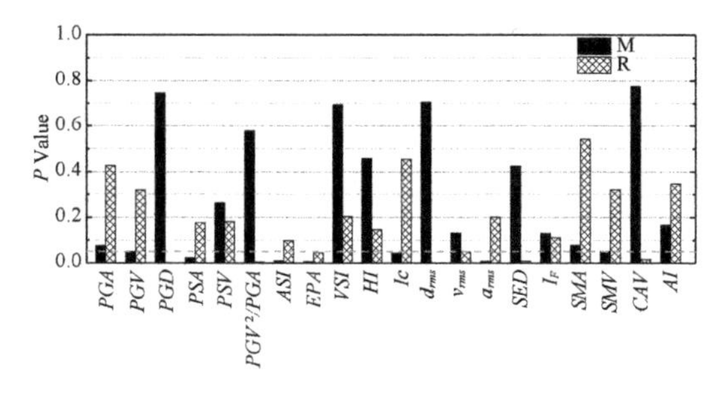

Figure 4 *P*-values of *IM*s-*OLDC* regressions

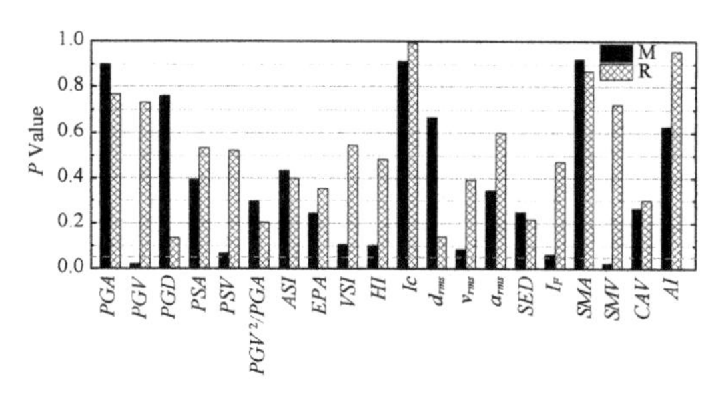

Figure 5 *P*-values of *IM*s-*OLDT* regressions

Based on the criteria of efficiency, practicality and sufficiency, Arias intensity measure is the optimal *IM* for the seismic performance assessment of mountain tunnels.

References

[1] HAMBURGER R O. The ATC-58 project: development of next-generation performance-based earthquake engineering design criteria for buildings[C]//Structures Congress 2006: Structural Engineering and Public Safety, 2006: 1-8.

[2] ZHAO M, GAO Z D, WANG L T, et al. Obliquely incident earthquake input for soil-structure interaction in layered half space[J]. Earthquakes and Structures, 2017, 13(6): 573-588.

[3] ZIENKIEWICZ O C, BIANIC N, SHEN F Q. Earthquake input definition and the transmitting boundary condition [M]// Advances in computational non-linear mechanics, Springer, Vienna, 1989: 109-138.

[4] FEMA P695. Quantification of building seismic performance factors [R]. Federal Emergency Management Agency, Washington, D.C, 2009.

[5] LUCO N, CORNELL C A. Structure-specific scalar intensity measures for near-source and ordinary earthquake ground motions[J]. Earthquake Spectra, 2007, 23 (2): 357-392.

[6] PADGETT J E, NIELSON B G, DESROCHES R. Selection of optimal

intensity measures in probabilistic seismic demand models of highway bridge portfolios [J]. Earthquake Engineering and Structural Dynamics, 2008, 37(5): 711-725.

[7] SHOME N, CORNELL C A. Probabilistic seismic demand analysis of nonlinear structures (Report No. RMS35) [D]. California: Stanford University, 1999.

[8] CORNELL C A, JALAYER F, HAMBURGER R O, et al. Probabilistic basis for 2000 SAC federal emergency management agency steel moment frame guidelines [J]. Journal of structural engineering, 2002, 128(4): 526-533.

Dynamic smart monitoring on deformation of initial lining of rock tunnel

M. Q. Chen & D. M. Zhang & M. L. Zhou & H. W. Huang
Key Laboratory of Geotechnical and Underground Engineering of Minister of Education and Department of Geotechnical Engineering, Tongji University, Shanghai 200092, China

Abstract

Monitoring the deformation of the tunnel can capture the surrounding ground conditions and dynamic changes of the supporting structure at each construction stage of the tunnel, judge the stability of the surrounding rock and the reliability of the supporting structure, and ensure the construction safety and long-term stability of the structure[1]. However, the traditional monitoring method is sometimes inefficient, time-consuming, greatly affected by the tunnel environment[2]. In order to overcome these limitations of traditional monitoring, this paper adopts a novel wireless monitoring system to monitor the deformation characteristics of surrounding rock and the content of gases (CO) in the process of tunnel construction in real time for a real project of Mopanshan tunnel in south-western of China. Through the analysis of the monitoring data and the actual operation of the construction site, the safety status of the tunnel during the construction process can be assessed in real time, and the variations of tunnel convergence over time can be characterized. At the same time, according to the results of data fusion analysis, the LED risk visualization equipment is used to transform the risk into the color change of the lighting equipment installed on the site, so as to support the construction safety of the tunnel.

As shown in Figure 1, during the tunnel excavation process, four sections were selected for the installation of the laser ranging dip fulcrum. The sensor can measure distance and inclination simultaneously. The sensors arranged in the four sections are used to monitor the convergence value of the tunnel section, and a gas concentration sensing fulcrum and a visual fulcrum are installed at section 2.

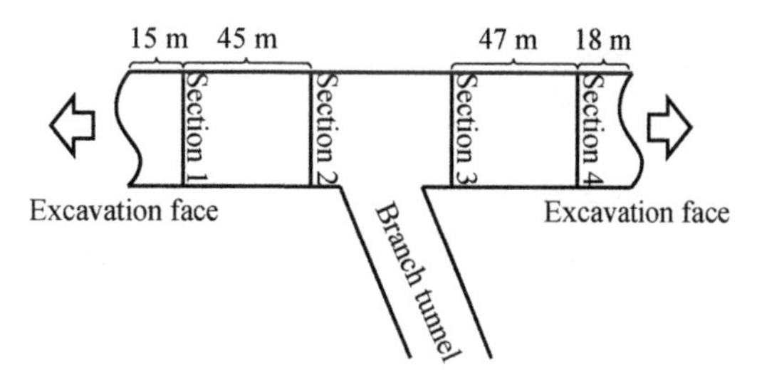

Figure 1 Section layout

The installation diagrams of Sections 1, 3, and 4 are shown in Figure 2, and the installation diagram of Section 2 is shown in Figure 3. The layout method of section sensor is referred to the related reference [3], [4]. Figure 4 shows an installed laser sensor.

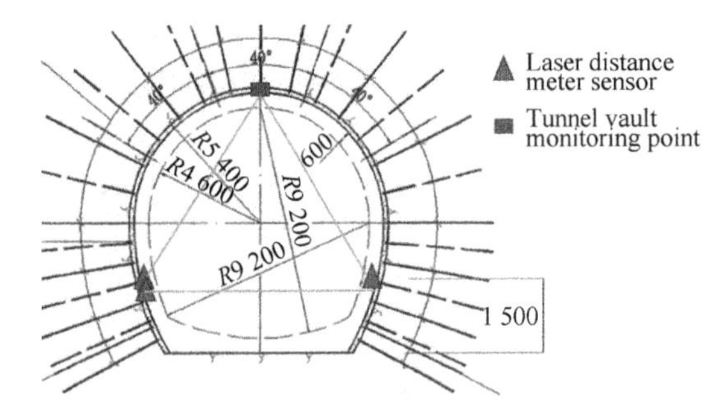

Figure 2 Layout of sections 1, 3 and 4 (unit: mm)

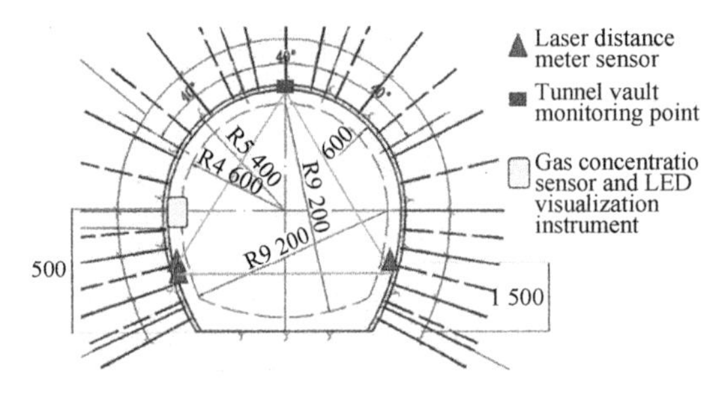

Figure 3 Layout of section 2 (unit: mm)

Figure 4 An installed laser sensor

After the sensor is installed, the monitoring data can be obtained from the web-based platform in real time, and the parameters can be set on the computer. During the blasting, due to the splash of the gravel and the large dust near the face of the face, the laser ranging function is limited, so the inclination value is mainly observed during this period. Figure 5 shows the face blasting operation at 8:40, and the CO content in the tunnel hole. It can be seen that a large amount of harmful gas (CO) is generated due to the blasting operation, and the harmful gas content in the tunnel rises sharply to 15.7 mg/m^3, which will damage people's health[5]. The content of harmful gases rapidly decreased under the action of ventilation equipment after the end of the explosion.

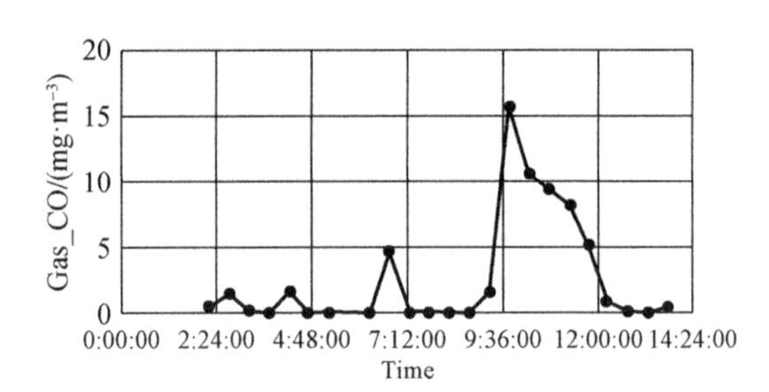

Figure 5 CO content change chart

Figure 6 and Figure 7 show the inclination value and vault settlement value of tunnel lining engraved near the tunnel face at the same time. The change value of inclination angle is less than

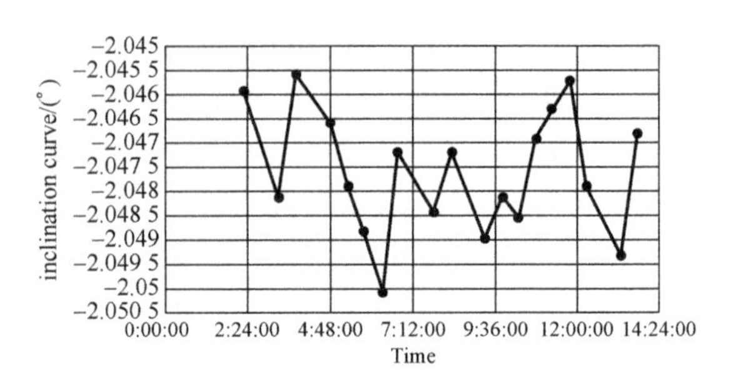

Figure 6 Dip change chart

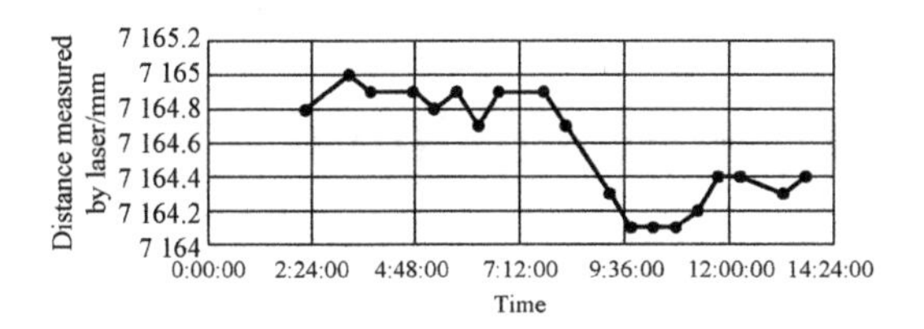

Figure 7 Vault to arch waist spacing change diagram

0.005° and the change of vault and arch waist distance is less than 1 mm, indicating that the structure is relatively stable and not affected by blasting[4], which is consistent with the monitoring results of other sensors. It shows that the monitoring system has good stability and reliability.

Acknowledgements

The research presented in this paper is supported by the Natural Science Foundation Committee Program of China (Grant No. 1538009, Grant No. 51778474), the Fundamental Research Funds for the Central Universities of China (Grant No. 0200219129) and Key Innovation Team Program of Innovation Talents Promotion Plan by MOST of China (No. 2016RA4059).

References

[1] FANG Q, SU W, ZHANG D L, et al. Study on deformation characteristics of tunnel surrounding rock based on field monitoring data[J]. Journal of rock mechanics and engineering, 2016, 35(9): 1884-1897.

[2] GAN Q Y, ZHOU J. Current Research on tunnel monitoring and measurement technology[J]. Journal of underground space and engineering, 2019 (S1): 400-415.

[3] ZHOU D H, CAO L Q, QU H F, et al. Monitoring and control of construction deformation of super large cross-section highway tunnel under different surrounding rock conditions [J]. Journal of rock mechanics and engineering, 2009, 12: 140-149.

[4] JTG F60. Technical Specification for Construction of Highway Tunnel[S]. China, 2009.

[5] LONG Z H, CHENG Z Q. Health effects of low concentration carbon monoxide [J]. Industrial health and occupational diseases, 1993, 2: 122-126.

Experimental study and three-dimensional numerical simulation analysis of nanoindentation process of concrete

H. Li & X. D. Ren

Department of Structural Engineer, Tongji University, Shanghai 200092, China

Abstract

Concrete is still the most widely used material in construction worldwide. With the development of concrete science, scholars are increasingly exploring the influence of concerete structures on macroscopic performance from the interior of concrete. In modern concrete science, concrete is considered a cement-based material with multiple scale microstructures from nanoscopic to macroscopic[1].

Nanoscale characterization of cement based materials has shown that nanoindentation is an adequate technique for extracting nanomechanical properties[2]. During a nanoindentation test, a diamond indenter tip is pressed into a material up to a specified force or depth and then withdrawn. At the same time, the indentation force and depth are recorded which can be extracted the mechanical properties such as hardness and the Young's modulus.

For a heterogeneous material such as concrete, how to obtain the random distribution characteristics of the properties of each phase medium is a fundamental and urgent task for researchers. In this paper, systematic concrete nanoindentation tests were performed at Shanghai Institute of Materials Research to obtain the random distribution characteristics of the mechanical properties of the nanoscale of concrete. On this basis, the indentation test data of different levels is modeled analysis by using the random field theory and mathematical statistics method as shown in Figure 1.

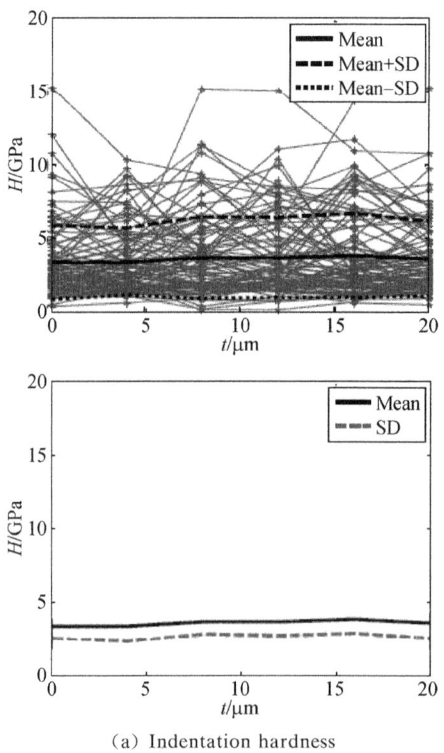

(a) Indentation hardness

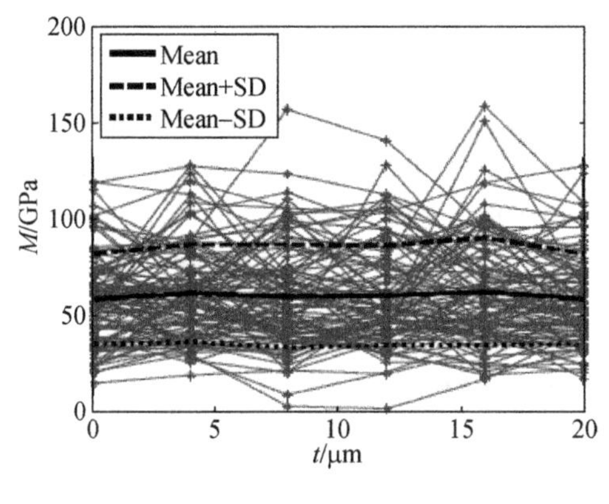

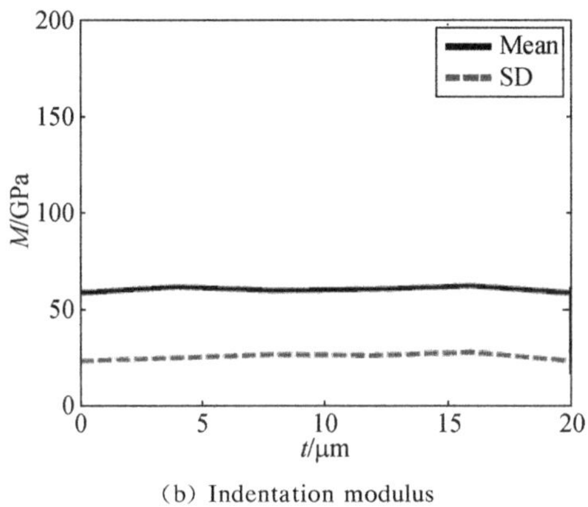

(b) Indentation modulus

Figure 1 Results of nanoindentation test for cement paste

The mechanical analysis of the contact process between the indenter and the material to be tested in the indentation test is very complicated, which need to consider the elastoplasticity of the material, in the meanwhile, the boundary conditions between the indenter and the material specimen are constantly changing. These factors make it difficult to analyze the problem by analytical means. So numerical simulation shows an advantage on this issue.

In this paper, the nanoindentation experiments performed on the cement paste were simulated using the finite element method (FEM). A three dimensional (3D) finite element (FE) model was created using ABAQUS finite element software as shown in Figure 2. The commonly used Berkovich geometric model was selected for the finite element model of the indenter, which allows a fine description of the indenter tip, namely an imperfection such as the one which occurs in the real geometry[3].

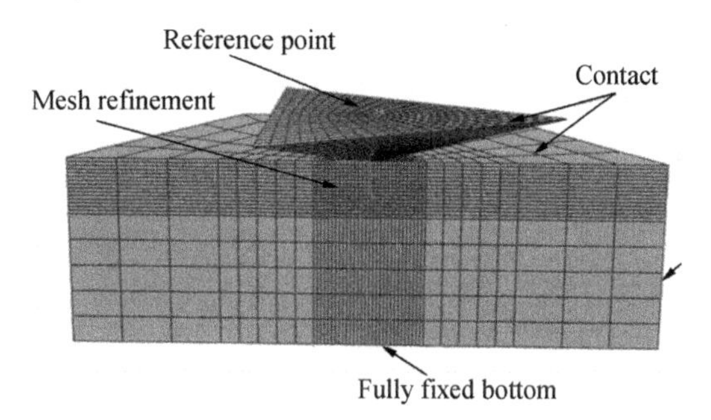

Figure 2 3D model of the cement paste and indenter developed in ABAQUS

An elastoplastic damage constitutive model based on irreversible thermodynamics was employed for to simulate the load-displacement data recorded during the nanoindentation test. It is shown that the load-ondentation curves can be well reproduced by the finite element model correctly in Figure 3.

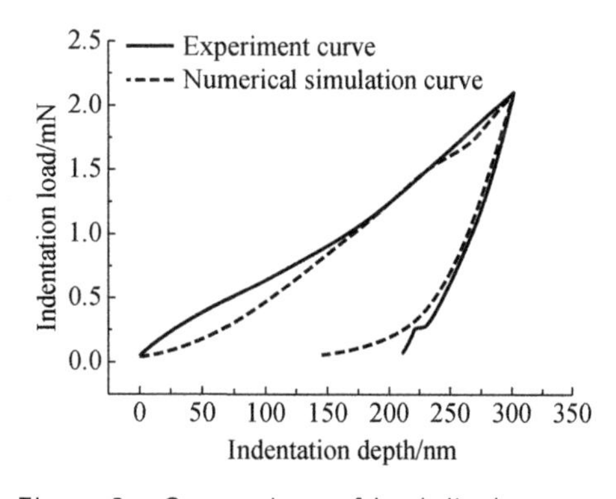

Figure 3 Comparison of load-displacement curves between nanoindentation test and numerical simulation

Acknowledgements

Financial support from the National Science Foundation of China (Grant No. 51678439) is appreciated.

References

[1] JENNINGS H M. A model for the microstructure of calcium silicate hydrate in cement paste [J]. Cement & Concrete Research, 2000, 30(1): 101-116.

[2] SAEZ DE IBARRA Y, GAITERO J J, ERKIZIA E, et al. Atomic force microscopy and nanoindentation of cement pastes with nanotube dispersions[J]. Physica Status solidi (A) Application and Materials, 2006, 203 (6): 1076-1081.

[3] ANTUNES J M, CAVALEIRO A, MENEZES L F, et al. Ultra-microhardness testing procedure with Vickers indenter[J]. Surface and Coatings Technology, 2002, 149 (1): 27-35.

Seismic retrofit in stages for residential buildings with soft and weak bottom story

T. C. Chiou

The Center for Research on Earthquake Engineering, Taipei, Taiwan, China

Y. H. Lin

Department of Civil Engineering, Taiwan University, Taipei, Taiwan, China

L. L. Chung & S. J. Hwang

The Center for Research on Earthquake Engineering, and Department of Civil Engineering, Taiwan University, Taipei, Taiwan, China

Q. Q. Yu

Department of Structural Engineering, Tongji University, Shanghai 200092, China

Abstract

Many commercial and residential mixed buildings collapsed in 1999 Chi-Chi Earthquake due to the soft and weak bottom story. In the traditional mixed buildings, the open area, such as shop stores and living rooms, were designed on the ground floor, however, the private area, such as residential bedrooms and bathrooms were usually designed on the upper floors. The open area is usually few partition walls, however, the private area needs lots of partition walls. Therefore, the structural system with soft and weak bottom story is a significant problem of the traditional mixed buildings.

Rebuild and seismic retrofit of the existing mixed buildings always has a big challenge because it is very difficult to come out with agreement from the multi-owners of the buildings. However, the occurrence of earthquakes is unpredictable, the seismic capacity of existing buildings should be improved as soon as possible. Therefore, seismic retrofitting in stages may be a feasible strategy[1]. The study proposed the seismic retrofit in stages can implement retrofit in the public area without affect into private area. That may be easier to come out with agreement of whole multi-owners. The proposed seismic retrofit in stages aims to improve the problem of soft and weak bottom story. The first phase of retrofit aims to reduce the collapsed probability of the buildings although that may not satisfy the seismic performance of the current seismic design code. The final phase shall still implement retrofit of whole buildings which can satisfy the seismic performance of the current seismic design code. This paper introduces a design procedure of the seismic retrofitting in stages for the existing reinforced concrete buildings with soft and weak bottom story. Moreover, a design example will be provided[2].

References

[1] Technical handbook for seismic retrofit in stages of a residential building [S]. The Center for Research on Earthquake Engineering, Construction and Planning Agency, Ministry of the Interior, 2018.

[2] LIN Y H. A Simple Method to Eliminate the Deficiency of the Soft Story for Reinforced Concrete Buildings [D]. Taipei: Taiwan University, 2018.

Comparative experimental study on seismic performance of different types of hollow piers

W. Wei

School of Civil Engineering, Southwest Jiaotong University, Chengdu 610031, China

C. J. Shao

School of Civil Engineering, Southwest Jiaotong University, Chengdu 610031, China & National Engineering Laboratory of Geological Disaster Prevention Technology in Land Transportation, Chengdu 610031, China

Abstract

Compared with the solid pier with the same external size, the hollow pier has the advantages of saving masonry and reducing the dead weight of the structure, and can effectively reduce the seismic inertia force to a certain extent, and achieve the purpose of protecting the overall safety of the bridge[1]. Although some achievements have been made in the study of seismic performance of hollow piers[2-5], these results are more derived from the research of rectangular hollow section piers, whether the research results can be widely used in other types of hollow piers is not clear[5].

In order to reveal the difference of seismic performance of reinforced concrete hollow piers with different cross-sections under repeated loads, low-cycle reciprocating loading tests were carried out on 6 hollow piers with three different cross-sections (Scheme of the hollow piers are seen in Table 1 and shown in Figure 1). The evolution mechanism of pier body crack of hollow pier was analyzed, and the differences of pier crack evolution mechanism, hysteretic energy dissipation capacity, plastic rotation capacity, damping and stiffness characteristics of bridge piers with different cross sections were discussed. The test and analysis results showed that the seismic performance of square hollow pier was consistent with that of rectangular hollow pier, but it was different from that of thin-walled hollow pier at circular end, which mainly performanced as the following aspects: (1) Compared with square hollow piers and rectangular hollow piers, the distribution of cracks in the pier body of thin-walled hollow pier with variable cross-section was denser and wider, grid oblique cracks appeared on the side of pier body, the distribution range of bending cracks in pier body was about 0.61 ~ 0.75 times of pier height, and the through cracks in pier body were scattered. There were certain vertical cracks at the top of the pier. (2) In the limit state, the tension / buckling and concrete collapse of the steel bar of the square hollow pier and the rectangular hollow pier were mainly concentrated in the chamfer position at the bottom of the pier, while the thin-walled hollow pier at the circular end was mainly concentrated near the variable cross-section chamfer at the bottom of the pier (Damage description of specimens are shown in Figure 2). (3) For the square hollow piers and the rectangular hollow piers, the plastic hinge position of the thin-walled hollow pier at the circular end moved up and the plastic area was enlarged, which showed a better ductile capacity (Hysteresis curves of specimens are shown in Figure 3). (4) Compared with square hollow piers and rectangular hollow piers, the stiffness of thin-walled hollow pier at circular end degraded faster, but the effective stiffness corresponding to the equivalent yield point of the three kinds of cross-section was consistent with the compression ratio,

Table 1 Design parameters of quasi-static test model for piers

Test ID	L /mm	$h \times b$ /mm^2	t_w /mm	Concrete f_c' /MPa	Longitudinal reinforcement ρ_l /%	Longitudinal reinforcement Layout	Stirrup reinforcement ρ_s /%	Stirrup reinforcement d_s /mm	Stirrup reinforcement s /mm	η_k	λ
SH1	2 950	500×500	120	42.5	2.12	20φ12+8φ16	3.10	10	65	0.05	5.9
SH2	2 950	500×500	120	42.5	2.12	20φ12+8φ16	1.34	10	150	0.05	5.9
RH1	2 950	800×500	120	42.5	1.63	26φ12+6φ16	2.13	10	100	0.05	5.9
RH2	2 950	800×500	120	42.5	2.69	32φ12+16φ16	2.13	10	100	0.05	5.9
CH1	5 000	803	7.4/11.3	32	0.906	24φ12	0.325	6	280	0.147	6.1
CH2	5 000	803	7.4/11.3	32	0.906	24φ12	0.91	6	100	0.107	6.1

Note: L is the height of the members, h is the height of the component section, b is the width of the component section, t_w is the thickness of wall of the hollow piers, f_c' is the compressive strength of concrete, ρ_l is the stirrup ratio, ρ_s is the stirrup volumetric ratio, λ is the aspect ratio.

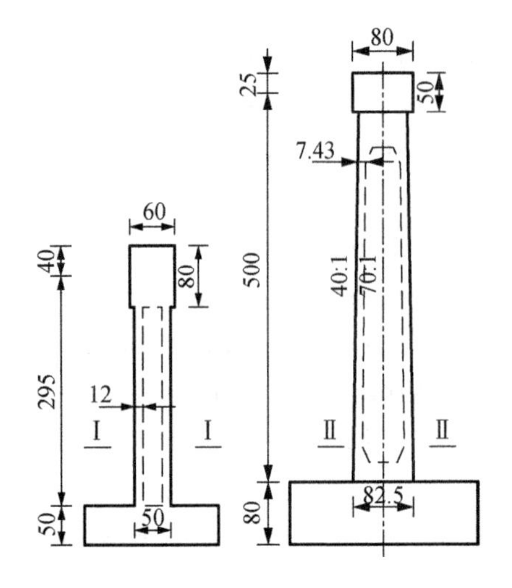

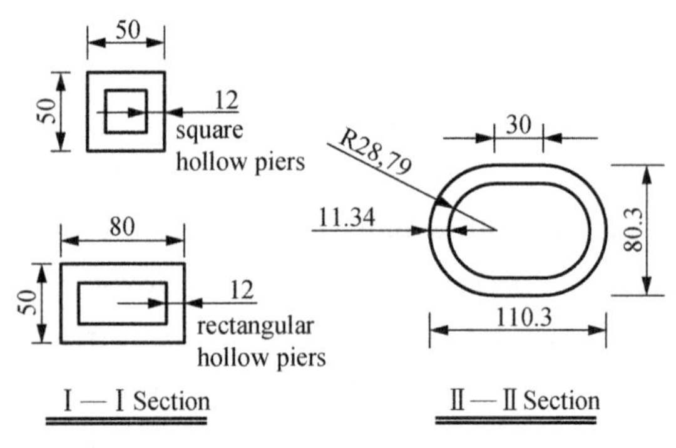

Figure 1 Scheme of the hollow pier (unit: cm)

shear span ratio and material characteristic parameters. (5) The equivalent damping ratio of thin-walled hollow pier with circular end before equivalent yield was slightly higher than that of hollow pier with square section and hollow pier with rectangular section. After equivalent yield, the equivalent damping ratio of hollow pier with square section and hollow pier with rectangular section changed faster than that of thin-walled hollow pier with circular end.

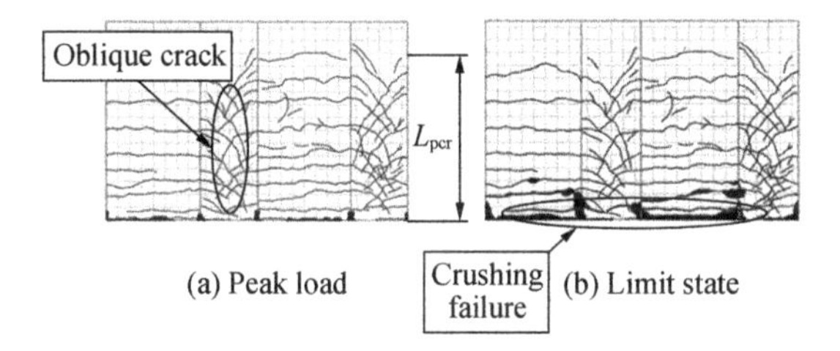

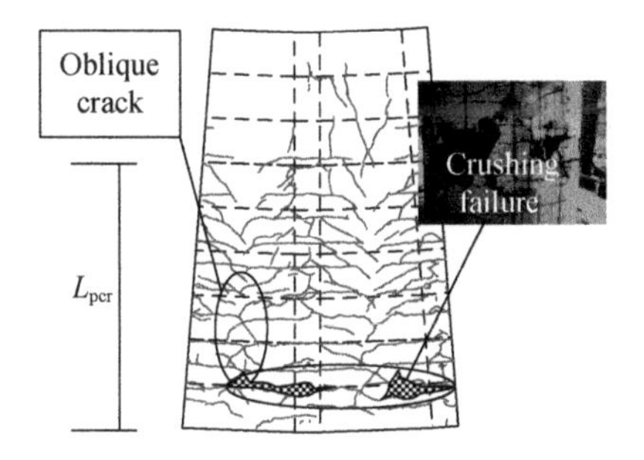

Figure 2 Damage description of specimens

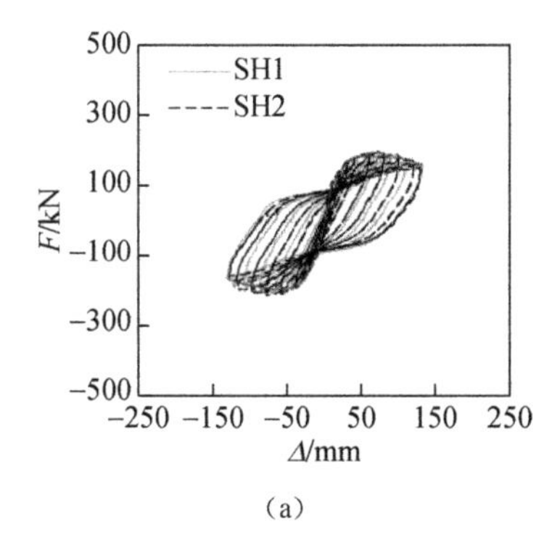

(a)

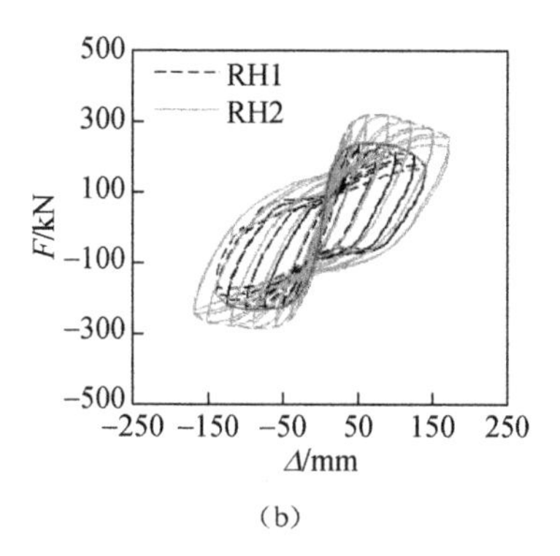

(b)

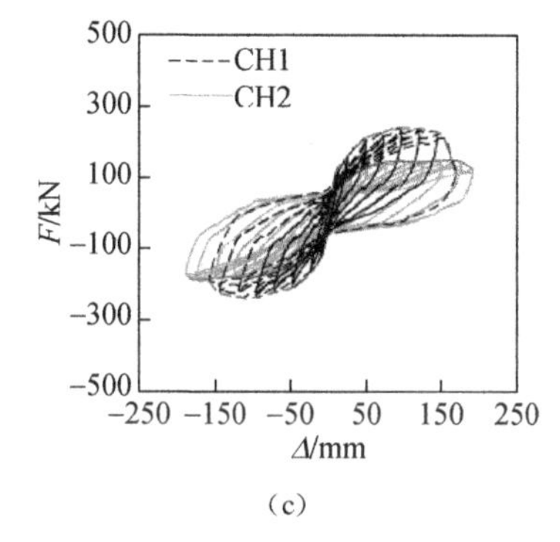

(c)

Figure 3 Hysteresis curves of specimens

Acknowledgements

The financial support provided by the National Natural Science Fund Committee of China (Grant No. 51178395) and the Applied Basic Research Plan are acknowledged gratefully.

References

[1] WEI W. Study on ductile and seismic behavior of concrete hollow pier based on quasi-static test [D]. Chengdu: Southwest Jiaotong University, 2019.

[2] SHAO C J, Qi Q M, WANG M, et al. Experimental study on the seismic performance of round-ended hollow piers[J]. Engineering Structures, 2019, 195:309-323.

[3] SHAO C J, QI Q M, WEI W, et al. Quasi-static test on seismic ductility of round-end

hollow-section railway piers[J]. China Civil Engineering Journal, 2019, 52(7):118-128.

[4] WEI W, SHAO C J, XIAO Z H, et al. Experiment on effective stiffness of reinforced concrete hollow piers[J]. China Civil Engineering Journal, 2019, 52(10): 101-110.

[5] SHAO C J, SUN N C, ZUO X, et al. Analyses on the length of plastic hinge of rec-tangular hollow-section concrete pier[J]. China Civil Engineering Journal, 2018, 51(11): 124-132.

Service limit/ultimate limit state ratio improvement for FRP concrete beams

E. J. Fernández González
Department of Disaster Mitigation for Structures, College of Civil Engineering, Tongji University, Shanghai 200092, China

Abstract

Currently discovers have proven the use of new materials to improve the mechanical properties in reinforced concrete (RC) structures rather than the common ones used mostly in the past century such as steel rebar and light weight concrete. One of those materials that can fulfill these expectations is the use of Fiber-Reinforced Polymer (FRP) reinforcement as a prime material instead of steel reinforcement.

FRP reinforcement is commonly made from basalt, aramid, glass or carbon. BENMOKRANE et al.[1-2] indicate some FRP reinforcement qualities like corrosion resistance, structure behavior and lightweight according to test results. The corrosion factor represents the main reason over steel reinforcement for choice selection. Hence, the FRP properties are being studied for the comprehension of its behavior in individual members as well as structural behavior.

Recently papers have been researching the FRP capacity under limit states design. FRP reinforced bars are checked for service limit state and ultimate limit state to control deflection, cracking and resistance. According to design, the ultimate limit is governed by the amount of reinforcement while the service limit is governed by the beam cross section and crack width respectively. International codes such as US codes[3-4] and Canadian codes[5] have similar approaches to identify service and ultimate limits state.

Due to its mechanical properties, FRP behaves as lineal elastic material up to its failure unlike steel reinforcement that yields. Also, because of its higher stress capacity (f_{fu}) joined with a lower elasticity modulus (E_f) causes the service limit state often controls the design of FRP rather than the ultimate state limit. According to international codes formulas for FRP design, the ratio state limit [R, Eq.(1)] will have lower values due to current parameters. This implies an overdesign member or overstressing induced by high loads.

Meanwhile, for conventional RC beams the ratio is rather higher and usually the ultimate limit state governs its designs. This study intends to investigate the parameters to increase the efficiency on service limit state of FRP concrete beams to prevent overdesigning for ultimate limit state.

$$R = \frac{\text{Service limit}}{\text{Ultimate limit}} \tag{1}$$

As a preliminary work, the study focuses on currently bibliographic resources (design codes, journey papers, reports) on the topic of service level on a first phase. Later on, it'll proceed to investigate the design parameters and establishes relationships between new and former variables from test specimens to develop prototype formulae for ratio improvement as a way to prevent overdesign member.

Acknowledgements

To my professor advisor Jiafei Jiang and partners from the Disaster Mitigation for Structures Department of Tongji University, Natural Science Foundation of Shanghai (19ZR1460900).

References

[1] BENMOKRANE B, el al. Recent Development on FBR as Internal Reinforcement & Applications in Buildings and Bridges[C]// Presentation of the 11th Conference on Application Technology of FRP in Infrastructure in China, Shanghai, 2019.

[2] BENMOKRANE B. Design of concrete structures internally reinforced with FRP bars[C]// Seminar on FRP Reinforcement for Sustainable Concrete Infrastructure, Changsha: Hunan University, 2019.

[3] ACI std 440.2R-17. Guide for Design and Construction of Externally Bonded FRP Systems for Strengthening Concrete Structures[S]. Farmington Hills: American Concrete Institute, 2017.

[4] ACI std 440.1R-15. Guide for Design and Construction of Structural Concrete Reinforced with Fiber-Reinforced Polymer (FRP) Bars[S]. Farmington Hills: American Concrete Institute, 2015.

[5] CAN/CSA std S806 - 12. Design and Construction of Building Components with Fibre-Reinforced Polymers [S]. Canada: Canadian Standards Association, 2012.

Rammed earth: a promising sustainable housing solution

B. Khadka & L. S. He

Department of Disaster Mitigation for Structures, College of Civil Engineering, Tongji University, Shanghai 200092, China

Abstract

With the increasing population and their housing demand, there has been tremendous use of energy and resources in several forms like during extraction of raw materials, manufacturing of building material, construction, operations and deconstruction of building due to which the level of greenhouse gases is increasing in earth's atmosphere causing the climate change. Buildings alone account for one-sixth of the world's freshwater withdrawals, one-quarter of its wood harvest, and two-fifths of its material and energy flows[1] resulting in massive side-effects in the form of global pollution, i. e. air pollution 23%, climate change gas 50%, drinking water pollution 40%, landfill waste 50% and ozone depletion 50%[2]. We need to be aware that our planet cannot always support the current level of resource consumption and global pollution associated with the building construction.

This miserable scenario shows the necessity of inventing new sustainable alternative materials and successfully implement it in modern building construction. In order to fulfill these necessities, this paper introduced soil (earth) in the form of rammed earth. Soil has been adopted and tested as a natural construction material for thousands of years, and in association with modern techniques (mainly stabilization), it can be easily used for modern ecological buildings. In the last decades, it has been found that many published papers are focused on earth construction (rammed earth), which shows that more and more research efforts are being dedicated to transform current building construction sector into a more sustainable one[3]. The majority of earth construction are seen in developing countries like Nepal, but currently, it has also increased substantially in developed countries like Australia, US, New Zealand, and UK due to global sustainable construction agenda. In order to match the present requirements (control quality, strength, mix proportion, etc.) for structurally stable sustainable housing, various countries like New Zealand, Australia, New Mexico, Zimbabwe, etc. have developed standards, national reference documents or guidelines like Standard Australia 2002, NZS 4297:

1998[4], NZS 4298: 1998[4], etc. for earthen construction. Inspite of being useful housing material and technique, rammed earth is still not familiar among local people, which has further made the housing sector more deviated towards environmental deterioration. So, to reveal environmental impacts of building, promote rammed earth construction as an affordable alternative eco-friendly sustainable housing (both in terms of environmental and structural stability), this paper is written.

1. Rammed Earth Construction

It is the soil-based ancient method of construction in which the continuous walls are built by placing correct proportions of damp or moist sub soils (Optimum moisture content: 8% ~ 12% of dry weight) consisting of clay/silt (15% ~ 25%), sand (50% ~ 60%) and gravel (15% ~ 20%) with or without stabilizers like cement (around 5%) in thin layers of about 100 mm to 150 mm into the temporary formwork and are compacted by hand or pneumatic tampers to the sufficient wall strength (Figure 1)[3]. The quality of soil, proper proportion of mixed ingredients, type of stabilizers, amount of water, effective/adequate compaction and curing are some of the essential criteria for good quality rammed earth construction.

Generally, there are two types of rammed earth construction, i.e. stabilized and unstabilized rammed earth. Unstabilized rammed earth is made from clay, silt, sand, and gravel, whereas stabilized rammed earth contains additives like cement or lime.

(a) Preparing mix

(b) Tamping mix

(c) Removing formwork

Figure 1 Rammed earth construction procedure

2. Mechanical Property

The compressive strength of rammed earth is in the range of 6 ~ 10 MPa for cement-stabilized and 1.5 ~ 3 MPa for unstabilized[3]. As per British[5] and New Zealand Standards[4], in single-storey construction, the characteristic wall strength required is about 0.5 N/mm^2. From the test results (Figure 2), the average characteristic strength obtained was much higher than 0.5N/mm^2, which shows that rammed earth can be used with full confidence.

3. Advantages of Rammed Earth

Currently, almost 50% of the world's population live in earth-based dwellings[6]. Many world heritage sites like Buddhist monasteries, some wall sections of Great

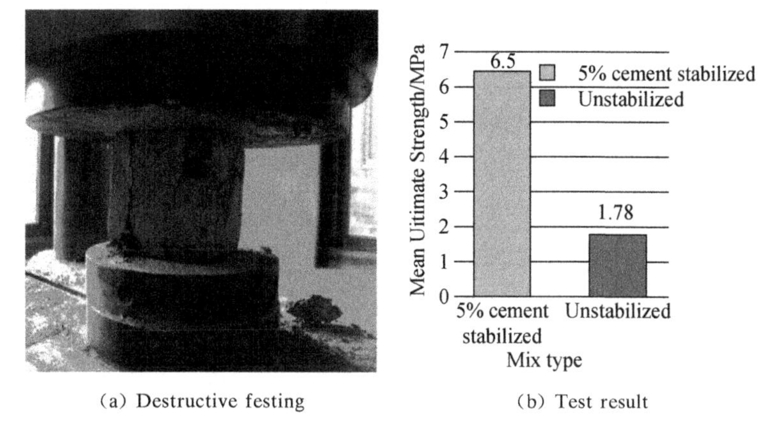

(a) Destructive festing (b) Test result

Figure 2 Compressive strength of rammed earth

Wall of China, Potala palace in Lhasa, etc. include earth construction. Recent interest in rammed earth is mainly due to its inherent green attributes. Table 1 shows the advantages of rammed earth.

Table 1 Advantages of rammed earth

S.N	Benefits	Description
1	Thermal Mass	Good thermal performance of rammed earth includes nearly ± 12 ℃ to 20 ℃ temperature difference between outside and inside room during summer and winter
2	Load Bearing	Strength of 5% cement-stabilized 100 mm cube is in the range of 6 ~ 10 MPa[3]
3	Strong and Durable	Investigation shows rammed earth as durable, weather-resistant, and strong enough to resist upcoming forces
4	Fire Proof & noise reduction	Good in fire resistance[7] and noise absorbent
5	Earthquake resistant	Investigation after 7.8 MW 2015 earthquake, Nepal, showed neglible damage
6	Versatile	Can be used a load-bearing and non-load bearing structure
7	Cost-effective	Earth being locally available can be used with reduced transport costs and without any secondary industrial transformation
8	Healthy & Environment friendly	Earth is recyclable, non-toxic, non-polluting. Minimum consumption of energy and resources like water

Compressed stabilized earth block (CSEB) is environmentally sustainable (less carbon-dioxide emission and less energy requirement as shown in Table 2) compared to the conventional building materials and would be appropriate in case of urban house construction.

4. Conclusions

Rammed earth can be considered boon in the building sector, in terms of structural stability and environmental sustainability. In context to Nepal, the rammed

Table 2 Energy requirement and carbon-dioxide emissions[6]

	CSEB	Fired brick	Hollow concrete blocks
Energy required/ ($MJ \cdot fu^{-1}$)	233	1 026	390
CO_2 Emission/ ($kg \cdot fu^{-1}$)	55	118	98

earth houses were used for office, residence, hospital, and school in the form of load-bearing and non-load bearing structure. The majority of cement stabilized rammed earth houses were in good condition and were able to resist the effects of weathering agents. Inspite of minimum utilization of energy and resources like water and stabilizer (cement) during the construction, the overall obtained results (strength, durability, versatility) are good, and it contributes towards decrement in CO_2 by building sector. Comparatively 5% cement stabilized samples had better results than unstabilized samples. Inspite of such good characteristics, it is still not widely accepted by most people, and is considered as an old material which has no existence in this modern world. So, necessary awareness regarding building impacts on the environment and importance of earth construction as an alternative sustainable construction material should be promoted to develop positive attitudes among local people towards earth construction.

References

[1] ROODMAN D M, LENSSEN N K, PETERSON J A. A building revolution: how ecology and health concerns are transforming construction [M]. Washington, DC: Worldwatch Institute, 1995.

[2] ALLINSON D, HALL M. Investigating the optimisation of stabilised rammed earth materials for passive air conditioning in buildings [C]// Proceedings for the International Symposium of Earthen Structures, Bangalore, India, 2007.

[3] KHADKA B, SHAKYA M. Comparative compressive strength of stabilized and un-stabilized rammed earth[J]. Materials and Structures, 2016, 49(9): 3945-3955.

[4] NZS 4297, Standard N Z. Engineering design of earth buildings[S]. Wellington, New Zealand: StandardNew Zealand, 1998.

[5] BS 5628, Code of practice for use of masonry [S]. United Kingdom: British Standards Institute, 1992.

[6] VROOMEN R. Gypsum stabilised earth: research on the properties of cast Gypsum-stabilised earth and its suitability for low cost housing construction in developing countries [D]. Eindhoven: Eindhoven University of Technology, 2007.

[7] AREZOUMANDI M, VOLZ J S, ORTEGA C A, et al. Effect of total cementitious content on shear strength of high-volume fly ash concrete beams[J]. Materials & Design, 2013, 46: 301-309.

Steel frame joints based on damage avoidance design

J. W. Liu & C. X. Qiu & X. L. Du
Beijing University of Technology, Beijing 100124, China

Abstract

In the conventional design of steel structure joint, the steel frame joint will achieve energy consumption through yield and plastic deformation under earthquake loads. This sacrificial design approach can lead to permanent, and often irreparable damage when interstory drifts are too large. A new kind of steel frame joint is proposed. By the superelasticity of shape memory alloy (SMA)[1], it can achieve damage avoiding and recentering capabilities[2]. In this new type of steel frame joint (Figure 1) using SMA, the beam and column are connected by steel angle and bolts on the upper flange of the beam.

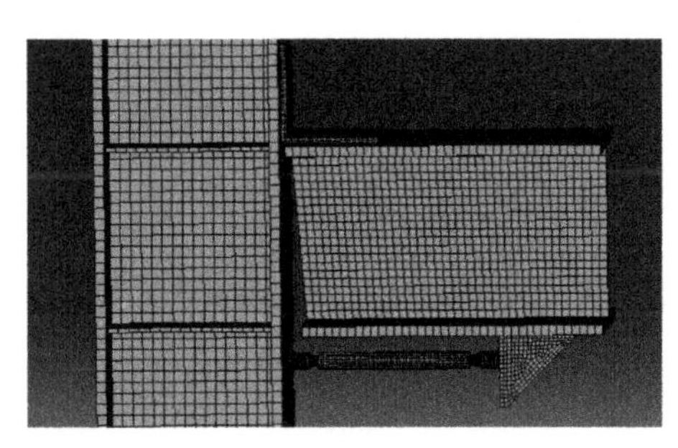

Figure 1 Recentering steel frame joint in ABAQUS

The buckling-restrained SMA member is located at the lower side of the beam, and is connected with the lower flange of the beam with steel angle and bolts. When the lateral displacement occurs, the rotation between the beam and the column around the upper steel angle occurs[3], and the buckling-restrained[4] SMA member deforms to achieve energy consumption. ABAQUS is used to simulate the buckling-restrained SMA member and the recentering steel frame joints. Simulation of the buckling-restrained SMA shows that when the inner core of SMA is subjected to the cyclic displacement loads, the external restraint sleeve restricts the buckling deformation of the inner core of SMA, and the buckling-restrained SMA member shows a stable flag-shape hysteretic curve (Figure 2), which preliminarily verifies the feasibility of the buckling-restrained SMA members.

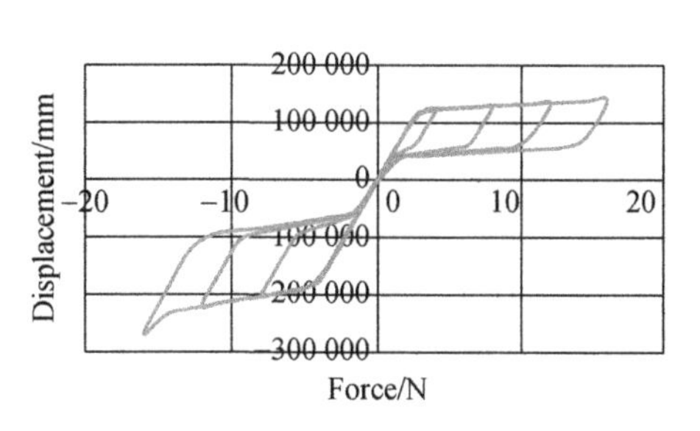

Figure 2 The flag-shape hysteretic curve of the buckling-restrained SMA member

Simulating results of the recentering steel frame joints indicate that the moment-rotation curve of the joint is obtained to verify the recentering performance of the joint. Therefore, this kind of steel frame joint can achieve recentering and energy dissipation without damage, in contrast to conventional sacrificial design.

References

[1] GRAESSER E J, COZZARELLI F A. Shape-memory alloys as new materials for aseismic isolation[J]. Journal of Engineering Mechanics, 1991, 117(11): 2590-2608.

[2] YAM M CH, FANG C, LAM A C C, et al. Numerical study and practical design of beam-to-column connections with shape memory alloys[J]. Journal of Constructional Steel Research, 2015, 104: 177-192.

[3] MANDER T, RODGERS G W, MANDER J B, et al. Damage avoidance design steel beam-column moment connection using high-force-to-volume dissipators [J]. Journal of Structural Engineering, 2009, 135 (11): 1390-1397.

[4] CLARK P W, AIKEN I, KIMURA I, et al. Large-scale testing of steel unbonded braces for energy dissipation [C]//Proceedings of the 2000 Structures Congress & Exposition, Philadelphia, PA, American Society of Civil Engineers, Reston, VA, 2000.

Experimental research on full-scale reinforced concrete beams with large openings

S. T. Zhao & Q. H. Huang & L. Lü

Key Laboratory of Performance Evolution and Control for Engineering Structures of Ministry of Education, Tongji University, Shanghai 200092, China

M. J. Tong & J. X. Yu

CIFI Holdings, Shanghai 200062, China

Abstract

Statistically, equipment pipelines installed under beams will occupy one-sixth to one-fifth of storey height of a building, reducing clear height of the building storey. If equipment pipelines cross beams through the openings in beam webs, they will not occupy the space under the beams. As a result, the storey height and the total height of the building will be reduced significantly, which will bring about great economic benefit.

A commercial and office building on the Middle Huaihai Road in Shanghai is designed as a reinforced concrete frame structure, with 3 floors below ground, 6 floors above ground, and a total height of 24 meters. The typical storey height is 3.65 meters. The typical beam height is 650 mm. And the typical beam span is about 9 meters. In order to satisfy the quality of Grade A, all equipment pipelines of the building are required to be installed through the openings in beam webs. Therefore, in each storey of the structure, more than 50% of the frame beams are designed to open a rectangular hole of 220 mm$H \times 400$ mm$W$ at both ends. The cross section of RC beam is weakened due to the openings, which may reduce the ultimate bearing capacity of the beam, causing the change of the failure mode of RC beam from ductile bending failure to brittle diagonal tension failure on hole side or shear failure of chord. Due to the stress concentration on the corners, rectangular holes may severely crack in advance, affecting the normal use of the beam and causing psychological anxiety to the user. The flexural stiffness of the beam with openings is reduced, resulting excessive deformation of the beam and the whole structure.

Based on test results of 61 simply supported RC beams with web openings, Cai J. et al.[1] summarized four typical kinds of failure modes. By regression analysis, the calculation method of shear bearing capacity of RC beams with web openings was proposed. Liu R. et al.[2] conducted six simply supported RC beams with web openings and four restrained ones. Openings in beam webs were divided into two different kinds, one subjected to bending and the other subjected to shearing, according to the position of openings on beams. Based on experimental results, calculation methods of shear bearing capacity and bending bearing capacity RC beams were proposed. Zhang S. et al.[3] studied the stress distribution characteristics of the circular, elliptical and rectangular openings of beams by photo-elastic experiment method. Huang T. et al.[4] analyzed effects of beam stiffness, opening height, opening width and opening eccentricity on deformation of beams by 42 RC beams' experiment. The relationship between crack width in the compressive corner of rectangular openings and the opening height was analyzed in the same study. Wu Y. et al.[5] conducted a pseudo-dynamic experiment to study the seismic performance of two single-storey RC frame structures with different sizes of beam opening.

Calculation methods of ultimate bearing capacity and deformation of RC beams proposed in above research were only applied to beams with small opening, whose height less than one third of beam height. The typical height of beam opening of the commercial and office building in this paper is 220 mm, higher than one third of the typical beam height, 650 mm. Considering the large number of beams with two big openings, the long typical span of beams, and the unfavorable location of openings at beam's shearing segments, it is significant to prove the mechanical performance of the typical beams with big openings, and the performance of the overall RC frame structure.

Two full-scale RC beams with openings (Figure 1 and Figure 2) and one without opening (Figure 3) are designed according the original RC frame structure. Loading experiments of the beams are conducted to study the mechanical performance of crack resistance, flexural stiffness, failure mode, ultimate bearing capacity and so on (Figure 4). The finite element numerical models for the beams are established and verified by the experimental results. Using the proved finite element model to analyze the bending stiffness of other beams with openings different from experimental beams. Considering the stiffness degradation of beams, the overall structure performance is analyzed, and compared with the analysis result without considering the stiffness degradation of beams due to big openings.

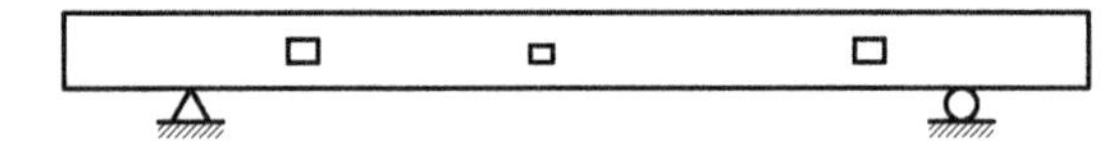

Figure 1 Experimental beam 1

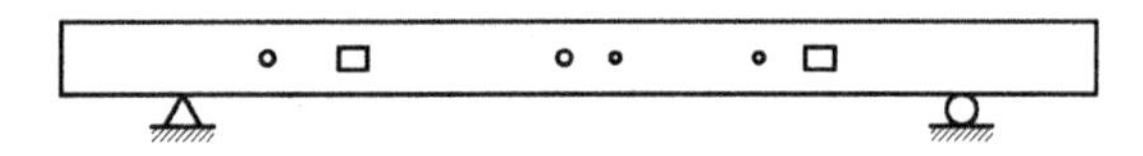

Figure 2 Experimental beam 2

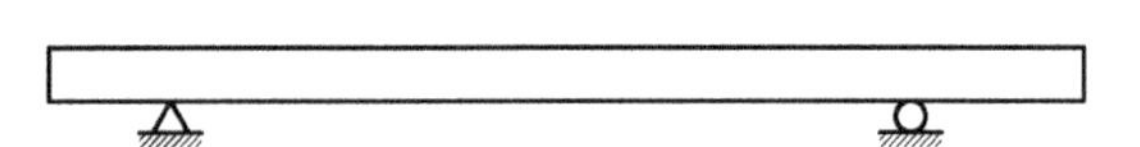

Figure 3 Un-opened experimental beam

Figure 4 The mechanical test of full scale member

Acknowledgements

The authors would like to acknowledge Chao Yin, Xinbo Xu and Kai Gao for their assistance in the experimental work.

References

[1] CAI J, WANG Y T, CHEN Q J, et al. Calculation of shear capacity of reinforced concrete simply supported beam with web openings[J]. Journal of Building Structures, 2014, 35(3): 149-155.

[2] LIU R G, LV Z T. Experimental research and theoretical analysis on reinforced and pre-stressed concrete beams with large openings [J]. China Civil Engineering Journal, 2004, 37(7): 29-34.

[3] ZHANG S Y, WU Y H, ZHENG B L, et al. Study of bearing behavior for simply-supported reinforced concrete beams with small opening-photoelastic test and power capacity test [J]. Chinese quarterly of mechanics, 2013, 34(1): 101-113.

[4] HUANG T B, CAI J. Experimental study on simply supported reinforced concrete beams with rectangular web openings[J]. China Civil Engineering Journal, 2009, 42(10): 36-45.

[5] WU Y H, CHENG H D. Experimental study on seismic behavior of reinforced concrete frame with openings [J]. Earthquake Engineering and Engineering Vibration, 2004, 24(2): 75-81.

Longitudinal loading of socket connection with different bottom plate depth

Z. L. Tong & Y. Xu & S. Shivahari

State Key Laboratory for Disaster Reduction in Civil Engineering, Tongji University, Shanghai 200092, China

Abstract

Socket connection, as a main method of accelerate bridge construction is especially suitable for the municipal bridge and other bridges which have the limited construction period. And it does attract a lot of attention to its horizontal bearing capacity such as its seismic performance. Marsh and Motaref used cyclic loading to investigate the horizontal performance of the socket connection in 2011[1,2]. But its longitudinal bearing capacity is neglected.

This paper totally studied 3 socket connection specimens with different thickness of bottom plate and compared with the cast-in-place specimen. Bonding failure condition is also considered between the grouting material and socket end. Model sketch is shown in Figure 1, gray part is prefabricated cap; green part is grouting material and red part is prefabricated column. Height of cap is 75 cm, inserted depth of the columns are 63 cm, 50 cm, 35 cm and the corresponding thickness of the bottom plate is 12 cm, 25 cm, 40 cm, respectively.

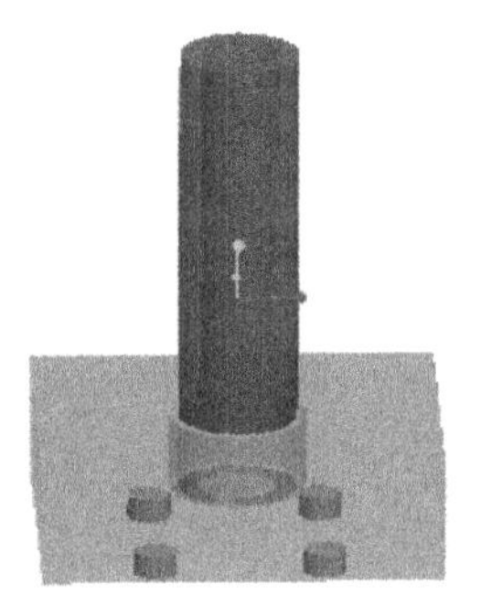

Figure 1 Model sketch

When bonding is well between grouting material and column end, the compress stress distribution of the whole structure is shown in Figure 2, longitudinal load is 5 000 kN, thickness of the bottom plate is 25 cm.

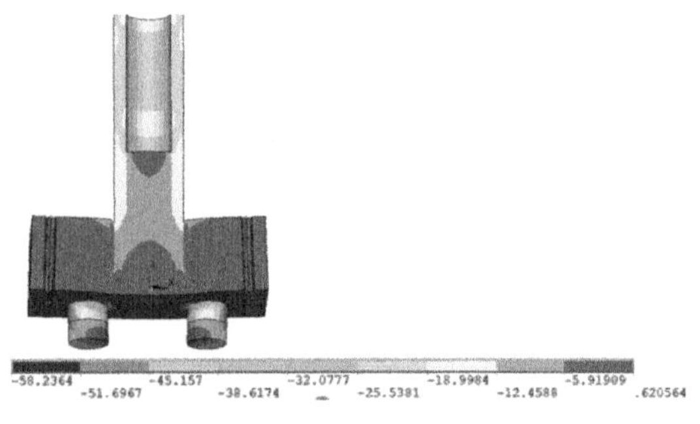

Figure 2 Compressive stress distribution under bonding well condition

When bonding failure between grouting material and column end under long service period, the distribution of compressive stress is shown in Figure 3.

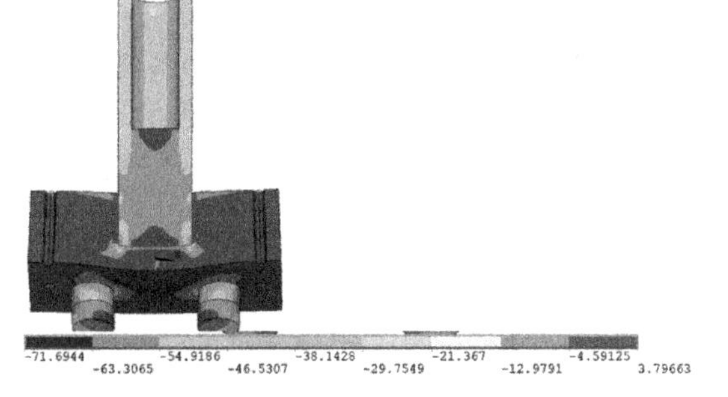

Figure 3 Compressive stress distribution under bonding failure condition

Comparing Figure 2 and Figure 3, after bonding failure between grouting concrete and column end, the main force is supported by the bottom plate of cap. After bonding failure, tensile stress of steel bars in the bottom plate increases from 78 MPa to 139 MPa. The tensile stress distribution is shown in Figure 4.

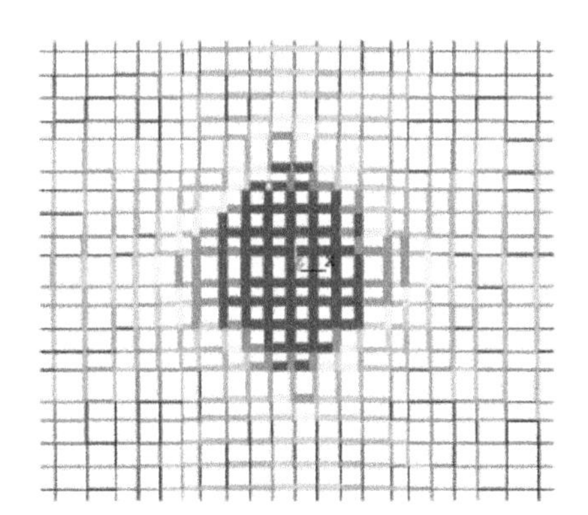

Figure 4 Bar stress distribution in the bottom plate of cap

About the longitudinal stiffness of each specimen, it can be reflected by the loading deformation, for the cast-in-place specimen, its longitudinal deformation is 0.52mm at the center of bottom plate. Deformation of other socket connection at the center of bottom plate is shown in Table 1. The longitudinal deformation of socket connection specimens is larger than the cast-in-place specimen and with the thickness of bottom plate decreasing, the longitudinal deformation increasing more and more.

Table 1 Deformation of socket connection at the center of bottom plate under bonding well condition

Thickness of bottom plate/cm	Longitudinal deformation/mm
40	0.74
25	0.76
12	0.87

When bonding failure between grouting material and column end, the longitudinal deformation is shown in Table 2.

Table 2 Deformation of socket connection at the center of bottom plate under bonding failure condition

Thickness of bottom plate/cm	Longitudinal deformation/mm
40	1.01
25	1.58
12	1.87

Comparing Table 1 and Table 2, after the cohesion invalid between grouting material and column end, the longitudinal deformation of the whole specimen increases about one time at the center of bottom plate.

Socket connection can have the comparative longitudinal bearing stiffness and capacity for the cast-in-place specimen. Bonding condition between grouting material and column end as an important factor influences the longitudinal performance of socket connection. When bonding is invalid, the longitudinal stiffness shows a discount.

References

[1] MARSH M L, WERNLI M, GARRETT B E, et al. Application of accelerated bridge construction connections in moderate-to-high seismic regions [R]. Washington, D. C.: National Cooperative Highway Research Program (NCHRP) Report No. 698, 2011.

[2] MOTAREF S, SAIIDI M S, SANDERS D. Seismic response of precast bridge columns with energy dissipating joints[R]. Center for Civil Engineering Earthquake Research, Department of Civil and Environmental Engineering, University of Nevada, Reno, Report No. CCEER-11-01, 2011.

Experimental study of the axial bearing capacity of prefabricated pier-cap with socket connection

Z. Zeng & Y. Xu
State Key Laboratory for Disaster Reduction in Civil Engineering, Tongji University, Shanghai 200092, China
Z. G. Wang
China Communications Second Highway Survey and Design Institute Co., Ltd., Wuhan 430058, China

Abstract

At present, Accelerated Bridge Construction (ABC) has been widely recognized and gradually promoted. This paper studies the mechanical behavior of a new type of pier-footing connection, namely socket connection, in the application of ABC. In the past decades, many scholars did a lot of research on socket connection. For example, Osanai et al.[1] studied the application of socket connection in the building structure and proposed that at least 1.5 times the column diameter (1.5*D*) could meet the requirements. M Mashal and A Palermo[2] studied the application of socket connection in bridge structure, not only verified the reliability of socket connection, but also suggested that the performance of 1.0*D* could be similar to that of cast-in-place structure. Haraldsson et al.[3] stated that the pier with 0.5*D* will be damaged by punching and shearing, and the footing cracks seriously, which means the vertical bearing capacity of the footing is insufficient. And then, scholars and engineers would like to believe that at least 1.0*D* was needed to ensure the performance of socket connection. In addition, Saiidi et al.[4-7] carried out not only quasi-static testing but also shaking-table testing in order to study socket connection and stated that socket connection may be a better choice for pier-footing connection or pier-cap beam connection.

As we all know, pile foundation is used extensively in China, so that the bottom plate is needed to be set under the pier when using socket connection, in order to meet the specification requirements of pile cap, that is, partial penetration. However, the configuration of socket connection above researches is usually not equipped with a bottom plate, that is, full penetration. Therefore, in order to apply the socket connection well in China, quasi-static testing of several specimens was conducted to study socket connection based on the existing research results, i.e. performance of socketed pier, reasonable embedment length and minimum reasonable embedment length were studied[8-10]. The results showed that the socket connection could achieve the similar effect of the cast-in-place, and the bottom plate could further reduce the embedment length. Obviously, it is necessary to study the performance of the structure with merely increasing vertical load to further demonstrate that the embedment length can be less than 1.0D. Therefore, this paper designs two specimens to study the vertical bearing capacity of the structure. Specially, in order to determine the vertical bearing capacity of the socket wall, the socket wall of specimen A is constructed with smooth surface and a bottom plate is set to study the vertical bearing capacity of the bottom plate, while the socket wall of specimen B is constructed with rough surface but no bottom plate is set to. The specimen is shown in Figure 1.

The maximum axial force is loaded to 10 000 kN, which is equal to section capacity of pier approximately and around 5 times of the designed axial load. In addition, the load increment of each grade is 500 kN. During the test of specimen A, when the axial force was loaded to 1 000 kN, cracks were found in the bottom plate and the maximum width was 0.062 mm, the width of 0.084mm when loaded to 1 935 kN, and the width of 0.101 mm when loaded to 2 500 kN. When loaded to 10 000 kN, the crack width reached 0.45 mm. However, no obvious damage was found in the pier and

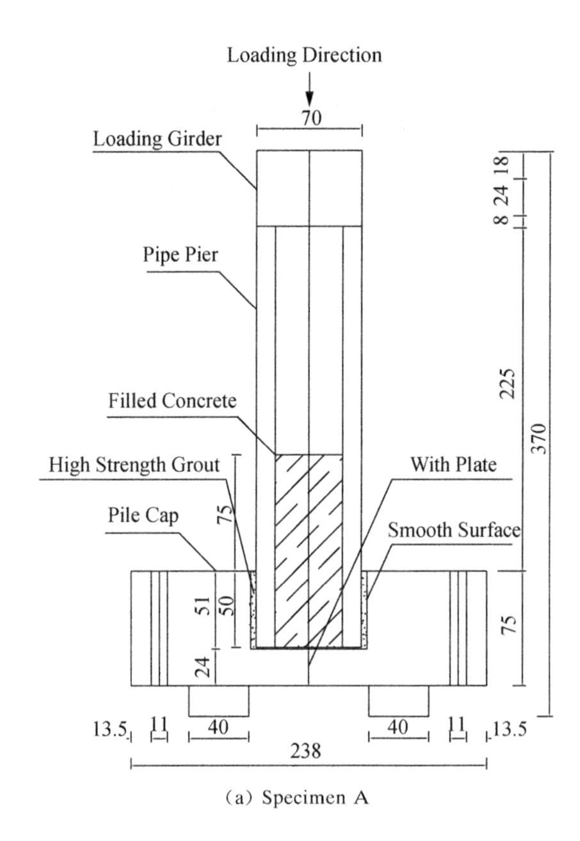

(a) Specimen A

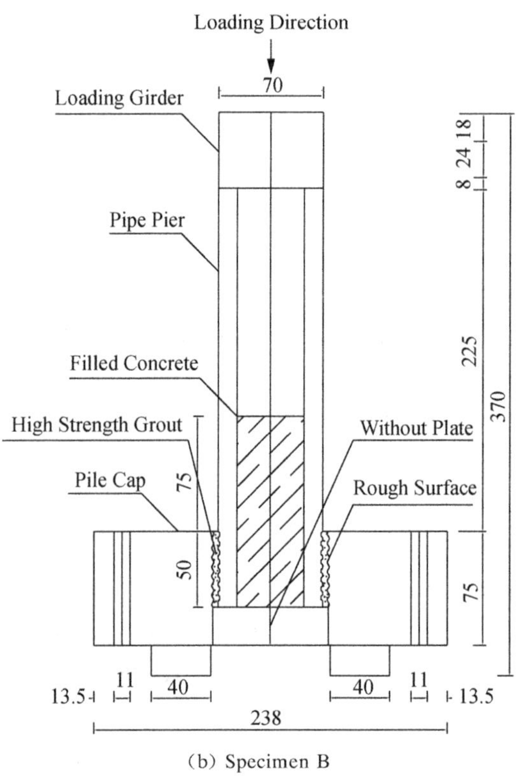

(b) Specimen B

Figure 1 Configuration of specimens (unit: cm)

the grouting material. For specimen B, when the axial force was loaded to 4 500 kN, a microcrack was found on the grouting material surface and its width was merely 0.033 mm, and when it was loaded to 10 000 kN, the maximum crack width of the grouting material was 0.08 mm. Simultaneously, the pier was not obviously damaged. Cracking is shown in Figure 2.

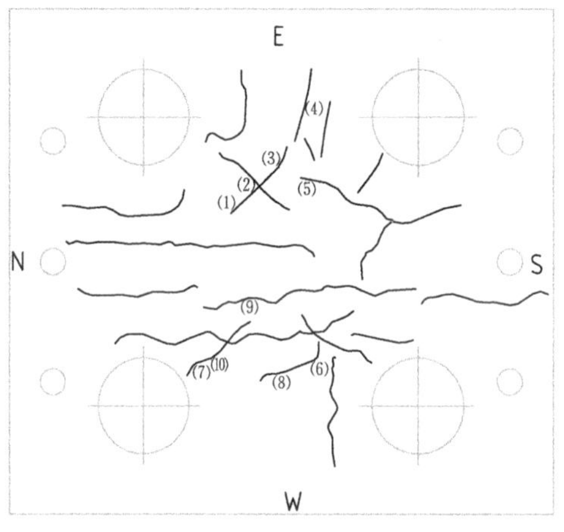

(a) Specimen A

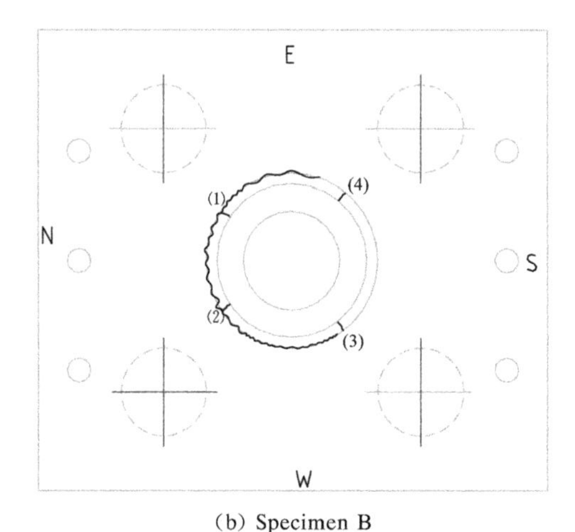

(b) Specimen B

Figure 2 Cracks of footing

In addition, the Figure 3 can be obtained according to the force and displacement monitored by the actuator. The curves show that the specimens did not break up until 10 000 kN. Interestingly, the curves are actually very close, which may mean that the vertical stiffness and strength of the bottom plate and the socket wall are relatively close. It is necessary to point out that the displacement may contain around 7 mm of actuator slip displacement so that there is an approximate flat line in the front section of the curve.

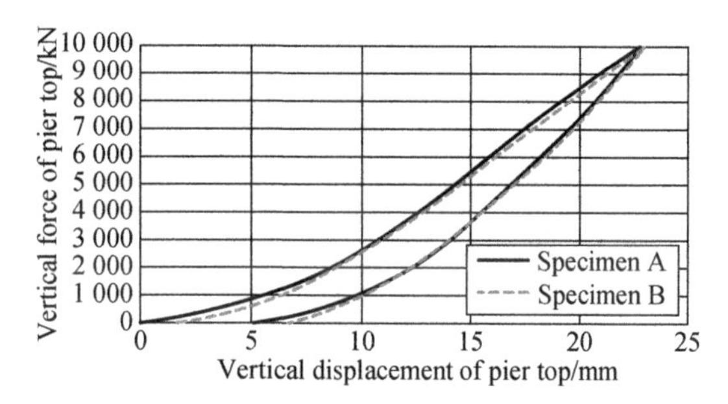

Figure 3 Vertical loading curve

According to the test results, the following conclusions can be obtained. (1) The vertical bearing capacity of both the bottom plate and the socket wall is more than 10 000 kN. (2) Based on the result of force-displacement relationship, the vertical stiffness and vertical strength provided by the bottom plate and the socket wall are little different. (3) Under the condition of only vertical loading, $0.7D$ can also meet the performance requirements. (4) It is reasonable to believe that if both the rough surface and the bottom plate are used in the socket connection, the vertical bearing capacity of the footing will be much larger than the section capacity of pier, so that only the pier can be damaged.

Acknowledgements

This research is funded by the National Science Foundation of China (Grant No. 51878492, Grant No. 51978511), and the National Key Research and Development Plan, China (Grant No. 2017YFC0806000).

References

[1] OSANAI Y, WATANABE F, OKAMOTO S. Stress transfer mechanism of socket base connection with precast concrete columns [J]. Journal of Structural & Construction Engineering, 1996, 93(3): 266-276.

[2] MASHAL M, PALERMO A. Quasi-static cyclic testing of half-scale fully precast bridge substructure system in high seismicity [C]// NZSEE Conference, 2014.

[3] HARALDSSON O S, JANES T M, EBERHARD M O, et al. Seismic Resistance of socket connection between footing and precast column [J]. Journal of Bridge Engineering, 2013, 18(9): 910-919.

[4] MEHRSOROUSH A, SAIIDI M S. Cyclic response of precast bridge piers with novel column-base pipe pins and pocket cap beam connections [J]. Journal of Bridge Engineering, 2016, 21(4): 04015080.

[5] MEHRAEIN M, SAIIDI M. Seismic performance of bridge column-pile-shaft pin connections for application in accelerated bridge construction. No. CCEER-16-01[R]. USA, Reno, NV: Department of Civil and Environmental Engineering, University of Nevada, 2016.

[6] MOHEBBI A, SAIIDI M S, ITANI A M. Shake table studies and analysis of a PT-UHPC bridge column with pocket connection [J]. Journal of Structural Engineering, 2018, 144(4): 04018021.

[7] MOHEBBI A, SAIIDI M S, ITANI A M. Shake table studies and analysis of a precast two-column bent with advanced materials and pocket connections[J]. Journal of Bridge Engineering, 2018, 23(7):04018046.

[8] XU Y, ZENG Z, GE J P, et al. The minimum reasonable pocket depth of precast pier-precast footing [J]. Journal of Tongji University (Natural Science), 2019. (Accepted)

[9] ZUO G H, HUANG Z Y, ZENG Y K, WANG Z G. Anti-seismic Performance Testing of Centrifugally Precast Pipe Pier with Socket Connection [J]. Structural Engineers, 2019. (Accepted)

[10] ZENG Z, WANG Z G, YU J, et al. Testing of Precast Pier with Different Socket Connection Construction [J]. Structural Engineers, 2019. (Accepted)

Dynamic performance analysis of Kaiyuan Pagoda based on ambient vibration test

F. Bai & N. Yang & S. Zhang

School of Civil Engineering, Beijing Jiaotong University, Beijing 100044, China

Abstract

The Kaiyuan Pagoda in Dingzhou has a pagoda height of 83.7 m and a perimeter of 128 meters. It was established in Song Dynasty and was the highest brick pagoda in the same period. As shown in Figure 1, the pagoda consists of a pagoda base, a pagoda body and a pagoda brake. It has eleven floor and is shrunk from bottom to top. In the tenth year of Guangxu reign (AD 1884), the northeast side of the pagoda collapsed (Figure 2(a)). In 2001, the collapsed part was repaired (Figure 2 (b)), while the first floor and the pagoda base were reinforced. Today, the Kaiyuan Pagoda stands on the land of China with its majestic style.

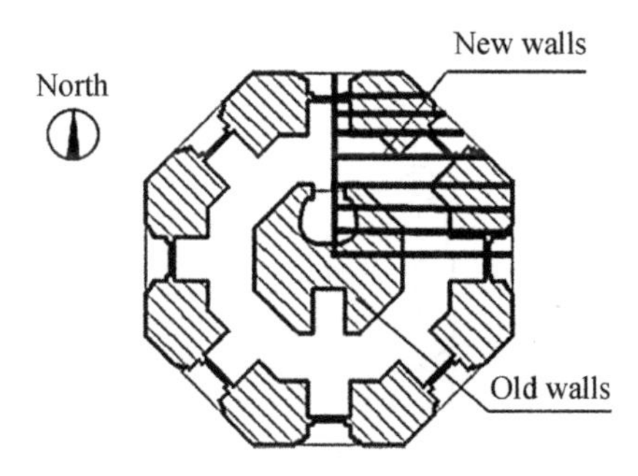

Figure 1 Plan of the top floor

The existing masonry pagoda reflects the essence of Chinese architectural art and is quite ingenious. It is not only an outstanding representative of ancient high-rise buildings, but also an important basis for studying ancient Chinese politics, religion and culture. However, due to earthquakes, winds, water disasters and man-made damage, a large number of masonry pagodas are seriously damaged over time. When the Pagoda is subjected to traffic vibration or earthquake, it is prone to damage or even collapse. Studying the damage of the existing masonry pagoda and assessing its safety limit is an urgent problem to be solved. The dynamic characteristics of the structure can effectively reflect the structural mechanical properties and can

(a) Before restoration

(b) After restoration

Figure 2 Photos of Kaiyuan pagoda before and after restoration

be also an important indicator to assess its health. In general, the dynamic characteristics of a structure can be obtained from established theoretical mechanical models or on-site measurements of structures. Yuan Jianli et al.[1] analyzed the dynamic characteristics of Suzhou Huqiu Pagoda under environmental excitation, and proposed a modeling method based on classical theory, test data and numerical simulation. Chen Ping[2, 3] conducted an analysis of the dynamic characteristics of Xi'an Da and Xiaoyan Pagoda, and conducted an analysis and evaluation of seismic capacity. Yang Qingshan[4], Li Tieying[5] and others researched the dynamic performance of the Yingxian wooden pagoda in Shanxi, and obtained the modal parameters of the wooden pagoda. Gao Yanan et al.[6] analyzed the dynamic response data of Feiyun Building using random subspace method improved by random decrement technique. Qin Shujie et al.[7] conducted a dynamic test on the Jingting wood structure with the characteristics of ancient Ming and Qing Dynasties in the Forbidden City in Beijing. The study also determined the type of damage of the structure and analyzed the cause of the damage[7].

Based on the ambient vibration test method, this paper proposes a new spatial test method for the ancient pagoda structure—the parallel test method. With this method, we can (1) obtain the spatial vibration mode and corresponding frequency of the Kaiyuansi pagoda. The vibration of each position of the structure is evaluated to discuss the application of the test scheme on similar ancient pagodas. (2) Based on the measured analysis results and structural dynamics theory, the inter-layer stiffness of the two orthogonal directions of the northeast and southeast sides of the ancient pagoda is reversed. The stiffness distribution of the structure is evaluated. (3) The finite element model of the ancient pagoda ABAQUS is established, and compared with the measured results, the numerical calculation model of the subsequent dynamic response analysis of the ancient pagoda is obtained.

The main results are shown in Table 1, Figure 3 and Figure 4.

Table 1 Dynamic characteristics of Kaiyuan Pagoda

Frequency /Hz	E-S	E-N	S-N	E-W	Torsional
First	0.844	0.906	0.844	0.844	3.188
Second	2.688	3.031	2.938	2.938	7.688
Third	3.313	3.250	—	3.125	13.06

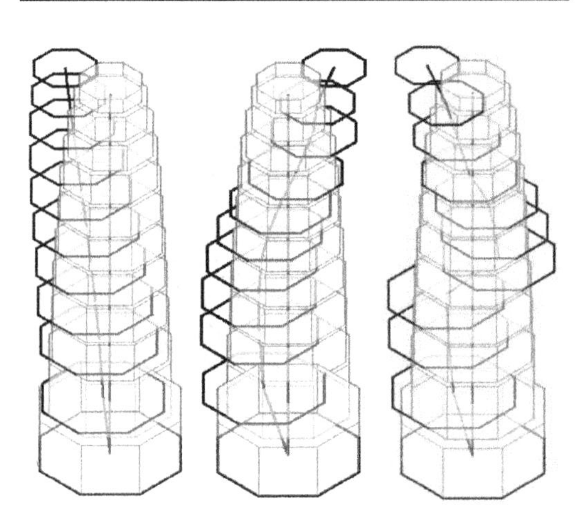

Figure 3 Three order mode shape in south-east direction

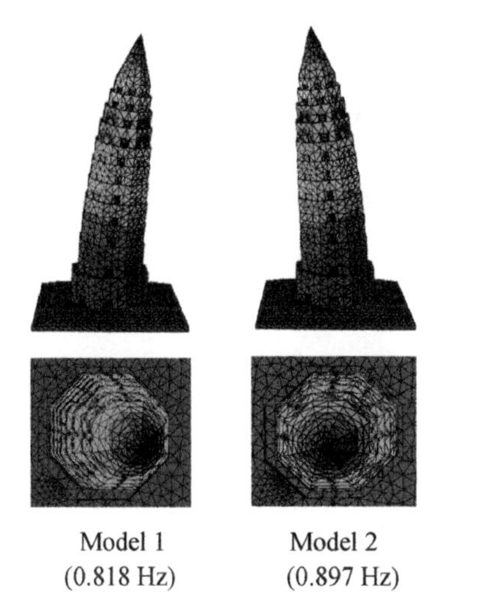

Figure 4 Finite element model of the Pagoda

Acknowledgements

The study presented in this paper was supported by National Natural Science Foundation of China for Excellent Young Scholars (NSFC 51422801), Beijing Natural Science Foundation of China (Key Program 8151003), National Natural Science Foundation of China (Key Program NSFC 51338001), National Natural Science Foundation of China (General Program NSFC 51178028), National Key Technology R & D Program (2015BAK01B02) and 111 Project of China (B13002).

References

[1] YUAN L J, et al. Experimental study of dynamic behavior of Huqiu Pagoda [J]. Engineering Mechanics, 2005(5):158-164.

[2] CHEN P, YAO Q F, ZHAO D. A study on the aseismic behavior of Xi'an dayan pagoda [J]. Journal of Building Structures, 1999, 20 (1):46-49.

[3] CHEN P, YAO Q F, ZHAO D. An exploration of the aseismic behavior of Xi'an xiaoyan pagoda[J]. Journal of University of Architecture and Technology, 1999, 31(2): 49-51.

[4] CHEN B, YANG Q S, WANG K, et al. Full-scale measurements of wind effects and modal parameter identification of Yingxian wooden pagoda [J]. Wind and Structures, 2013, 17(6): 609-627.

[5] LI T Y, et al. Experiment and analysis of vibration characteristics of yingxian wooden pagoda[J]. Engineering Mechanics, 2005, 22(1): 141-146.

[6] GAO Y A, YANG Q S, et al. Dynamic performance of the ancient architecture of Feiyun pavilion under the condition of environmental excitation [J]. Journal of Vibration and Shock, 2015(22): 144-148.

[7] QING S J, YANG N, et al. Study on dynamic characteristics of damaged wooden structures in Ming and Qing Dynasties[J]. Journal of Building Structure, 2018(10):130-137.

Model for galloping behavior analysis of iced conductors involving plunge, twist and swing

X. Guo & J. Chen & Y. Peng
Tongji University, Shanghai 200092, China

Abstract

As to the transmission line accreted with ice, it is very likely to arise a kind of vibration with large amplitude and low frequency, which is the so-called galloping. The amplitude ranges typically from 0.1 to 1.0 times the sag of the line[1], and might result in break of conductors, damage of connectors or even collapse of the tower. Therefore, galloping of iced-conductors has been attached extensive attention in recent years.

The first galloping model was proposed in 1932 by Den Hartog[2], which revealed the classical vertical galloping mechanism. According to field measurement, however, the twist of the transmission line may also contribute to the initiation of galloping. The importance of twist was later emphasized by Nigol and Buchan, and the torsional galloping mechanism was then proposed[3]. Besides, the horizontal oscillation along wind direction was proved to be essential to initiate the vertical galloping when it is coupled to the plunge[4]. In this paper, a simplified three-degree-of-freedom model (3DoF model) for describing the galloping behavior involving plunge, twist and swing is developed, as shown in Figure 1.

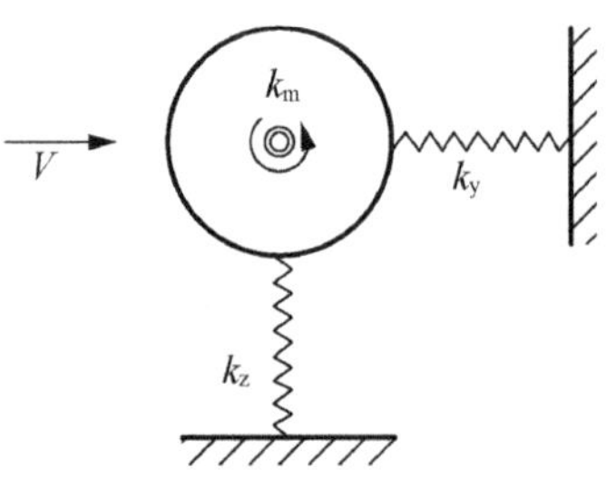

(a) Mechanical elements

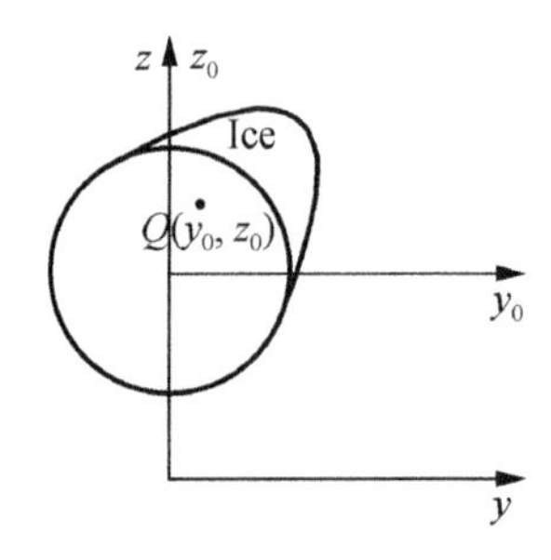

(b) Galloping motion

Figure 1 Three-degree-of-freedom model representing iced conductors

In this model, the cross-section eccentricity caused by the coated ice is considered, and the interactions between the plunge, twist and swing of the conductor along wind direction are included as well. Based on Lagrange's equation, the equation of dynamic motion of the 3DoF model is derived as follows.

$$\begin{bmatrix} m_0 & & S_{y0} \\ & m_0 & -S_{z0} \\ S_{y0} & -S_{z0} & I_0 \end{bmatrix}\begin{Bmatrix} \ddot{y} \\ \ddot{z} \\ \ddot{\varphi} \end{Bmatrix}+\begin{bmatrix} 2m_0\xi_y\omega_y & & \\ & 2m_0\xi_z\omega_z & \\ & & 2I_0\xi_m\omega_m \end{bmatrix}\begin{Bmatrix} \dot{y} \\ \dot{z} \\ \dot{\varphi} \end{Bmatrix}+$$

$$\begin{bmatrix} k_y & & \\ & k_z & \\ & & k_m \end{bmatrix}\begin{Bmatrix} y \\ z \\ \varphi \end{Bmatrix}=\begin{Bmatrix} \frac{1}{2}\rho_0 DV_r^2 C_y(\alpha) \\ \frac{1}{2}\rho_0 DV_r^2 C_z(\alpha) \\ \frac{1}{2}\rho_0 D^2 V_r^2 C_m(\alpha) \end{Bmatrix} \tag{1a}$$

$$\begin{cases} C_y(\alpha)=C_L(\alpha)\sin\beta+C_D(\alpha)\cos\beta \\ C_z(\alpha)=C_L(\alpha)\cos\beta-C_D(\alpha)\sin\beta \\ C_m(\alpha)=C_M(\alpha) \\ \alpha=\alpha_0+\varphi-\beta,\ \beta=\arctan[\dot{z}/(V-\dot{y})] \\ V_r=\sqrt{(V-\dot{y})^2+\dot{z}^2} \end{cases} \tag{1b}$$

where m_0 denotes the mass per unit length of iced conductor; S_{y0} denotes the first mass moment of area with respect to y_0 axis; S_{z0} denotes the first mass moment of area with respect to z_0 axis; I_0 denotes the mass moment of inertia per unit length; ξ_y, ξ_z and ξ_m denote the damping ratio along y, z and φ direction, respectively; ω_y, ω_z and ω_m denote the natural frequency along y, z and φ direction, respectively; k_y, k_z and k_m denote the stiffness along y, z and φ direction, respectively; α_0 denotes the static torsional angle; φ denotes the dynamic torsional angle; V is the wind velocity; ρ_0 denotes the density of air; D denotes the diameter of conductor; C_L, C_D and C_M denote the aerodynamic coefficient of lift, drag and torsional moments, respectively.

The galloping behavior of a D-shaped iced conductor is simulated using the proposed 3DoF model, which was experimentally investigated in the previous study[5], as shown in Figure 2.

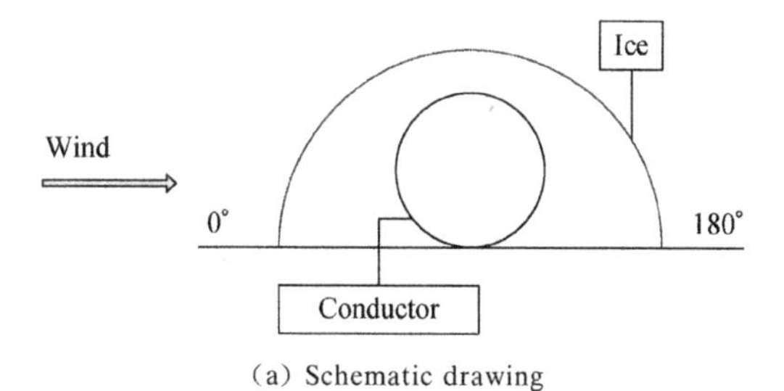

(a) Schematic drawing

(b) Aerodynamic coefficient against wind attack angle

Figure 2 D-shaped iced conductor

Using the Runge-Kutta algorithm, the equation of motion of the model can be solved efficiently. The galloping displacements of the D-shaped iced conductor along directions of plunge, twist and swing are shown in Figure 3, respectively. It is seen that the amplitude of galloping is increasing with time going, and it eventually achieves at a steady-state value.

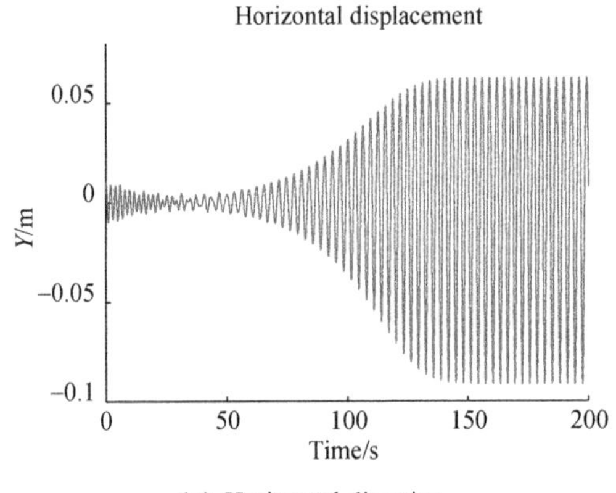

(a) Horizontal direction

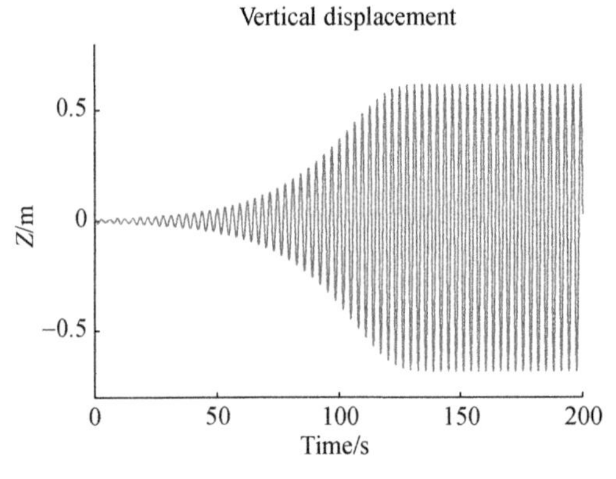

(b) Vertical direction

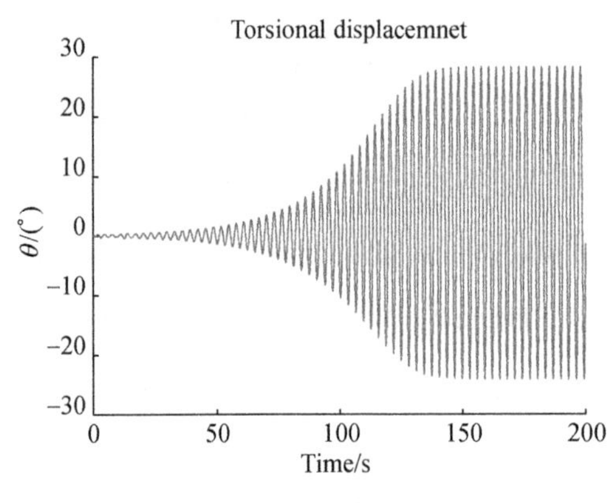

(c) Torsional direction

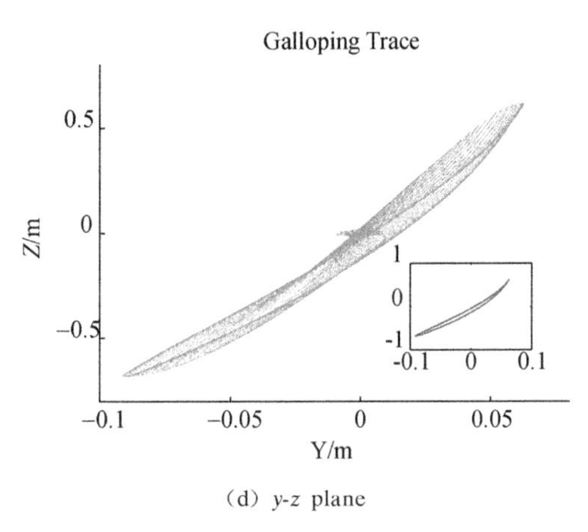

(d) y-z plane

Figure 3 Galloping displacement along

Table 1 presents the numerical result and its error in comparison with the data from wind tunnel test. It is seen that the vertical galloping amplitude by the proposed 3DoF model shows a consistence with that from the experimental test. One might recognize that the proposed model remains the aerodynamic nonlinearity although it ignores the geometrical and material nonlinearities. Therefore, the proposed 3DoF model provides a reliable mean to understand the galloping behaviors of iced conductors as well as to underlie the refinement control of galloping motion.

Table 1 Comparison between numerical result and wind tunnel test

	Maximum vertical displacement /m	Maximum torsional angle (°) and relative error
Wind tunnel test	0.67	—
Proposed method	0.681 3 (0.2%)	28.30

Acknowledgements

The supports of the National Key R &

D Program of China (Grant No. 2017YFC0803300) and the National Natural Science Foundation of China (Grant Nos. 11672209, 51725804, and 51878505) are highly appreciated.

References

[1] WANG J W, LILIEN J L. Overhead electrical transmission line galloping. A full multi-span 3-DOF model, some applications and design recommendations [J]. IEEE Transactions on Power Delivery, 1998, 13 (3): 909-916.

[2] DEN HARTOGs J P. Transmission line vibration due to sleet[J]. Transactions of the American Institute of Electrical Engineers, 1932, 51(4): 1074-1076.

[3] NIGOL O, BUCHAN P G. Conductor galloping-part II Torsional mechanism [J]. IEEE Transactions on Power Apparatus and Systems, 1981, PSA-100(2): 708-720.

[4] JONES K F. Coupled vertical and horizontal galloping [J]. Journal of engineering mechanics, 1992, 118(1): 92-107.

[5] STUMPF P, NG H C M. Investigation of aerodynamic stability for selected inclined cables and conductor cables[D]. Manitoba: University of Manitoba, Canada, 1990.

Finite element simulation of flexural behavior of storage rack beam-to-upright connections

L. S. Dai

Department of Civil Engineering, Shanghai University, Shanghai 200444, China

Abstract

Substantial experimental investigations[1, 2] have been carried out to study the flexural behavior of steel storage rack beam-to-upright connections. However, the experimental procedures are expensive and time-consuming. Recently, the fracture theory has been developed for ductile metals[3, 4], and is gradually being used to predict the strength and post-ultimate behavior of hot-rolled steel beam-to-upright connections[5]. However, very limited research can be found on the fracture simulation of cold-formed steel members and structures. Therefore, high accuracy numerical analysis, incorporating the geometrical nonlinearity, material nonlinearity and fracture, is required to predict the full-range moment-rotation behavior of beam-to-upright connections in steel storage racks.

This paper presents a finite element (FE) analysis of a typical steel storage rack beam-to-upright boltless connection. A three-dimensional numerical model is established via the finite element (FE) package ABAQUS considering the realistic interactions among different structural components, as illustrated in Figure 1. A calibrated fracture model is employed to research the possible fracture of the tab and the tearing of the upright wall.

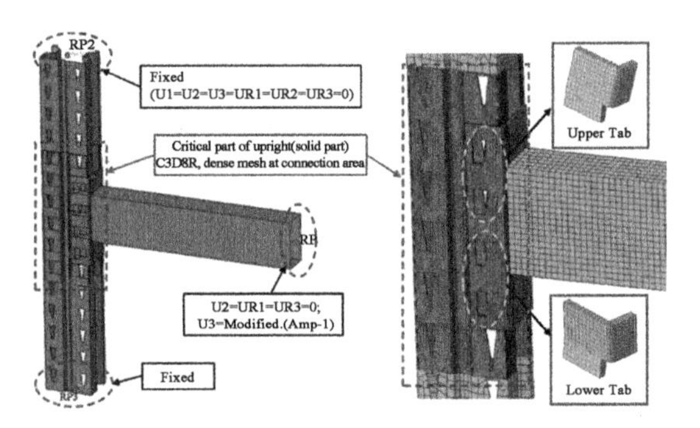

Figure 1 FE model for a typical boltless connection

The test results of the connections presented in reference[6] are used to verify the established models. Comparisons of connection behaviors between the finite element models and the test results are performed in terms of the failure mode and the moment-rotation curve, as shown in Figure 2. The comparison results demonstrated that the established numerical model can successfully simulate the typical failure mechanisms and satisfactorily predict the full-range moment rotation behavior of connections.

Finally, based on the developed model, the influences of the gap between

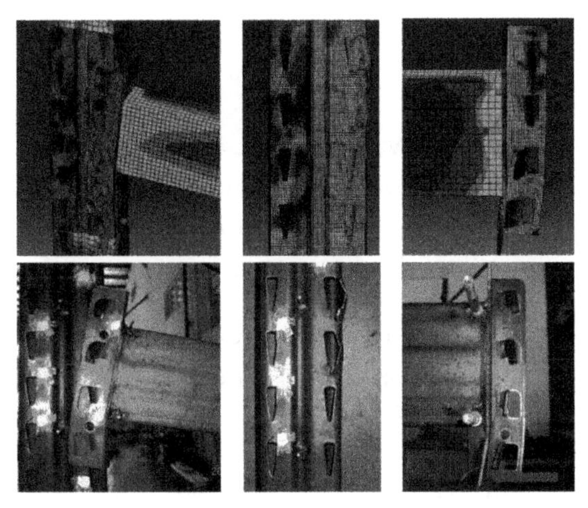

(a) Failure mechanism — Tearing of upright wall

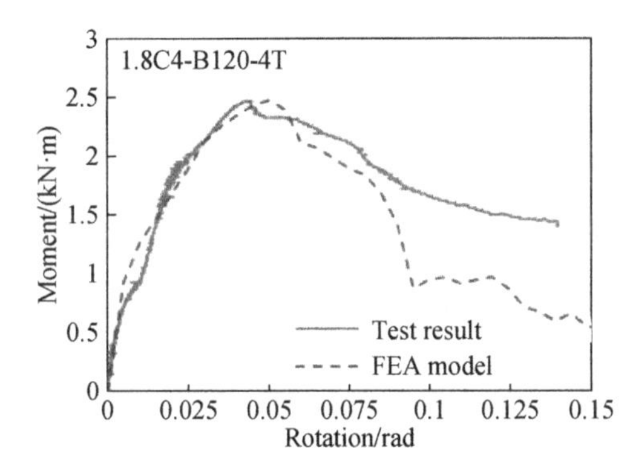

(b) Moment-rotation curve

Figure 2 Comparation between FEA results and the test data for a typical boltless connection "1. 8C4-B120-4T"

end-connector and upright flange are examined. Six values of the gap, i.e. 0. 0 mm, 0. 5 mm, 1. 0 mm, 1. 5 mm, 2. 5 mm and 5. 0 mm, were selected to investigate the effects of the gap on the connection behavior. The connection type "1. 8C4 - B105-4T" is used as an example. Based on the established FE model, six finite element models with the varied gap values were established and analysed. The numerical moment-rotation relationships of these connections are illustrated in Figure 3. It can be seen from the figures that the gap values, to varying extent, influence the connection behavior.

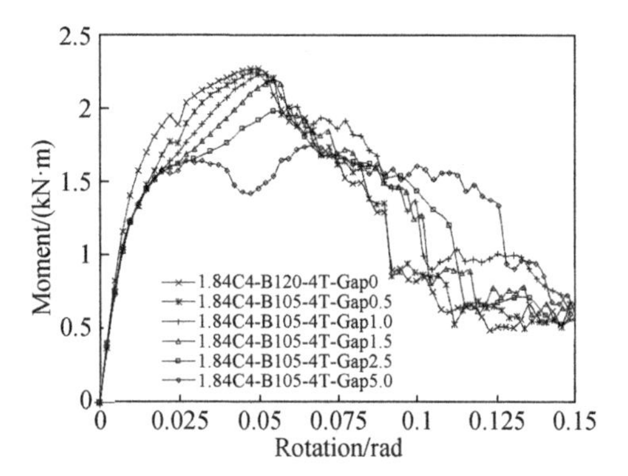

Figure 3 Moment-rotation relationships of connections with varied values of the gap

This study presents a refined numerical investigation of storage rack beam-to-upright boltless connections. The FE model is validated against the test results in terms of the possible failure mechanisms and the full-range moment-rotation curves. The results show that the FE model can successfully simulate the failure mechanisms of boltless connections, i.e. tab crack and the tearing of the upright wall, and can satisfactorily predict connections' full-range moment-rotation behavior including the post-ultimate range. Finally, the proposed FE model is employed to evaluate the effects of the gap between end connector and upright face. The developed FE model can provide a numerical method to be used in the evaluation and design of mechanical connections in steel storage racks.

References

[1] DAI L, ZHAO X, RASMUSSEN K J R. Flexural behavior of steel storage rack beam-

to-upright bolted connections [J]. Thin-Walled Structures, 2018, 124: 202-217.

[2] DAI L, ZHAO X, RASMUSSEN K J R. Cyclic performance of steel storage rack beam-to-upright bolted connections [J]. Journal of Constructional Steel Research, 2018, 148: 28-48.

[3] WIERZBICKI T, BAO Y B, LEE Y W, BAI Y L. Calibration and evaluation of seven fracture models[J]. International Journal of Mechanical Sciences, 2005, 47(4-5): 719-743.

[4] YAN S, ZHAO X. A fracture criterion for fracture simulation of ductile metals based on micro-mechanisms [J]. Theoretical and Applied Fracture Mechanics, 2018, 95: 127-142.

[5] WANG W, FANG C, QIN X, et al. Performance of practical beam-to-SHS column connections against progressive collapse[J]. Engineering Structures, 2016, 106: 332-347.

[6] ZHAO X H, WANG T, CHEN Y, et al. Flexural behavior of steel storage rack beam-to-upright connections [J]. Journal of Constructional Steel Research, 2014, 99: 161-175.

Numerical investigation on the tensile behavior of carbon fabric reinforced cementitious mortar using discrete element method

C. Q. Zeng & J. H. Zhu

Guangdong Province Key Laboratory of Durability for Marine Civil Engineering, College of Civil and Transportation Engineering, Shenzhen University, Shenzhen 518060, China

Abstract

Carbon-Fabric Reinforced Cementitious Matrix (C-FRCM) uses carbon fiber fabric as reinforcement, cement mortar as filler material, which has not only excellent mechanical strength, but also good electrical conductivity and formability. Researches have shown that the use of C-FRCM composites as anode material to retrofit reinforced concrete structures, which has the function of both anti-corrosion and structural strengthening[1], therefore significantly improves the durability of coastal infrastructure. However, several factors affect the durability of the C-FRCM anode. One of the major factors is the degradation of the bond strength caused by anodic polarization[2]. Anodic polarization would cause decrease of the bonding strength between the carbon fiber fabric and the cement matrix, and also the tensile strength of the C-FRCM would be affected.

In order to investigate the effect of anodic polarization on the tensile properties of the C-FRCM, Discrete element method (DEM) base model was developed. DEM is a numerical technique initially introduced by Cundall in 1971 for the study of rock mechanics[3]. Compared to traditional finite element method (FEM), DEM has some advantages especially for modelling fracture of materials. There are no limits in displacement continuity or compatibility conditions. Along with the development of DEM and its advantages, nowadays researchers begin to extend DEM to model other materials such as soils, hot mix asphalt, Portland cement concrete, and polymer composites.

The DEM model developed in this article was based on geometric configuration of C-FRCM. Voronoi meshes were used to discretize the mortar matrix. Rectangular meshes were used for carbon fiber fabric. The constitutive behavior of mortar matrix and carbon fiber fabric was calibrated based on experimental results. The material properties of the cement matrix and carbon fiber fabric are shown in Table 1.

Numerical uniaxial tensile tests were performed to study the decrease of the tensile strength of C-FRCM due to the decrease of bonding strength between carbon fiber fabrics and cement binder due to anodic polarization.

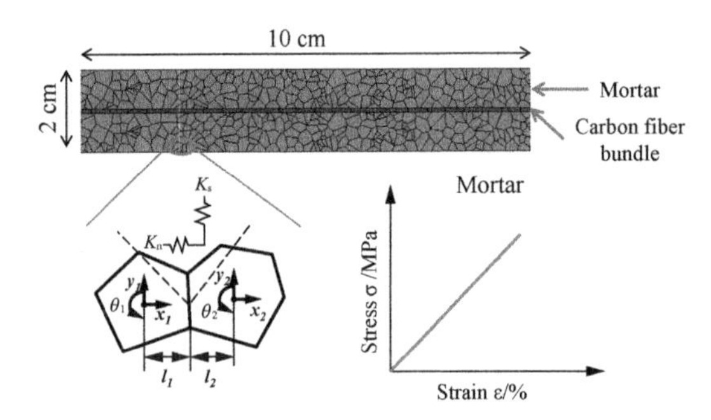

Figure 1 DEM model of C-FRCM composite and the corresponding material model

Table 1 Material properties of C-FRCM

Material	Properties	Symbol	Value
Carbon fiber bundle	Tensile strength/MPa	f_{cf}	2 125
	Elastic modulus/GPa	E_{cf}	196.4
	Strain-to-failure/%	ε_{cf}	1.1
	Cross-sectional area/mm^2	A_{cf}	0.462
Cement matrix	Tensile strength/MPa	f_{cm}	4.2
	Elastic modulus/GPa	E_{cm}	17.6
	Strain-to-failure/%	ε_{cm}	0.01
	Cross-sectional area/mm^2	A_{cm}	520

The numerical simulation was performed with LMGC90 simulation software. LMGC90 is an open source platform developed by French lab LMGC using Fortran 90 language[4]. It is capable of simulating composite materials using both FEM and DEM. The chosen numerical parameters of the simulation were listed in Table 2. An implicit integration scheme was used to ensure the convergence of the numerical scheme.

Table 2 Parameters of the numerical simulation

Parameters	Symbol	Value
Mesh number	N	2 304
Time step	$\mathrm{d}t$	10^{-4} s
Total steps	N_t	10^5
Implicit	—	Yes
Interaction law	—	IQS_CZM
Velocity	c	10^{-4} m/s

The results of the DEM simulation are shown in Figure 2. It can be seen from the figure that a weaker bonding strength would lead to a slippage of the carbon fiber bundle after cracking of the mortar. Therefore the tensile strength is greatly reduced. This is because the load is applied on the matrix. In order to transfer the applied load to the reinforced material, the carbon fiber fabric, the interface should be strong. When the strength of the interface decreases due to anodic polarization, this transfer mechanism becomes weaker. For very weak bond strength, the applied load cannot be transferred to the reinforced material; the failure mode is the cracking of the mortar matrix and slippage of the carbon fiber fabric. The low strength range at large deformation observed in Figure 2 referred to the slippage of the carbon fiber fabric. This strength depends on the friction coefficient between the carbon fabric and mortar matrix.

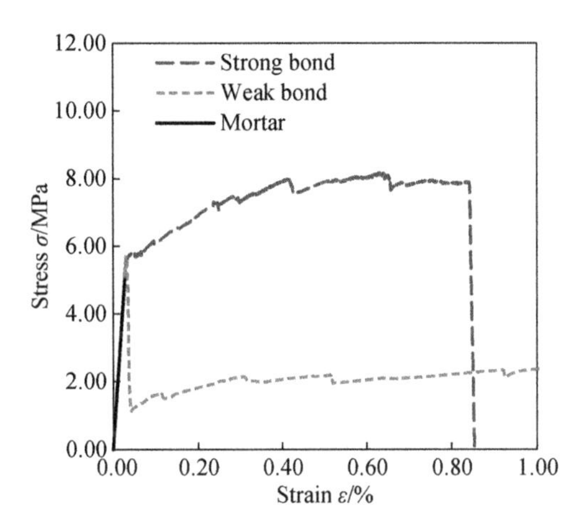

Figure 2 Effect of bonding strength on the tensile behavior of C-FRCM

Since anodic polarization would lead to a reduction on the bonding strength between carbon fiber fabric and mortar matrix, it would also affect the tensile strength of C-FRCM. In order to quantify this influence, a bond degradation parameter β is defined:

$$\beta = 1 - \frac{\sigma_d}{\sigma_0} \quad (1)$$

where σ_0 is the initial bond strength, σ_d is the bond strength after damage caused by anodic polarization. With this definition, the degradation of the bond strength is normalized by the initial bond strength. Figure 3 shows the relationship between the tensile strength of C-FRCM and bond degradation parameter β. It could be seen from the figure that the tensile strength decreases with increasing damage of the interface.

With these results above, it's now possible to characterize the effect of the anodic polarization on the tensile strength of the C-FRCM when applied as both anode material and structure strengthening material for RC structures. Another numerical model was developed using Comsol Multiphysic software to compute the degradation of the bond strength regarding to input electric current density and duration of energization using FEM model[5]. Therefore combining the results of FEM model and DEM model, it is possible to predict the tensile behavior of C-FRCM with input current density and duration of energization, which is critical for the design of ICCP-SS method using C-FRCM composite[6].

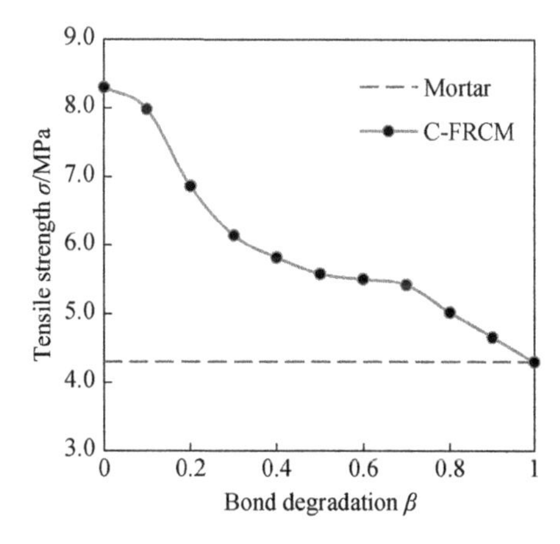

Figure 3 Effect of bond degradation on the tensile strength of C-FRCM

Acknowledgements

We would like to appreciate the support from the Chinese National Natural Science Foundation (Grant No. 51778370, Grant No. 51538007), Natural Science Foundation of Guangdong (Grant No. 2017B030311004), the Shenzhen science and technology project (JCYJ20160308104259253, JCYJ2017081809-4820689).

References

[1] ZHU J H, ZENG C, SU M, et al. Effectiveness of a dual-functional intervention method on the durability of reinforced concrete beams in marine environment[J]. Construction and Building Materials, 2019, 222(7): 633-642.

[2] ZHANG E Q, ABBAS Z, TANG L. Predicting degradation of the anode-concrete interface for impressed current cathodic protection in concrete [J]. Construction and Building Materials, 2018, 185(6): 57-68.

[3] CUNDALL P A, STRACK O D. A discrete numerical model for granular assemblies[J]. Geotechnique, 1979, 29(2): 47-65.

[4] DUBOIS F, JEAN M. LMGC90: une plateforme de développement dédiée à la modélisation des problèmes d'interaction[M]// In M. Pottier-Ferry, M. Bonnet et A. Bignonnet, éditeurs. sixième Colloque National en Calcul des Structures, Giens, 2003, 1: 111-118.

[5] ZENG C, ZHU J, WEI L. FEM model of bonding strength decrease caused by anodic polarization [J]. Construction and Building Materials. (Submitted)

[6] WEI L, ZHU J, UEDA T, et al. Tensile behavior of carbon fabric reinforced cementitious matrix composites as both strengthening and anode materials [J]. Composite structures. (Submitted)

Study on mounting site of micro-wind turbines at tall building's flat roof

L. Chen & H. B. Xiong

Department of Disaster Mitigation for Structures, Tongji University, Shanghai 200092, China

Abstract

Growing public awareness of utilizing renewable energy has led to popularity to adopt wind energy in building environments[1]. Distributed wind systems, which are located near the consumption center, can provide clean, renewable power for on-site use and have gained rising attention in the world recently[2]. There is an excellent potential for wind energy utilization on tall buildings rooftop for micro-wind turbines installation. Nevertheless, the installation of micro-wind turbines in urban areas is limited because the lack of accurate estimation of wind potential. Some researchers have used Computational Fluid Dynamic (CFD) method to address wind potential of different roof shapes on the same cubic building, including flat, domed, and pyramidal shapes[3]. Roof mounting site analysis for micro-wind turbines has found that flat roofs are likely to yield higher and more consistent power[4]. The study about wind energy on a flat roof with staircase is limited.

This paper presented the CFD simulation of the wind flow over a tall building's roof with two staircase out of the roof. The optimum mounting sites of turbines were researched by evaluating the vertical distribution of wind speed and turbulence intensity. In addition, the influence of non-staircase on wind flow characteristics was compared.

The cross-section of target building is shown in Figure 1. The cuboid domain is used with the size settings refer to reference [5]. The distance from the windward side of the building to the inlet is $5H$, from both sides of the calculation domain to the building's lateral is $5H$, from domain's top to the building's roof is $5H$, and from the outlet to the back of the building is $15H$. Where H is the height of the building. The block rate of the calculation domain is less than 5%, which meets the requirement of numerical calculation.

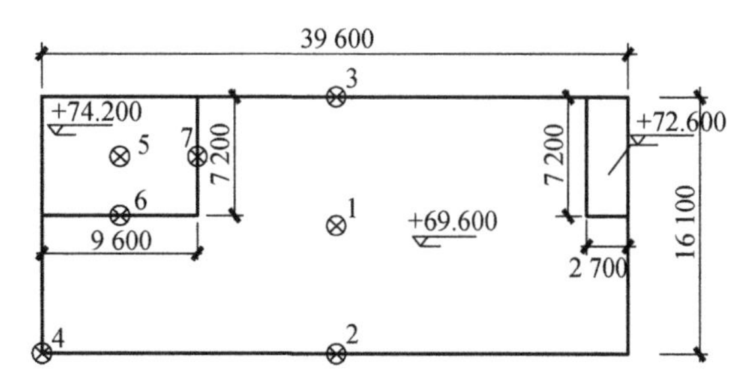

Figure 1 Potential locations layout

Unstructured prism grids are divided in ANSYS ICEM. Boundary layer grids are set around building with 0.1 meters initial height and 1.15 times growth rate. The grid has a total amount of about 120 000 cells. Fluid dynamics simulation was carried in ANSYS FLUENT. The boundary conditions are presented in Table 1. The velocity inlet satisfied Eqs. (1) to (3), which were stream wise velocity profile, turbulent kinetic energy (TKE), and turbulent dissipation rate (TDR) respectively.

Table 1 Boundary conditions for the simulation domain

Boundary	Boundary setup
Inlet	Velocity inlet as depicted in Eq. (1) with user defined function(UDF)
Outlet	Outflow conditions are adopted
Top and lateral sides	Symmetry boundary conditions are applied
Ground and building surface	Wall boundary with standard wall function

$$U(z) = \frac{1}{\kappa} u^* \ln\left(\frac{z + z_0}{z_0}\right) \quad (1)$$

$$k = \frac{u^{*2}}{\sqrt{C_\mu}} \sqrt{C_1 \ln\left(\frac{z + z_0}{z_0}\right) + C_2} \quad (2)$$

$$\varepsilon = \frac{u^{*3}}{\kappa(z + z_0)} \sqrt{C_1 \ln\left(\frac{z + z_0}{z_0}\right) + C_2} \quad (3)$$

Where κ is the von Karman constant, and taken as 0.42. u^* is the friction velocity, z_0 is the aerodynamic roughness length. And $C_\mu = 0.028$, $C_1 = 1.5$, $C_2 = 1.92$. The bottom boundary is modeled as a rough wall and standard wall functions are used.

A critical precondition affecting the consistency of the CFD simulation is the horizontal homogeneity of domain. The vertical profiles of wind speed, turbulent dissipation rate (TDR) and turbulent kinetic energy (TKE) at the inlet, 1/4 stream wise length, 1/2 stream wise length, 3/4 stream wise length, outlet were shown in Figure 2.

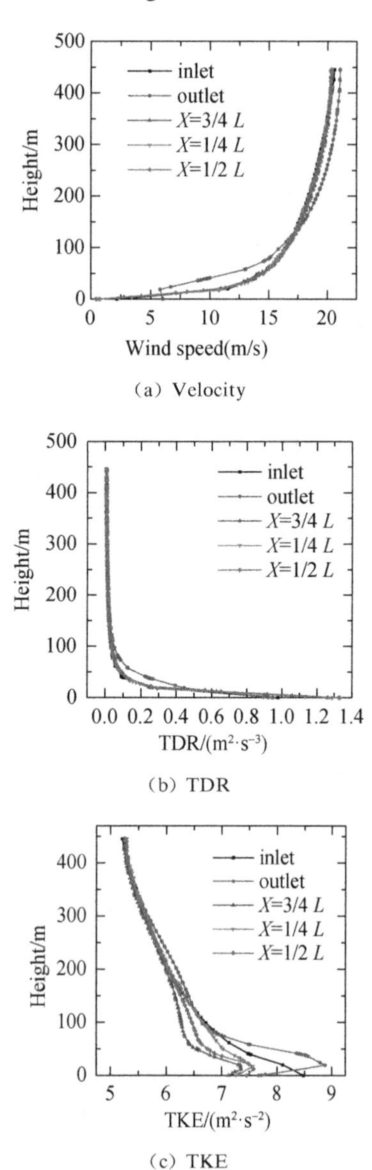

Figure 2 Comparison of velocity, TDR and TKE at different location of stream wise

Turbulence intensity affects the normal operation and service life of the micro-turbines, and service life would be shortened if turbulence intensity exceeds the threshold of 16% ~18%. Wind speed influences the efficiency of wind power generation. The variations of turbulence intensity and wind speed with height were shown in Figure 3.

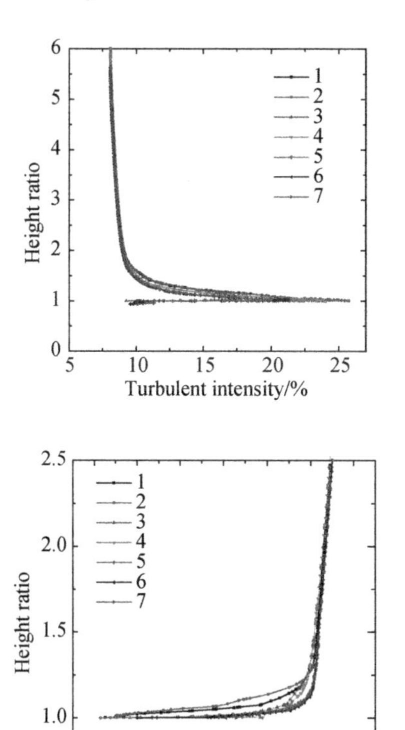

Figure 3 The variations of turbulence intensity and wind speed with height

The turbulence intensity increases sharply with height at first and reaches its peak value following with slowly decrease, and wind speed increases sharply at first with the increase of height ratio, which indicates that a small rise in height will bring a significant increase in wind energy yielding, and then the wind speed increases flatly. The minimum installation height ratio of turbines was shown in Table 2.

Table 2 Height ratio limits for installation

Wind direction	Location						
	1	2	3	4	5	6	7
North	1.15	1.15	—	1.11	1.06	1.07	1.07
Southwest	1.22	1.21	1.21	1.14	1.12	1.12	1.13

Results showed that normal operation of micro-wind turbines can be satisfied without affecting its service life as long as the installation height is higher than $1.15H$. Figure 4 showed the distribution of turbulence intensity and wind speed at $Z = 1.15H$.

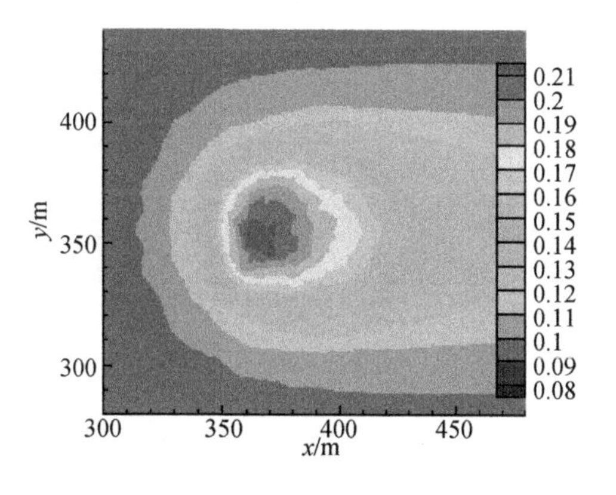

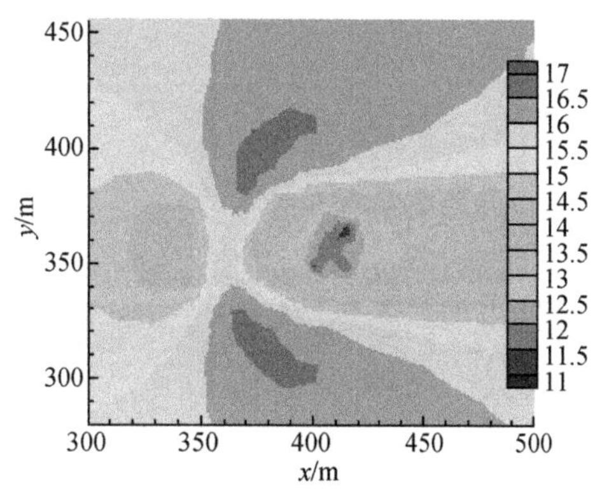

Figure 4 The distribution of turbulence intensity and wind speed at $Z = 1.15H$

High-rise buildings generally have staircases out of the roof for the necessity of vertical traffic inside, which would cause differences of wind flow on the roof compared with the ordinary flat roof of the low-rise and multi-story buildings. To compare the influence of roof staircase on the utilization of wind energy, two staircases out of the roof of the target building are removed and re-simulated. Take the cross section $Z = 71.6$ m through the stairwell of the roof as an example. The distribution of wind speed of this section was shown in Figure 5.

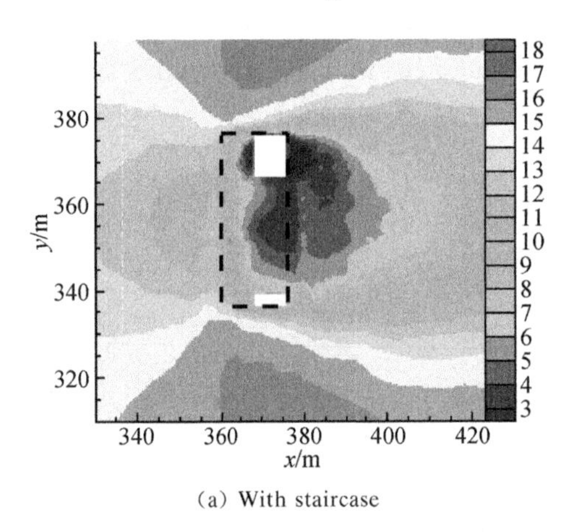

(a) With staircase

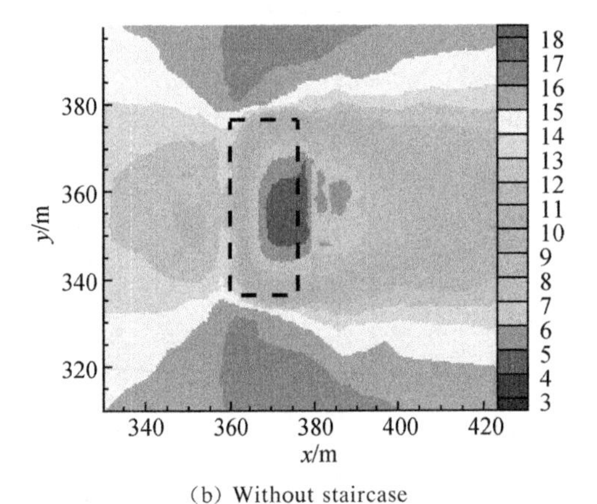

(b) Without staircase

Figure 5 The distribution of wind speed of section $Z = 71.6$ m

It can be seen that the presence of staircases outside the roof has a great impact on the distribution of wind speed. Specifically, the area of the low wind speed area of the roof is enlarged and the wind speed is lower, which means obvious blocking effect and makes against to the wind energy utilization. The windward turbulence intensity of the staircase increases and the area of the high turbulence area decreases slightly.

The simulation results show that the suitable mounting site for wind turbines is on the top of the staircase with higher wind speed and lower turbulence intensity. The optimum installation height ratio is $1.15H$ with the acceleration factor of 1.5, which means good potential for wind energy application.

Acknowledgements

This work was supported by the project of Development of Energy Efficiency Technologies for Tall Buildings in Megacities Combined with Structural Safety Solutions under Multi-Hazard Impacts (2016YFE0105600).

References

[1] OHUNAKIN O S. Wind resource evaluation in six selected high altitude locations in Nigeria[J]. Renewable Energy, 2011, 36: 3273-3281.

[2] STANKOVIC S, CAMPBELL N, HARRIES A. Urban wind energy [M]. London: Earthscan, 2009.

[3] ABOHELA I, HAMZA N, DUDEK S. Effect of roof shape, wind direction, building height and urban configuration on

the energy yield and positioning of roof mounted wind turbines [J]. Renewable Energy, 2013, 50: 1106-1118.

[4] LEDO L, KOSASIH P B, COOPER P. Roof mounting site analysis for micro-wind turbines[J]. Renewable Energy, 2011, 36: 1379-1391.

[5] FRANKE, JÖRG, et al. The best practice guideline for the CFD simulation of flows in the urban environment: an outcome of COST 732 [C]// The Fifth International Symposium on Computational Wind Engineering. Chapel Hill, North Carolina, USA, 2010.

Joint estimation for time-varying wind load and structural model parameters from spatially sparse structural output measurements

J. X. Cao & H. B. Xiong
Department of Disaster Mitigation for Structures, Tongji University, Shanghai 200092, China
F. Ghahari & E. Taciroglu
Department of Civil and Environmental Engineering, University of California, Los Angeles (UCLA), U.S.
S. Spence
Department of Civil and Environmental Engineering, University of Michigan, Ann Arbor, U.S.

Abstract

Wind load is one of the important later loads for civil engineering, such as high-rise buildings[1] and long span bridges. Currently the methods to explore the wind-induced response of structure can be classified into computational fluid dynamics, wind tunnel test and field measurement[2]. Recently many researchers focus on the field measurement as the real-life data provides valuable information to investigate the wind load characteristic and wind-induced response of structure. However, direct measurements of the wind load on large-scale buildings are unfeasible and difficult due to wind-structure interaction and the limitation of measurement equipment. In comparison, measurement of acceleration and displacement responses is easier and more accurate than the measurement of force. It would be meaningful if the wind loads on structures can be estimated from limited measurements of responses.

This paper proposes an input-state estimation method that identifies the unknown time-varying wind load and unknown model parameters together from spatially-spare output response measurements. It employs the Unscented Kalman Filter (UKF) to estimate the mean vector and covariance matrix of the argument state vector by using a set of deterministic sampling points. As the argument state vector includes the unknown wind load and model parameters, the size of the state vector increases with the time increasing. To reduce the size of argument state vectors, the wind load is estimated with a finite time interval referred to as a time window and an overlapping time window technique is utilized to improve the accuracy of the estimation result. Figure 1 shows the procedure of the proposed method, where k is the time step.

$\widetilde{\boldsymbol{X}}_{k|k}$ and $\widetilde{\boldsymbol{P}}_{k|k}$ are the posterior mean and covariance matrix of the state vector

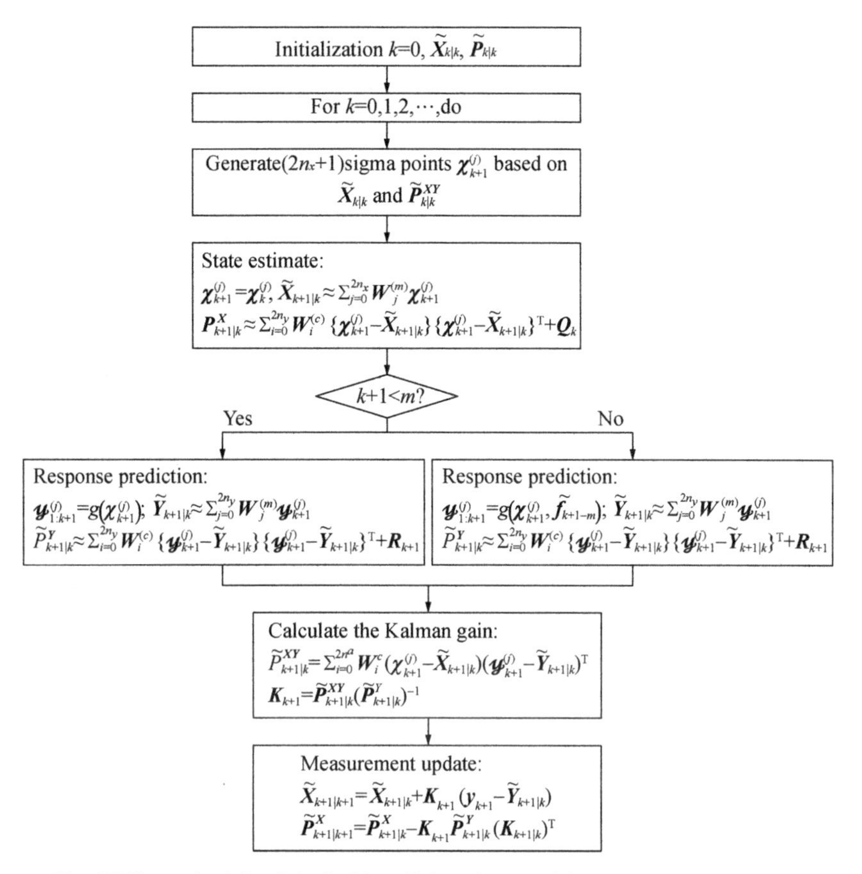

Figure 1 The UKF method for jointly identifying the model parameters and wind load using the roll time window

at time k; n_x is the number of state vector; $\boldsymbol{\chi}_{k+1}^{(j)}$ is the sigma point, $j = 0, 1, 2, \cdots, 2(n_x + 1)$; $W_{\mathrm{m}}^{(j)}$ and $W_{\mathrm{c}}^{(j)}$ are the weight for mean and covariance of the state vector, respectively; $\boldsymbol{K}_{k+1}$ is the Kalman gain; $\boldsymbol{y}_{k+1}^{(j)}$ is the predicted response.

To verify the proposed wind load estimation framework, a numerical model is created, which is a 40-story 4-bay steel frame structure (Figure 2). The wind loads are generated from a random wind model, which are applied to each floor as point loads. The structural responses are generated according to the 40-story building model, which are regarded as true values in the estimation stage after being artificially contaminated by Gaussian noise.

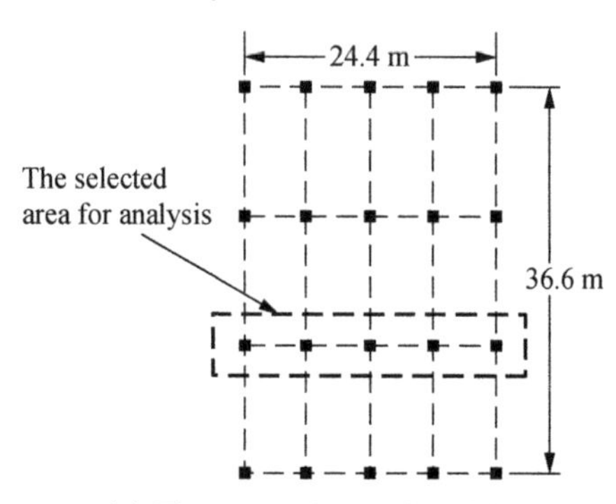

(a) Floor plan of the tall building

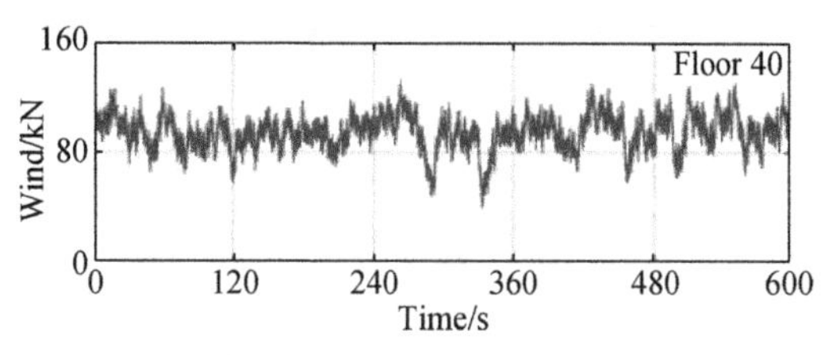

(b) Wind load of the 40th floor

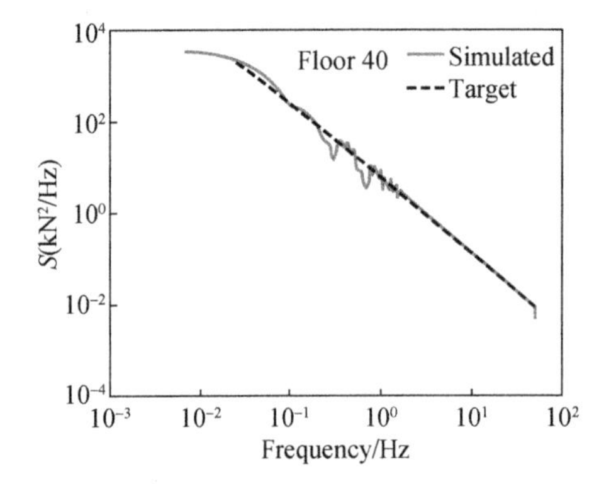

(c) Comparison of the target spectrum and simulated result

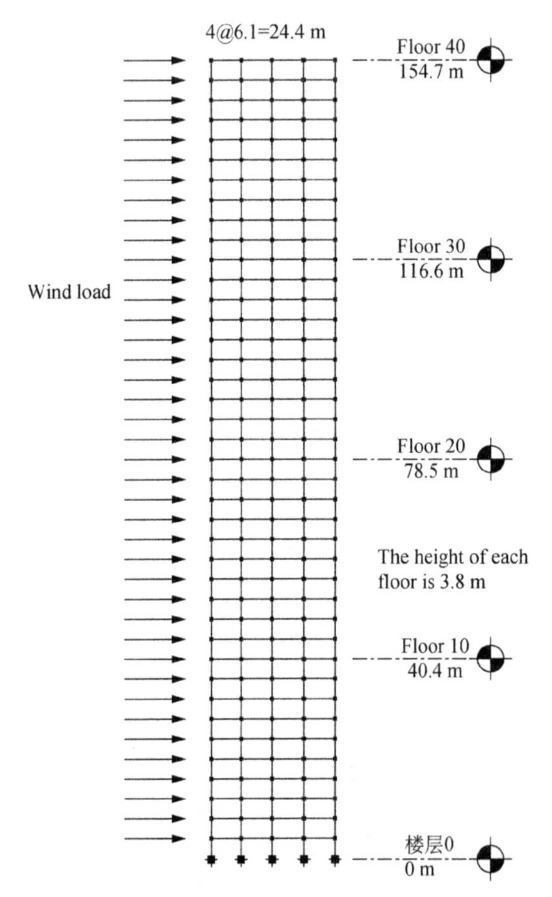

(d) Vertical information of the tall building

Figure 2 Basic information of the 40-story building

In the estimation process, the true displacement of floor 5, 7, 10, 12, 15, 17, 20, 22, 25, 27, 30, 32, 35, 37 and 40 is regarded as the known displacement, which is then used to jointly identify the model parameters and wind load. Figure 3 shows the comparison between the estimated wind load and true values. Due to the limit length of paper, only the 40th floor estimated time-history wind load is presented in Figure 3 (a), which has a good agreement with the true value. To quantify the accuracy of the estimated quantities, the relative root mean square error (RRMSE) between the true and estimated results is defined as follows:

$$RRMSE = \sqrt{\frac{1}{N_t}\sum_{k=1}^{N_t}(R_k^T - R_k^E)^2} \Bigg/ \sqrt{\frac{1}{N_t}\sum_{k=1}^{N_t}(R_k^T)^2} \tag{1}$$

where N_t denotes the total number of data sample; R_k^T denotes the true response; R_k^E denotes the response from estimation. The RRMSE of the wind load for floor 1~20 and floor 21~40 is shown in both Figure 3 (b) and Figure 3 (c), respectively. The values of *RRMSE* show large in the floor 1~7, whereas the *RRMSE* decreases as the number of floor increase.

The estimated displacement of the floor 40 is presented in Figure 4 (a), which is consistent the true displacement. The *RRMSE* between the estimated and true results are also compared in Figure 4 (b), with the value less than 1%. The estimated damping ratio is convergence to the true value after a few seconds.

In the above synthetic example, both the wind load and model parameters can be identified with relatively precision. This case study demonstrates that the proposed method can be used to estimate

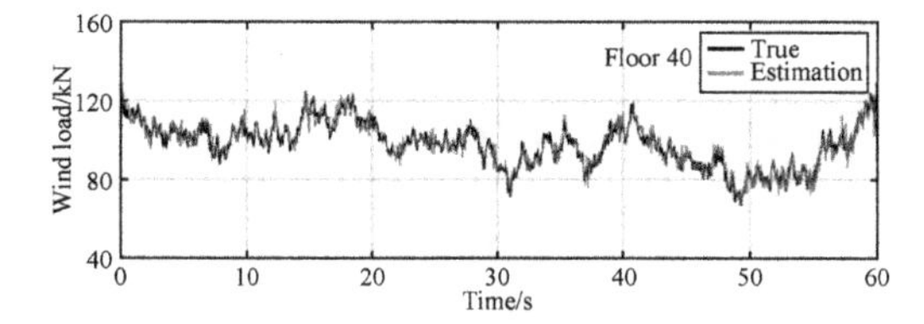

(a) The estimated time-history wind load of the 40th floor

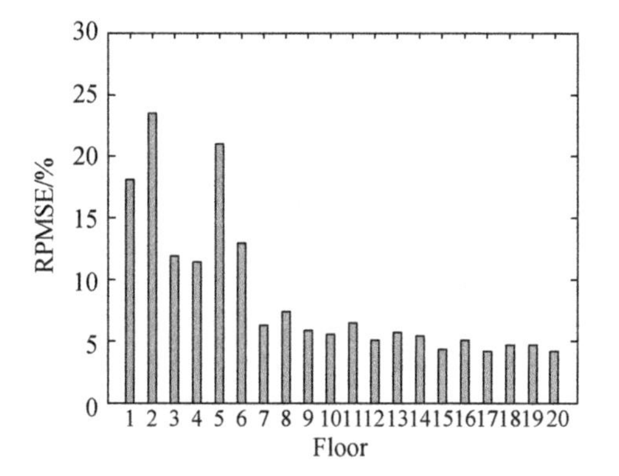

(b) The value of RRMSE from floor 1 to floor 20

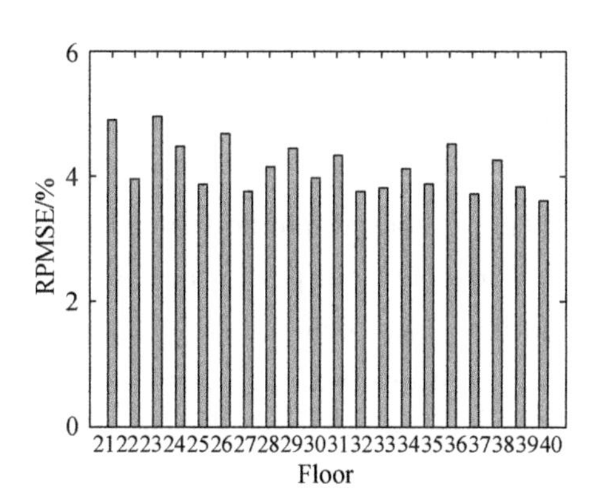

(c) The value of RRMSE from floor 21 to floor 40

Figure 3 Comparison of the true wind load and estimated results

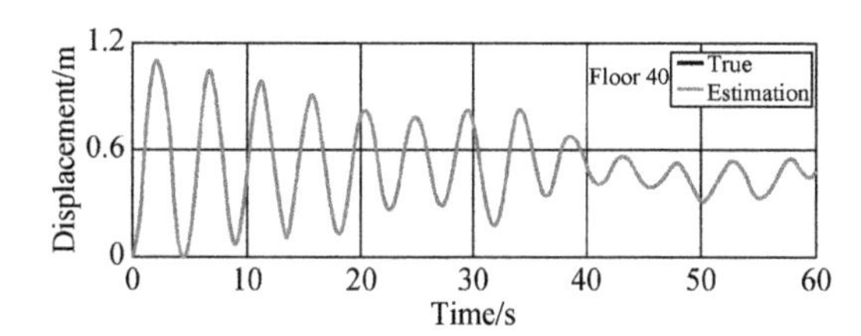

(a) Comparison of the true displacement and estimated results in the 40th floor

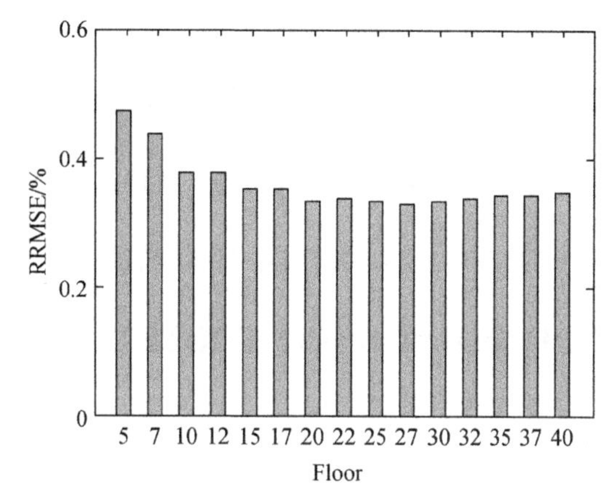

(b) *RRMSE* of the displacement between the true values and estimated results

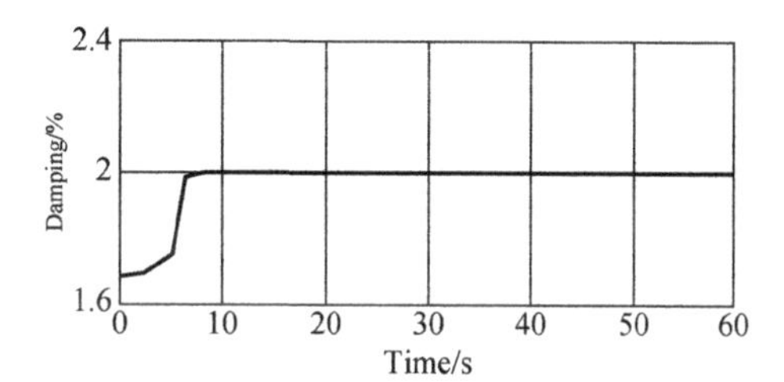

(c) Time-history of the estimated damping

Figure 4 The estimated and true results

the unknown dynamic input excitation and unknown model parameters jointly from limited number measurement. This tool provides a promising route for wind loading identification.

References

[1] XIONG H B, CAO J X, ZHANG F L. Inclinometer-based method to monitor displacement of high-rise buildings [J]. Structural Monitoring and Maintenance, 2018, 5(1): 111-127.

[2] KUBOTA T, MIURA M, TOMINAGA Y, et al. Wind tunnel tests on the relationship between building density and pedestrian-level wind velocity: Development of guidelines for realizing acceptable wind environment in residential neighborhoods [J]. Building and Environment, 2008, 43(10): 1699-1708.

Insight into polyvinyl alcohol stabilized graphene dispersion based on molecular dynamics and its modification effect on cement-based materials

C. Pei & X. Y. Zhou & J. H. Zhu

Guangdong Province Key Laboratory of Durability for Marine Civil Engineering & School of Civil Engineering, Shenzhen University, Shenzhen 518060, China

Abstract

One-step combined modification using graphene and polyvinyl alcohol (PVA) to improve mechanical properties of cement-based materials and overcome limitations of other modification methods by polymer and nanomaterials is presented in this paper. The complementary and synergistic effect of graphene and PVA can solve key problems such as high cost, low production of graphene and poor dispersion of graphene in cement-based material.

An important advantage of PVA is that it functions as a stabilizer, preventing graphene aggregation by its adsorption on graphene surfaces. Figure 1 shows pictures of a dispersion that is freshly prepared and allowed to sit for 6 h, 12 h, 7 d, 14 d and 28 d, respectively. As desired, no difference can be found indicating no agglomeration even after long durations. The concentration of graphene in the resulting PVA dispersion was measured using absorbance, vacuum filtration and TGA measurements. Figure 1(a) presents the measured absorbance relative to the concentration of graphene/PVA dispersions. The linear relationship also suggests the stability of graphene dispersion in PVA. The calculated graphene concentration is consistent with the results obtained from vacuum filtration measurement and TGA test. The proposed method of graphene/PVA aqueous dispersions by shear exfoliation has high stability and can directly substitute water in cement casting. The TEM image of the prepared graphene free from PVA is shown in Figure 1(b).

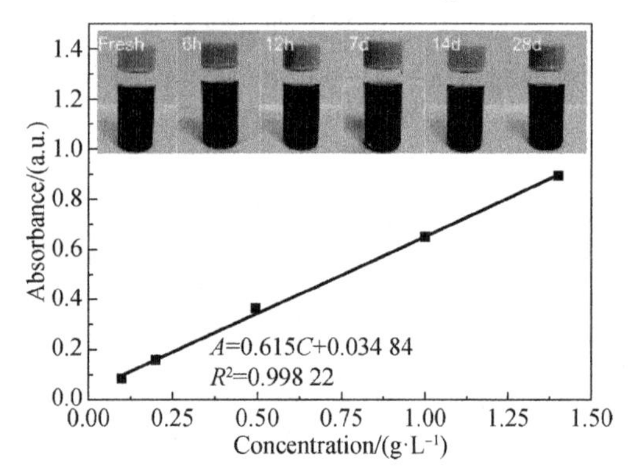

(a) Absorbance relative to the concentration of graphene/PVA dispersions. The inset shows pictures of graphene/PVA dispersions

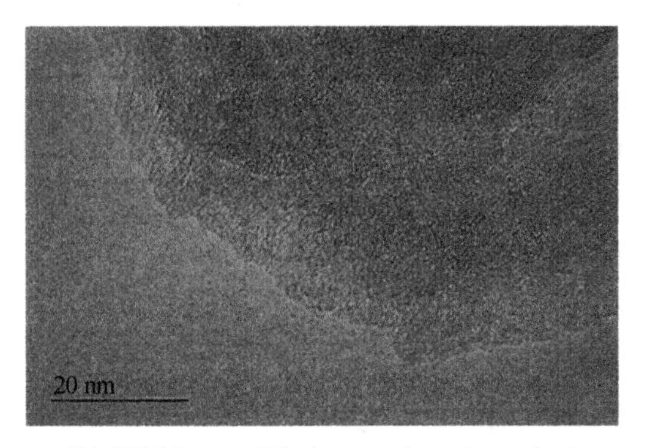

(b) TEM image of the prepared graphene sheet

Figure 1 Absorbance relative to the concentration and TEM image of the prepared graphene

A series of MD simulations were performed to provide complimentary information at a molecular level on the role that PVA plays in stabilizing graphene dispersion. Firstly, adsorption of a single chain of PVA ($N = 20$) from the aqueous phase onto graphene sheet was simulated. The graphene sheet was modeled as a square in the $x-y$ plane with a lateral length of 2.8 nm^2. Geometry optimization was first performed to relax the system to a local energy minimum. Afterwards, the position of the graphene sheet was fixed and dynamic simulation was performed until reaching equilibrium. Figure 2(a) displays the time evolution of the distance between the center-of-mass of the PVA chain and the graphene sheet. This distance is slightly larger than the interlayer distance of graphite (0.34 nm), which means graphene can be stabilized by PVA in water without aggregation. Figure 2(b), (c) show snapshots of PVA absorption onto graphene surface, which reflects the time evolution of PVA motion towards the graphene surface. Initially, the PVA chain and the graphene sheet are separately solvated in water. PVA chain gradually approaches the graphene surface and attaches itself to the graphene surface. At 2 ns, the PVA chain is completely absorbed on the graphene surface. These simulation results show that even though graphene is hydrophobic, water does not hinder PVA absorbance onto graphene. In the second step of MD simulation, the possibility of graphene aggregation after shear mixing was considered. As shown in Figure 2(e), (f), the two graphene sheets that were initially separated would aggregate spontaneously in water. This result is consistent with observations of poor graphene dispersion in water[1, 2]. From the time evolution of center-of-mass distance between the top and bottom graphene sheets (Figure 2(d)), the distance between the two graphene sheets varies little with time after 0.8 ns, indicating an equilibrium state. The equilibrium distance between the two graphene sheets was 0.34 nm, which is the interlayer distance of graphite. A model with PVA chain ($N = 20$) placed between two graphene sheets was built. The time evolution of center-of-mass distance of the PVA chain from the top graphene sheet and between the top graphene sheet to the bottom graphene sheet is shown in Figure 2(g), with Figure 2(h), (i) showing snapshots at 0 ns and 2 ns. The two distances remain stable (Figure 2(g)), indicating a geometric equilibrium of the system. The equilibrium distance between the two graphene sheets is 0.75 nm, while the distance from PVA to the bottom graphene sheet is 0.37 nm. It can be observed that due to presence of the PVA chain, the top graphene sheet cannot attach to the bottom graphene sheet (Figure 2 (h), (i)). This result confirms that the adsorption of PVA onto the graphene surface can prevent aggregation of graphene in water[3].

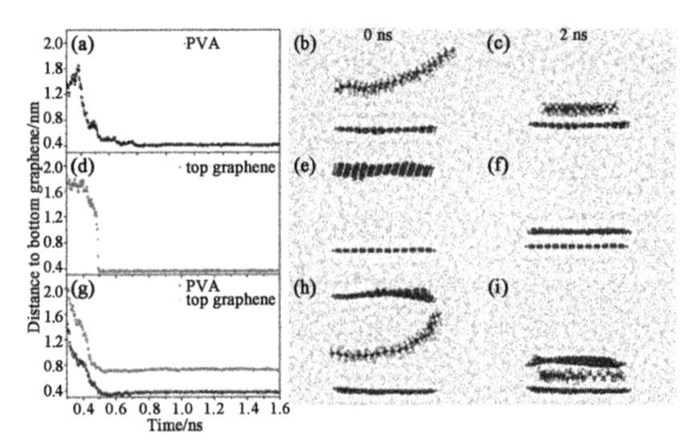

Figure 2 MD simulations of PVA stabilized graphene dispersion

The properties of hydrated cement depend on the characteristics of its microstructure. On the fracture surface of plain cement (Figure 3 (a)), needle-shaped crystals and flocculation structures can be seen, representing cement hydration products ettringite and calcium silicate hydrate (C—S—H) gel, respectively[4, 5]. Distinctly different morphology (Figure 3 (b)) can be observed when graphene was introduced into the cement. The size and shape of the pores of well-dispersed cement and the crystalline hydration products were completely different. With graphene in cement, clusters of rod-like ettringite crystals and granular C—S—H can be observed. The shape of ettringite changed from a needle to a rod, indicating that the incorporation of graphene into cement promoted the hydration process. With the assistance of PVA, graphene dispersed uniformly in cement paste and functioned as a nucleation-inducing agent[6]. In the cement hydration stage, as the amount of water decreased, PVA was gradually restricted in capillary pores, forming a gel membrane on the surface of the hydrated cement[7]. The membrane wrapped the hydrated cement gel, the unhydrated cement particles and graphene together to form a three-dimensional interpenetrating network structure, thereby enhancing the mechanical properties of the cement, especially its flexural strength.

Figure 3(c) compare compressive and flexural strengths of cement pastes with different graphene concentrations. Compared to cement with 1.8% PVA by weight, the modified cement with graphene and PVA has increases of 71.1% and 106.5% in compressive and flexural

(a) Plain cement

(b) Graphene/PVA modified cement

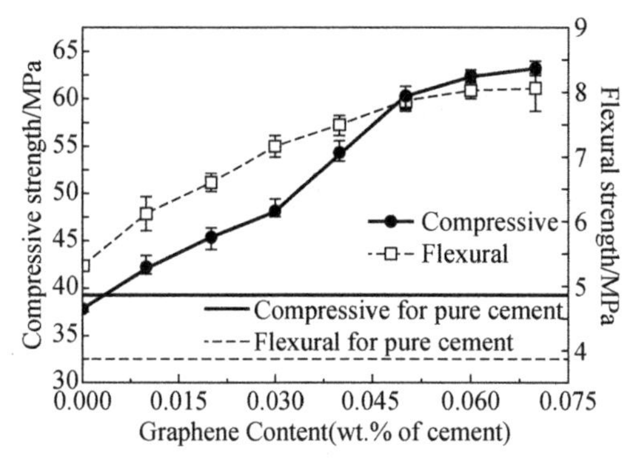

(c) mechanical properties of plain cement, PVA modified cement ($C_g = 0$) and graphene/PVA synergistically-modified cement-based materials

Figure 3 SEM images and mechanical properties of graphene/PVA synergistically-modified cement-based materials

strength, respectively. The corresponding values are - 4% and 36.2% respectively when the cement is modified by 1.8% PVA only. This strongly demonstrates the synergistic effects of graphene/PVA in improving cement properties.

The graphene processing approach of this paper provides a new method of synergistic modification of cement-based materials for practical application of graphene in mass produced concrete in large scale projects.

Acknowledgements

We would like to appreciate the funding support from the Chinese National Natural Science Foundation (Grant No. 51778370, Grant No. 51538007), Natural Science Foundation of Guangdong (Grant No. 2017B030311004) and the Shenzhen science and technology project (JCYJ 20170818094820689).

References

[1] LIN S, SHIH C J, STRANO M S, et al. Molecular Insights into the Surface Morphology, Layering Structure, and Aggregation Kinetics of Surfactant-Stabilized Graphene Dispersions [J]. Journal of the American Chemical Society, 2011, 133(32): 12810-12823.

[2] STUKOWSKI A. Visualization and analysis of atomistic simulation data with OVITO—the Open Visualization Tool[J]. Modelling and Simulation in Materials Science and Engineering, 2010, 18(1): 2154-2162.

[3] ARUNACHALAM V, VASUDEVAN S. Probing Graphene—Surfactant Interactions in Aqueous Dispersions with Nuclear Overhauser Effect NMR Spectroscopy and Molecular Dynamics Simulations[J]. Journal of Physical Chemistry C, 2017, 121(30): 16637-16643.

[4] LV S H, MA Y J, QIU C C, et al. Effect of graphene oxide nanosheets of microstructure and mechanical properties of cement composites [J]. Construction and Building Materials. 2013, 49: 121-127.

[5] PARVEEN S, RANA S, FANGUEIRO R, et al. Microstructure and mechanical properties of carbon nanotube reinforced cementitious composites developed using a novel dispersion technique[J]. Cement and Concrete Research, 2015, 73: 215-227.

[6] Hou D S, Lu Z Y, Li X Y, et al. Reactive molecular dynamics and experimental study of graphene-cement composites: Structure, dynamics and reinforcement mechanisms[J]. Carbon. 2017, 115: 188-208.

[7] SINGH N B, RAI S. Effect of polyvinyl alcohol on the hydration of cement with rice husk ash [J]. Cement and Concrete Research, 2001, 31: 239-243.

Experimental study on the electrochemical deposition method for the repair of the cracked underground structures

Z. Y. Zhu & Q. Chen

Department of Civil Engineering Materials, College of Materials and Science, Tongji University, Shanghai 201804, China & Department of Structural Engineering, College of Civil Engineering, Tongji University, Shanghai 200092, China

Abstract

Cracks deteriorate concrete's strength, performance and durability[1], particularly for underground structures, which will reduce extended service life of the structure. In light of the limitations of traditional repairing methods in an aqueous environment[2, 3], the electrochemical deposition method (EDM), as a new technique for repairing concrete cracks in a water environment, has been developed in the past 20 years and applied to marine structures and other situation[4-8].

This paper focuses on the experimental analysis of the asymmetric restoration of the cracked underground structures using EDM.

1. Experimental Program

To consider the asymmetry of underground structure restoration, the device of the EDM can be summarized as follows. The embedded rebars of the concrete structure and the titanium plateare connected to the negative terminaland positive terminalof the power supply, respectively, as displayed in Figure 1. The cracked concrete specimen is put between the water and electrolyte, while titanium plates are immersed in alkaline electrolyte.There is a continuous electrochemical deposition in the concrete member or structure when an appropriate direct current (DC) is maintained in the system. Therefore, the cracks and pores in the concrete members and structures can be healed or filled gradually by the deposition products.

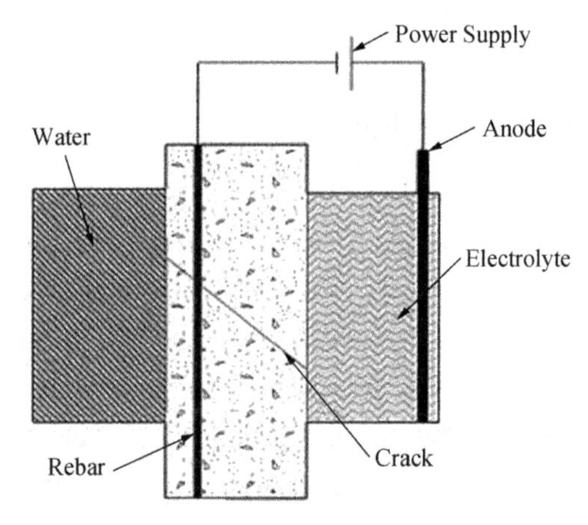

Figure 1 Experimental device

The main electrochemical reactions

during the electrodeposition are as follows:

Cathodic reaction:

$$2H_2O + 4e^- \rightarrow H_2 \uparrow + 2OH^- \quad (1)$$

Athodic reaction:

$$Y^{2+} + 2OH^- \rightarrow Y(OH)_2 \downarrow \quad (2)$$

where Y^{2+} is the ion in the solution, such as the Mg^{2+} and Ca^{2+}. In the experiment, the influence of electrolytes, electrodes, modes of cathode connection are considered. Meanwhile, the surface coverage and the microstructures are adopted to assess the repairing effects.

2. Experimental Results

Figure 2 shows the deposition effectunder different proportions of electrolyte solution. Compared with the mix solution of Zn^{2+} (0.05 mol/L) and Ca^{2+} (0.01 mol/L), the one with Mg^{2+} (0.05mol/L) and Ca^{2+} (0.01mol/L) can improve the deposition amounts. In addition, when the proportion of Mg^{2+} and Ca^{2+} is (0.050 mol/L):(0.016 7 mol/L) or (0.050 mol/L):(0.010 mol/L), the surface coverage is better than those with Mg^{2+} (0.050 mol/L) and Ca^{2+} (0.007 mol/L).

Figure 3 displays the different materials used as anodes in the experiment. From left to right, they are respectively the ruthenium-iridium titanium plates, the ordinary titanium plates, the graphite rods and the graphite plates. Compared with the other materials, ruthenium-iridium titanium is more stable after the repairing process. The other materials are all damaged to some extend after the electrochemical depositon process.

(a) Mg^{2+}: 0.050 mol/L
Ca^{2+}: 0.016 7 mol/L

(b) Mg^{2+}: 0.050 mol/L
Ca^{2+}: 0.010 mol/L

(c) Mg^{2+}: 0.050 mol/L
Ca^{2+}: 0.007 mol/L

(d) Zn^{2+}: 0.05 mol/L
Ca^{2+}: 0.010 mol/L

Figure 2 Deposition effect with different electrolyte solution

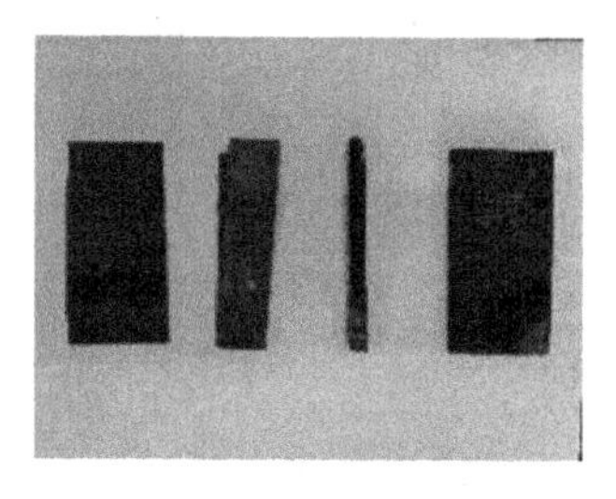

Figure 3 Electrodes used in the experiment

To investigate the influence of the connecting modes of the cathode on the repairing effectiveness, apart from connectingthe negative terminal of the power supply to the rebars (Mode a), the other approach by using the titanium plate (Mode b), which sticks to the side of the cracked specimen, is also employed, as shown in Figure 4.

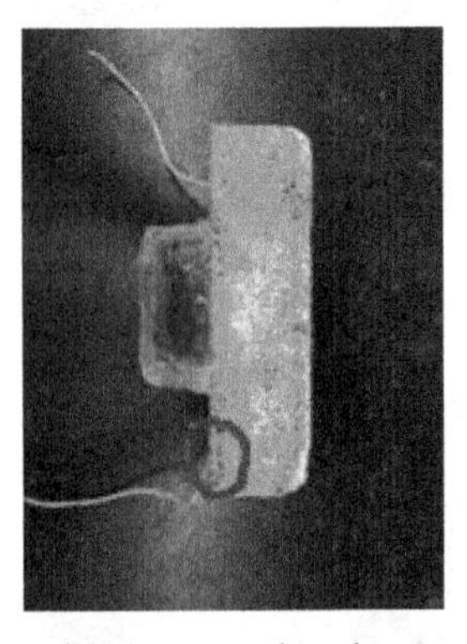

(a) Access to the rebars

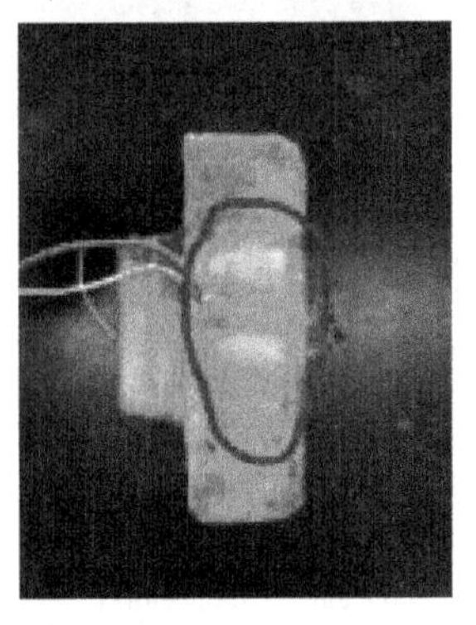

(b) Stick to the side

Figure 4 Modes of cathode connection

Figure 5 shows the coverage the deposition products using different connecting modes. It can be observed that the coverage using Mode a is better than those with Mode b.

Figure 6 shows the microstructures of different deposition products. According to the solution adopted in the experiment, the deposition products are mainly the $CaCO_3$ and $Mg(OH)_2$, as Figure 6(a), (b) display.

(a) Mode a

(b) Mode b

Figure 5 Deposition effect with different connection mode

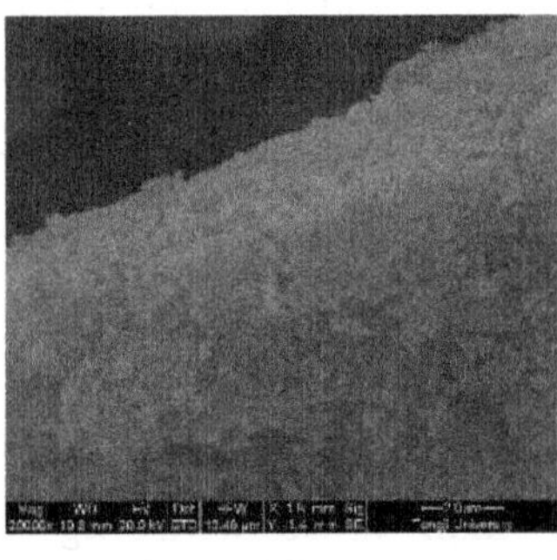

(a)

(b)

Figure 6 Microstructure of different deposition products

3. Conclusions

To model the repair of the cracked underground structures using the EDM, the asymmetric restoration device is employed in the lab. Meanwhile, the different factors are investigated on the healing effects. It can be realized that the cracks can be well repaired by electrodeposition with the proper electrolyte solutions and electrodes.

References

[1] LIMY M, LI V C. Durable repair of aged infrastructures usingtrapping mechanism of engineering cementitious composites [J]. Cement and Concrete Composites, 1997, 19: 373-385.

[2] VAGELIS G P, et al. Physical and chemical characteristics affecting the durability of concrete[J]. ACI Materials Journal, 1991, 88 (3): 234-239.

[3] SUN LZ, LIU H T, JU J W. Effect of particle cracking on elastoplastic behavior of metal matrix composites [J]. International Journal for Numerical Methods in Engineering, 2003, 56(14): 2183-2198.

[4] CHU H Q, JIANG L H, YOU L S, et al. Influence of mineral admixtures on the electro-deposition healing effect of concrete cracks[J]. Journal of Wuhan University of Technology, 2014, 29(6): 1219-1224.

[5] CHEN Q, ZHU H H, JU J W, et al. A stochastic micromechanical model for multiphase composites containing spherical inhomogeneities[J]. Acta Mechanica, 2015, 226(6): 1861-1880.

[6] ZHU HH, CHEN Q, JU JW, et al. Maximum entropy based stochastic micromechanical model for a two-phase composite considering the inter-particle interaction effect [J]. Acta Mechanica, 2015, 226(9): 3069-3084.

[7] YAN Z G, CHEN Q, ZHU H H, et al. A multi-phase micromechanical model for unsaturated concrete repaired by electrochemical deposition method [J]. International Journal of Solids and Structures, 2013, 50(24): 3875-3885.

[8] ZHU H H, CHEN Q, YAN Z G, et al. Micromechanical models for saturated concrete repaired by electrochemical deposition method [J]. Materials and structures, 2014, 47(6): 1067-1082.

Experimental study on FRP-RAC-steel hybrid columns under axial compression

Y. M. Liu & Z. H. Lan

School of Civil and Transportation Engineering, Guangdong University of Technology, Guangzhou 510006, China

G. M. Chen

School of Civil Engineering & Transportation, South China University of Technology, Guangzhou 510641, China

M. X. Xiong

Protective Structures Center, School of Civil Engineering, Guangzhou University, Guangzhou 510006, China

Abstract

The performance of recycled aggregate concrete (RAC) is usually inferior compared with the normal aggregate concrete (NAC). Using RAC in composite member is an easy and promising way to avoid the unfavourable effects (e. g. reduced strength and durability) of including recycled aggregates. In this study, RAC was in combination with steel and fibre-reinforced polymer (FRP) to form an FRP-RAC-steel hybrid column. In such hybrid column, the FRP tube and steel tube were placed outside and inside, respectively, with the concrete filling in-between. Existing studies already showed that if NAC was used, the said hybrid column possessed many good properties such as enhanced strength and deformation capacity. The compressive behavior of both DTCCs and DTSCs with RAC (under axial compression loading) was tested. For comparison purpose, plain concrete columns (CC), concrete filled steel tubes (CFST) and concrete filled FRP tubes (CFFT) were also tested. The test results showed favourable effects existed due to the beneficial interaction among the difference components (FRP, steel and RAC) of the said hybrid column members. As a result, FRP-RAC-steel hybrid column has a similar compressive behavior as FRP-NAC-steel hybrid column although including recycled aggregates which had a certain adverse effects.

1. Introduction

Due to the existence of interfacial transition zone (ITZ) and pre-cracks/damage in the parent concrete, recycled aggregate concrete (RAC) usually suffers from some drawbacks, such as reduced strength, inferior durability compared with the normal aggregate concrete (NAC). As a result, the use of RAC is mainly limited to non-structural applications to date. Using RAC in composite member is an easy and promising way to avoid the unfavourable effects (e. g. reduced strength and durability) of including recycled aggregates. Following this approach, the RAC was in combination with steel and fibre-reinforced polymer (FRP) in this study to form FRP-RAC-steel hybrid column so as to compensate/suppress the unfavourable effects of the said drawbacks. In such hybrid column, FRP tube was placed outsides and steel tube inside with the concrete filling in-between. If the steel tube is filled with concrete, the hybrid column is termed as FRP-concrete steel double-tube concrete column (DTCC) and if there is no concrete filling in the steel tube, it is termed as FRP-concrete steel double-skin tubular column (DSTC). Existing studies already showed that if NAC was used, the said hybrid column possessed many good properties such as enhanced strength and deformation capacity. This is because the following favourable interactions exist in the said hybrid columns: (1) steel tube and FRP tube provide effective confinement to the concrete between the tubes; (2) the FRP confinement together with the confined concrete supresses the local outwards buckling of the steel tubes; (3) the FRP tube can provide protection to the concrete and steel tubes against harsh environment[1]. Furthermore, both the steel tube and FRP tube can work as formwork for casting concrete which leads to ease in construction. In this study, the compressive behavior of both DTCCs and DTSCs with RAC (under axial compression loading) were tested. For comparison purpose, plain concrete columns (CC), concrete filled steel tubes (CFST) and concrete filled FRP tubes (CFFT) were also tested. The test results showed that beneficial interaction also existed in the difference components (FRP, steel and RAC) of the said hybrid column members. As a result, it should be possible to use RAC in the said hybrid columns.

2. Experimental Program

The cross-sectional sizes of specimens are shown in Figure 1 where CC, CFST, CFFT, DSTC and DTCC stands for the concrete column, concrete filled steel tubular column, concrete-filled FRP tube, double-skin tubular column, and the double-tube concrete column, respectively. The DSTCs and DTCCs are a promising type of FRP-RAC-steel hybrid columns, and the DTCCs may be taken as a variation of DSTCs[2]. Totally 34 column specimens were tested subjected to axial compression considering the effects of replacement ratio of RCAs and the

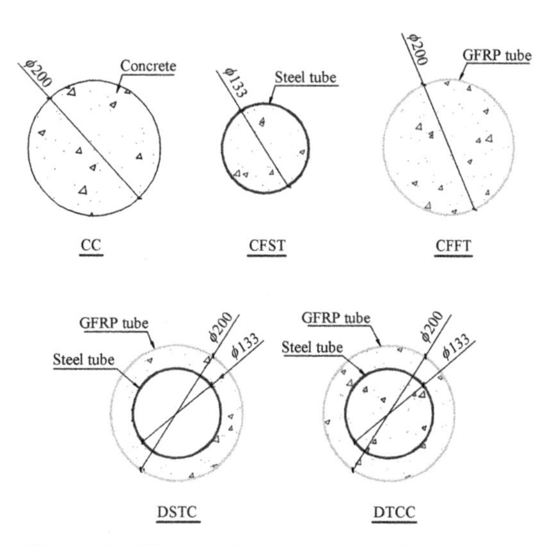

Figure 1 Sizes and designations of specimens

thickness of GFRP tube. All specimens had a height of 400 mm.

The RCAs were taken from waste concrete after demolition. The replacement ratios of 0, 50% and 100%, respented by R0, R50 and R100, were adopted. The compressive strength of R0, R50 and R100 on 28 days are respectively 45.6 MPa, 41.6 MPa and 40 MPa. The ultimate strength, yield strength, elastic modulus, and Poisson's ratio of steel section were 504 MPa, 315 MPa, 205 GPa and 0.3, respectively. The hollow steel columns with a height of 400 mm were also tested under axial compression to obtain the compressive or buckling strength of steel. The solid GFRP tubes were fabricated in a filament-wound process with fiber winding angles of $\pm 80^{\circ}$ to the longitudinal axis of the tube. Two types of GFRP tube, 4-ply and 8-ply, were prepared, having a thickness of 1.96 mm and 3.69 mm, respectively. The axial strength of 4-ply and 8-ply is 77.7 MPa and 83.4 MPa, respectively. The compressive strength and elastic modulus of the 8-ply tubes were higher than those of the 4-ply tubes, but the Poisson's ratio was lower. The hoop strength of 4-ply (G4) and 8-ply (G8) is 40.3 MPa and 44.2 MPa, respectively. The 8-ply tubes also exhibited higher elastic modulus and lower Poisson's ratio, when compared with the 4-ply tubes.

3. Results and Discussion

3.1 Failure Observations

The maximum diameter over thickness value for local buckling provided by EC4[3] indicates the local buckling would not occur in the steel tube. However, it should be mentioned that the EC4 maximum value corresponds to the value of classification for Class 3 section in EC3[4]. This means the EC4 can only tell that the steel tube would not have local buckling in the elastic stage. Whether the local buckling would happen in the plastic stage should be based on the classification according to EC3. It is then found that the steel tube falls into Class 2 but very close to Class 1 according to EC3, implying that the local buckling would possibly happen in the plastic stage but may experience a large plastic deformation firstly. Figure 2 shows the failure modes of the steel tubes under axial compression. The local buckling could be found in the CFST and DSTC specimens, outwardly and inwardly respectively, and it is severer for the CFST specimens since the outward buckling is more likely to happen than the inward buckling. The local buckling was not noticeable for the DTCC specimens as the steel tubes were restrained from both sides.

Figure 3 shows the failure modes of GFRP tubes. No significant difference was found when the replacement ratios of recycled coarse aggregates varied. All

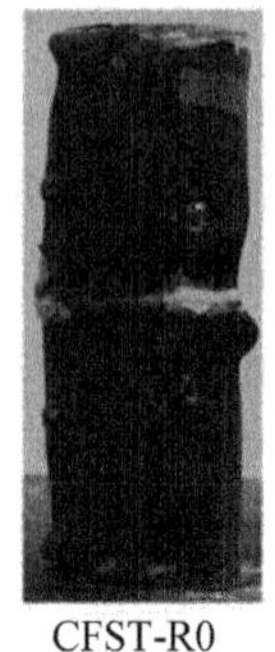

CFST-R0

CFST-R50

CFST-R100

(a) CFSTs

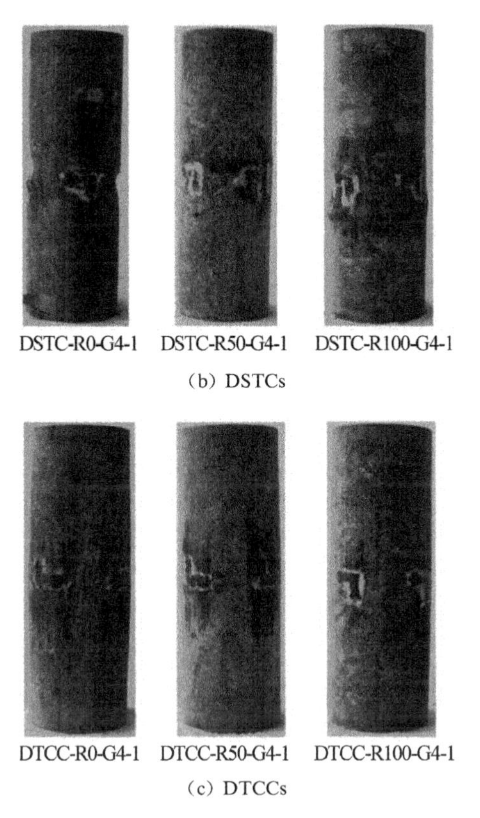

(b) DSTCs

(c) DTCCs

Figure 2 Failure observations on steel tubes

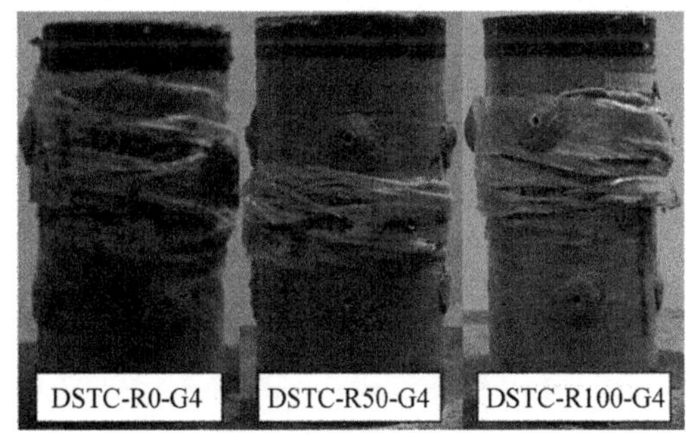

(a) DSTCs

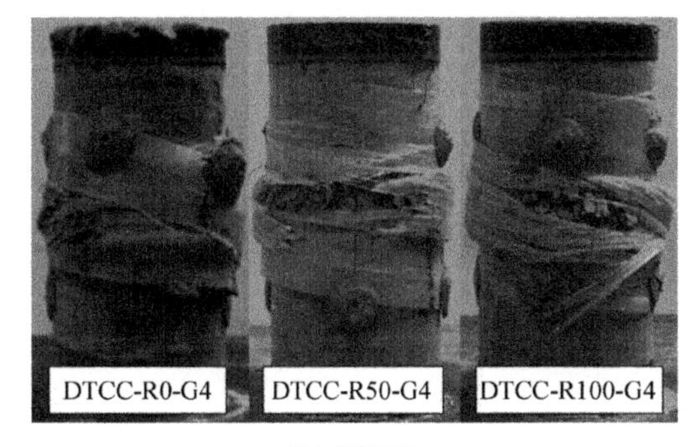

(b) DTCCs

Figure 3 Failure observations on GFRP tubes

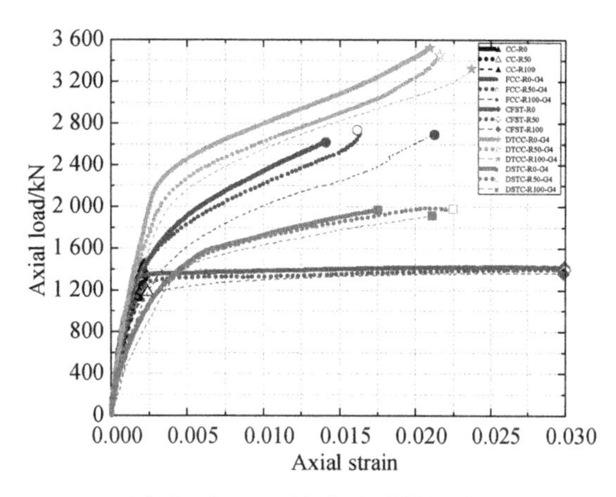

(a) Specimens with 4-ply GFRP tubes

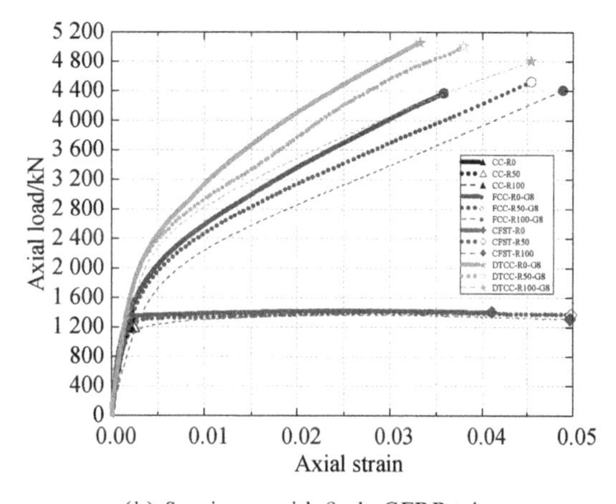

(b) Specimens with 8-ply GFRP tubes

Figure 4 Axial load-axial strain curves: effect of replacement ratio

DSTCs and DTCCs specimens failed in rupture of the GFRP tubes.

3.2 Load-Strain Response

Figure 4 shows the typical load-axial strain curves considering the effect of replacement ratio. It can be found that the load-strain curves of the specimens confined by the GFRP tubes presented a monotonically ascending bi-linear shape where there were no load drops and the loads kept increasing until the GFRP tubes ruptured[5]. The second ascending portion of the bi-linear curve is defined as the hardening range. For DSTCs and DTCCs, the ultimate load capacity decreased with an increase in the replacement ratio of the recycled coarse aggregates and the thickness of the GFRP tubes. However, the ultimate axial strain shows the opposite trend.

4. Conclusions

This paper has been concerned with an experimental study on the compressive behavior of FRP-RAC-steel hybrid columns. It has been found that the replacement ratio of recycled coarse aggregate and thickness of GFRP tubes had significant effects on the load-strain responses of such hybrid columns. Increasing the replacement ratio of recycled coarse aggregates and the thickness of the GFRP tubes reduced the ultimate load capacity but increased the ultimate axial strain.

Acknowledgements

The authors would like to acknowledge the financial support by the National Natural Science Foundation of China (Project No. 51978281, 51678161), and the Guangdong Natural Science Foundation under project No. 2018A030313752.

References

[1] TENG J G, YU T, WONG Y L, et al. Hybrid FRP-concrete-steel tubular columns: concept and behavior[J]. Construction and building materials, 2007, 21(4): 846-854.

[2] TENG J G, WANG Z H, YU T, et al. Double-tube concrete columns with a high-strength internal steel tube: Concept and behavior under axial compression [J]. Advances in Structural Engineering, 2018, 21(10): 1585-1594.

[3] EN 1994 - 1 - 1, Eurocode 4: Design of Composite Steel and Concrete Structures — Part 1-1: General Rules and Rules for Buildings [S]. European Committee for Standardization, 2004.

[4] EN 1993-1-1. Eurocode 3: Design of steel structures — Part 1-1: General rules and rules for buildings[S]. European Committee for Standardization, 2005.

[5] LAM L, TENG J G. Design-oriented stress-strain model for FRP-confined concrete[J]. Construction and Building Materials, 2009, 17(6-7): 471-489.

Coarse-grained simulations on creep behavior of polypropylene

C. Wu & R. D. Wu & L. H. Tam
School of Transportation Science and Engineering, Beihang University, 37 Xueyuan Road, Beijing, China

Abstract

Polypropylene (PP) polymer has considerable practical usages in various engineering applications, such as food packaging, coating, glass panel for automobiles, and filler for concrete[1-4]. During long-term service life, PP-based materials are usually subjected to sustained external loads (e.g. wind and structural loading), causing a time-dependent deformation (i.e. creep) of the polymer material. The continuously increasing deformation during creep leads to damage and fracture of the material even at a loading level much lower than the designed material strength[4-7]. Therefore, understanding the creep behavior of PP is crucial to the knowledge of long-term performance of polymer materials. In this work, atomistically informed coarse-grained (CG) technique is employed to model the PP polymer system, which is demonstrated to possess similar structural and mechanical properties as the experimental sample. Afterwards, CG molecular dynamics (MD) simulations are conducted to investigate the creep behavior of the PP molecule model under different stress levels. According to the obtained strain-time curves, it is revealed that there exists a threshold stress, above which the maximum strain of PP within the simulation timespan increases dramatically. Meanwhile, by observing the changes in polymer chains and potential energy, it is learned that the conformational changes of the polymer chains, including chain stretching, unfolding, and sliding accounts for the creep at different stages. Our study provides physical insight into the creep behavior of PP at a fundamental molecular level.

Acknowledgements

C.W. acknowledges the support from the National Sciencefoundation of China (grant number 51608020) and the Thousand Talents Plan (Young Professionals) in China. L. T. acknowledges the support from the National Science foundation of China (grant number 51808020) and the China Postdoctoral Science Foundation (grant number 2017M620015 and 2018T110029).

References

[1] LEE J W, SON S M, HONG S I. Characterization of protein-coated polypropylene films

as a novel composite structure for active food packaging application [J]. Journal of Food Engineering, 2008, 86(4): 484-493.

[2] ALARIQI S A S, KUMAR A P, RAO B S M, et al. Effect of γ-dose rate on crystallinity and morphological changes of γ-sterilized biomedical polypropylene [J]. Polymer Degradation & Stability, 2009, 94(2): 272-277.

[3] CABRERA N O, ALCOCK B, PEIJS T. Design and manufacture of all-PP sandwich panels based on co-extruded polypropylene tapes [J]. Composites Part B: Engineering, 2008, 39(7-8): 1183-1195.

[4] VRIJDAGHS R, PRISCO M D, VANDEWALLE L. Uniaxial tensile creep of a cracked polypropylene fiber reinforced concrete [J]. Materials and Structures, 2018, 51: 5.

[5] SINCLAIR J E, EDGEMOND J W. Investigation of creep phenomena in polyethylene and polypropylene [J]. Journal of Applied Polymer Science, 1969, 13(5): 999-1012.

[6] CHANG B B, SCHNEIDER K, HEINRICH G. Microstructural Evolution of Isotactic-Polypropylene during Creep: An In Situ Study by Synchrotron Small-Angle X-Ray Scattering [J]. Macromolecular Materials and Engineering, 2017: 1700152.

[7] XU Z Y. Modeling and Prediction of Creep Behavior of Polypropylene Packaging Belt [J]. Applied Mechanics and Materials, 2011, 117-119: 1168-1171.

Probability density function-informed reliability analysis of deteriorating structures

H. Y. Guo

Department of Structural Engineering, College of Civil Engineering, Tongji University, Shanghai 200092, China & Department of Civil and Environmental Engineering, Hong Kong Polytechnic University, Hong Kong 999077, China

Y. Dong

Department of Civil and Environmental Engineering, Hong Kong Polytechnic University, Hong Kong 999077, China

Abstract

In general, there exist two types of deterioration scenarios: continuous deterioration[1] and sudden damage[2]. The continuous deterioration usually refers to corrosion-induced deterioration which could cause cumulative damage during the structural service life. The sudden damage may endanger structural safety in a short period. Ignoring the relevant deterioration may overestimate the safety and service life of engineering structures. Thus, it is significant to establish a comprehensive framework for life-cycle assessment and systematically consider different deterioration scenarios. To date, reliability-informed assessment is one of the most popular approaches to consider the condition and safety evolution in life-cycle engineering comprehensively.

In this paper, a novel probability density function (PDF)-informed method (PDFM) is proposed to compute the system-level time-dependent reliability of deteriorating structures suffering from various deterioration scenarios. For a general deteriorating system, its performance function g can be written as

$$g = G(\boldsymbol{\Theta}, t, \boldsymbol{b}) \qquad (1)$$

where t is the time parameter; $\boldsymbol{\Theta} = [\boldsymbol{\Theta}_1, \boldsymbol{\Theta}_2, \cdots, \boldsymbol{\Theta}_d]^{\mathrm{T}}$ is a d-elements random input vector; and $\boldsymbol{b} = \{b_k, k = 1, 2, \cdots, n_d\}$ is the random vector of n_d critical time instants. Two types of deterioration modes are supposed: derivable deterioration and un-derivable deterioration. For derivable deterioration, the performance function $G(\boldsymbol{\Theta}, t, \boldsymbol{b})$ is derivable to time t; and un-derivable deterioration refers to the continuous $G(\boldsymbol{\Theta}, t, \boldsymbol{b})$ but discontinuous $\dot{G}(\boldsymbol{\Theta}, t, \boldsymbol{b})$ at the time instant $\boldsymbol{b}$.

In this study, the PDFs of performance function under different types of deterioration modes are obtained to compute the time-dependent failure probability $p_{\mathrm{f}}(t)$. For derivable deterioration, the PDF of performance function $p_{Y\boldsymbol{\Theta}\boldsymbol{B}}(y, \boldsymbol{\theta}, \boldsymbol{b}, t)$ can be solved by using existing algorithms of PDFM[3]. For other deterioration scenarios, the "two-step translation method" is developed to calculate the $p_{Y\Theta B}(y, \boldsymbol{\theta}, \boldsymbol{b}, t)$. Applying the point evolution method[4], n_{sel} selected points are acquired and the principal processes of "two-step translation method" are listed as following.

(1) For each representative point $\boldsymbol{\theta}_a$, $a=1, 2, \cdots, n_{sel}$, the critical time instants $\boldsymbol{b}_a = \{b_{a,1}, b_{a,2}, \cdots, b_{a,nd}\}$ can be identified and surrogate model $Y(t) = G(\boldsymbol{\theta}_a, t, \boldsymbol{b}_a)$ is translated to $\tilde{Y}(t) = \tilde{G}(\boldsymbol{\theta}_a, t, \boldsymbol{b}_a)$ with $\Delta g(t)$.

$$\Delta g(t) = \sum_{s=1}^{nd} \left\{ [g_{s+1}(b_{a,s}^{+}) - g_s(b_{a,s}^{-})] \cdot H(t - b_{a,s}) \right\} \tag{2}$$

where $b_{a,s}^{-}$ and $b_{a,s}^{+}$ are the time instants just before and after $b_{a,s}$; and $g_s(b_{a,s}^{-})$ and $g_{s+1}(b_{a,s}^{+})$ are $G(\boldsymbol{\theta}, b_s^{-}, \boldsymbol{b}_a)$ and $G(\boldsymbol{\theta}, b_s^{+}, \boldsymbol{b}_a)$.

(2) Applying traditional PDFM then obtaining $p_{\tilde{Y}\Theta B}(\tilde{y}, \boldsymbol{\theta}, t, \boldsymbol{b})$, $p_{\tilde{Y}\Theta B}(\tilde{y}, \boldsymbol{\theta}, t, \boldsymbol{b})$ is translated to $p_{Y\Theta B}(y, \boldsymbol{\theta}, \boldsymbol{b}, t)$ with $\Delta g(t)$ at the inverse direction.

For illustrative purpose, a stochastic deteriorating system with one single sudden damage of random occurred time instant t_{drop} is assumed.

$$f(t) = (1 - 6 \times 10^{-6} t^3) \cdot f_0 - 5 \cdot H(t - t_{drop}) \tag{3}$$

where both f_0 and t_{drop} are Gauss random variable [$f_0 \sim N(20, 1)$ and $t_{drop} \sim N(25, 2)$].

Figure 1 compares the differences between the two-step translation method and traditional PDFM by the PDFs after 23, 25, and 27 years. It could be noticed that the PDFs obtained by the proposed method are smoother than traditional ones, which demonstrates the efficiency of the proposed method.

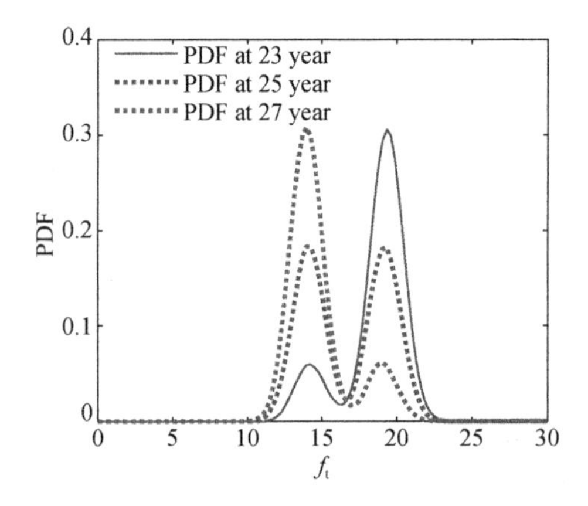

(a) Two-step translation method

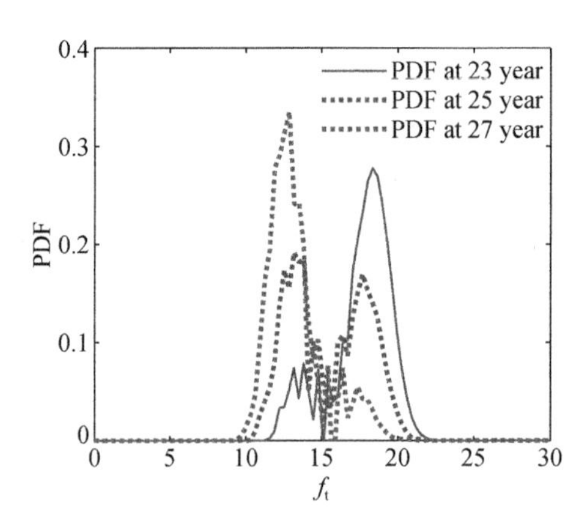

(b) Traditional PDFM

Figure 1 Comparison of PDFs at 23, 25 and 27 years

Besides, supposing a simply supported beam with a cover thickness of 25 mm with a dimension of 6 000 mm × 200 mm × 500 mm which is subjected to uniform load. More detailed information refers to reference [5]. Considering the ductile and brittle failure of an RC beam, the results of the PDF are presented in Figure 2. Figure 3 compares the reliability index obtained by PDFM and 1 million trials of Monte-Carlo simulation (MCS), which

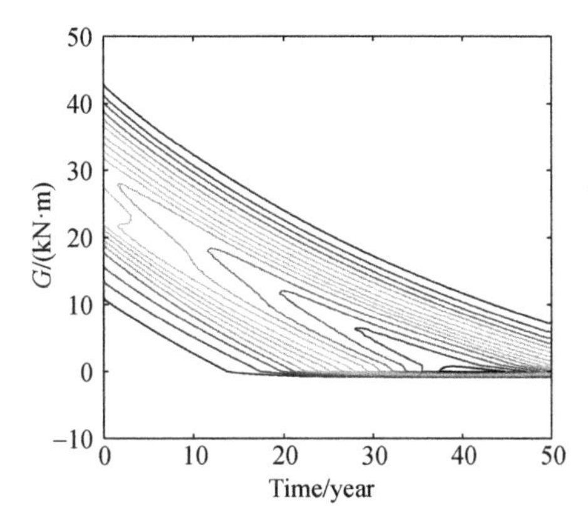

(a) Ductile failure

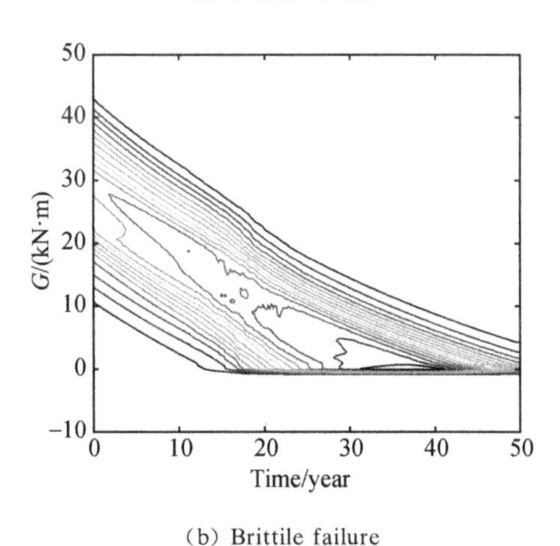

(b) Brittile failure

Figure 2 PDF contour of RC beam

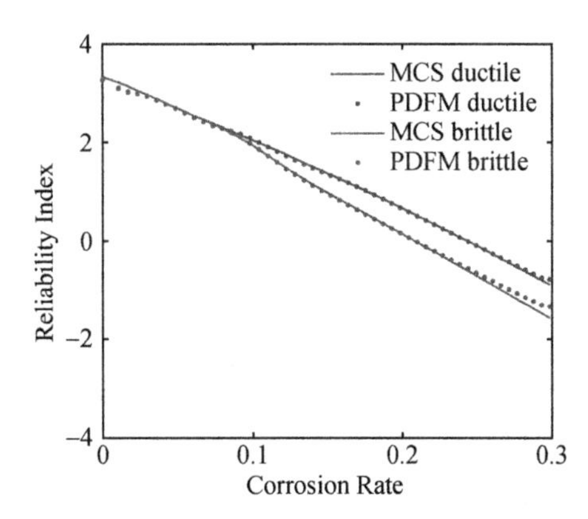

Figure 3 Comparison of time-dependent reliability index

demonstrates the proposed method also suit for durability engineering.

Overall speaking, the PDFM based reliability analysis framework is developed and successfully applied in the general form of deterioration system. Numerical cases demonstrated that the proposed method could overcome the shortage of existing PDFM and maintain high computational accuracy comparing with MCS.

Acknowledgements

The authors also gratefully acknowledge the financial support of the National Natural Science Foundation of China (Grant No. 51320105013 and Grant No. 51808476), the National Basic Research Program of China (973 program) (Grant No. 2015CB655100), and Research Grants Council of the Hong Kong Special Administrative Region, China (Grant No. PolyU 252161/18E).

References

[1] WANG C. Time-dependent reliability of aging structures: From individual facilities to a community [D]. Sydney: University of Sydney, 2019.

[2] KUMAR R, CLINE D B H, GARDONI P. A stochastic framework to model deterioration in engineering systems [J]. Structural Safety, 2015, 53: 36-43.

[3] LI J, CHEN J B. Stochastic dynamics of structures[M]. John Wiley & Sons, 2009.

[4] TAO W, LI J. An ensemble evolution numerical method for solving generalized density evolution equation[J]. Probabilistic Engineering Mechanics, 2017, 48: 1-11.

[5] GU X, GUO H, ZHOU B, et al. Corrosion non-uniformity of steel bars and reliability of corroded RC beams [J]. Engineering Structures, 2018, 167: 188-202.